PAINTER'S HANDBOOK

by

WILLIAM McELROY

Craftsman Book Company
6058 Corte del Cedro P.O. Box 6500 Carlsbad, CA 92008

Acknowledgements

The author wishes to thank the following for providing materials and information used in preparing portions of the text.

Binks, Inc., 9201 West Belmont Avenue, Franklin Park, IL 60131

The DeVilbiss Company, 300 Phillips Avenue, P. O. Box 913, Toledo, OH 43692

International Conference of Building Officials, 5360 South Workman Mill Road, Whittier, CA 90601

Star Bronze Company, Box 2206, Alliance, OH 44601

Titan Tool Inc., 107 Bauer Drive, Oakland, NJ 07436

Figures 8-2 and 8-3 are reproduced from the ***Uniform Building Code***, 1985 edition, copyright © 1985, with the permission of the publisher, the International Conference of Building Officials.

Library of Congress Cataloging-in-Publication-Data

McElroy, William.
Painter's handbook.

Includes index.
1. House painting. I. Title.
TT320.M55 1987 698'.1'068 87-30484
ISBN 0-934041-28-8

contents

chapter one

An Introduction to Painting

Painting is as old as the history of man. In fact, the first history was recorded with crude paint on rock in caves. Both paint and painting have come a long way since then. Today's paints are synthetics based on complex technology that was unknown as recently as the 1940's. The paint you dip a brush into tomorrow is a high-tech by-product of research done for military and aerospace applications.

Any painter worth his pay needs both an understanding and some appreciation of paints and coatings. That's the starting point of this book and the primary subject of this chapter. Once we've covered what's in a paint can, we'll turn to painting itself. My intention is to give you all the information needed to handle nearly any residential, commercial or industrial painting job — quickly, efficiently and profitably.

There's more to painting than just slapping color on a wall. There are good painters and bad painters, profitable painting companies and unprofitable painting companies. Anyone who reads and understands what's written here is well on the way to becoming a knowledgeable, skilled, professional painter or paint contractor.

The History of Paint

The next few paragraphs provide a brief history of paint and painting. But before getting into that, let me define a few terms. For ease of understanding, I'll use the term *paint* to mean all coatings that are liquid at room temperature and are applied either to protect or color a surface. This includes dyes, stains, clear coatings and paint. For convenience, I'll use the term *he* to mean painter — whether male or female. I don't mean to imply that this book is written for men only. Where you see ''he'' in this book, please understand that I mean ''he or she.''

The first paints— These were made from clays and plants ground into powders, then mixed with water. Some were exceptionally good: they're still around today. Pottery and cave-wall paintings exist which date back to prehistoric times, some 50,000 years ago.

The Egyptians probably invented the paintbrush. They were also the first to manufacture what we would call paint, some 8,000 years ago. Around 1500 B.C., both Crete and Greece produced paints. Somewhere between 400 B.C. and 400 A.D. the first metallic pigment was developed: The Romans used white lead. But when the Roman Empire declined, so did paint technology.

In the Middle Ages, English priests and monks began to use paints on their churches. In the late Middle Ages, around 1500 A.D., the artists of Italy developed excellent paints. Some formulations were closely guarded secrets — and remain a mystery to this day.

American Indians used paints for both decorative and religious purposes centuries before Europeans began to settle the New World. The Indians, too, found that rocks could be ground into pigment and that the leaves and bark of some trees produced stains. Many paints made in the U.S. in the late 1700's and early 1800's were based on formulas that came from American Indians. As late as the 1860's many painters carried bags of powder which they mixed into paint as needed. The first liquid, premixed paints were manufactured and sold in this country in 1867.

While all this was going on in Europe and the United States, China and Japan were also developing paints. The Orientals had an advantage: The Tung and Lac trees grow only in the Far East. Tung oil is still used to make varnish; insects from the Lac produce the base for shellac. The Japanese also found that sap from the Sumacs made an excellent tinted varnish.

Modern-day paints— The first modern paints were made during World War II. The old ways of protecting the surface of equipment and weapons had to be improved. The quantities of paint needed were too large. Materials that had been used previously weren't available. The quality just wasn't good enough. The answer was synthetic paints — synthetics that went on faster, lasted longer and could be produced in volume.

In the 1950's and 1960's, nearly all paints were improved beyond anything that had been known before. The number of pigments, dyes, additives, carriers, and application methods multiplied. Research and testing resulted in lower cost, better ways to protect and decorate every type of surface.

As paints evolved, the methods and equipment for applying them evolved. Cavemen used fingers and sticks; the Egyptians and the Romans used paintbrushes. We use everything from brushes to electrostatic sprays. This book will cover them all.

Some safety history— Lead has been used in paint for centuries — poisoning millions of people. Finally, in 1972, the U.S. government restricted lead content in most paint to less than one-half of one percent. Many paint manufacturers now use no lead at all in their paints.

Today, hundreds of brands and types of paints are being marketed by dozens of companies. Many of these products are toxic, flammable, poisonous, or reactive when used incorrectly. Local and state governments, as well as building departments, are becoming more aware of the dangers associated with painting. Some communities have adopted regulations that require empty paint cans to be separated from regular garbage — so they can be disposed of at hazardous dump sites. Some building codes restrict the use of some paints to certain types of buildings or products. Other building codes require that special facilities such as paint booths and fume/dust scrubbers be used when painting.

You can expect that both paint manufacturers and painters will be more heavily regulated in the future. That's just one good reason to read the section in this manual that deals with paint safety and paint chemistry. Every professional painter should be a safety-conscious painter. Safety is just good business. It pays! That's all you need to say.

Where Do You Fit in the Picture?

You, like every painter in the trade, help protect and beautify property. You communicate style and mood through design and color. You're dealing with modern materials, competitive conditions and human emotions. It's not easy to survive in this business. And many don't. But painters who know how to make the most of the equipment and materials available, who price their services fairly, who deal intelligently and honestly with clients and fellow tradesmen, and who take pride in their work, will thrive in this business. That's true today and always will be.

That's what this manual is all about. It's meant to help you establish and build a career in painting. It shows how to stay out of trouble while building a reputation for quality and professionalism.

Setting up your business— If you don't already have a degree in business administration — and I know a few paint contractors who have graduate degrees — this book can be your introduction to setting up and running a paint contracting business. You need some goals and should know what to expect. The first section covers all this, and more practical details like insurance, taxes, and estimating . . . everything you need.

Chemistry and color— You need to know something about the chemistry of paint if you're going to recommend the right paint for the job. Will it cover? Does it have the weather-resistance you need? Will it react with what's already there? Don't be like a painter I know who did a nice job on a huge airplane hangar. His client was a communications company and had sensitive antennas in the hangar. Eager to do a lasting job, my friend used top quality lead-based paint. He clean forgot that lead reflects radio waves. His next step was to remove all the paint he had applied so carefully the week before.

But paint does more than protect exposed surfaces. It can also change the way form and texture are perceived. Color is the finishing touch for nearly every building. That's why this book has a section on color and its psychological and physical effects on people.

Doing the work— The last section of this manual explains how experienced painters apply coatings: do's and don'ts, the problems and solutions. There are plenty of creative ideas and tips here for your use, even if you've been painting for years.

If you're an experienced painter, congratulations. But I'll bet the price of this book that there's plenty of information that you can use between this page and the back cover — practical tips that even an old master painter could use.

If you're just starting on a career in painting, I can offer some encouragement. The field is wide open for enthusiastic, eager young men and women willing to work hard, deal honestly and develop the skills needed for success. You've made the right choice, both in following a career in painting and in picking up this book.

Having covered these important preliminaries, let's get down to business. How do you start and build a profitable paint contracting company? That's the subject of Chapter 2.

chapter two

Creating a Profitable Paint Contracting Company

There are at least as many ways to run a profitable paint contracting company as there are profitable painters. I won't claim otherwise. Nor will I insist that doing it your way is wrong. But there are some general business guidelines that nearly all painters should follow. That's the subject of this chapter.

I'll also offer at least a hundred tips and ideas that you should consider. Pick and choose among the ideas in this chapter. All have worked for me or paint contractors I know — including some painters who became millionaires in the business. Finally, I'll explain why some paint contracting companies fail and what you can do to avoid their fate.

First, ask yourself some questions: Why am I in the paint contracting business? How long do I plan to stay in this business? What are my long-range business goals? Am I willing to study, work, and sacrifice to make this business a success? You must know the answers to these questions. To run any business, you need a dream, a goal, and the commitment to make it happen.

It's *you* who'll make it happen — no one else. Some people will support you and your ideas. Some will work with you or for you. Some will offer ideas. Others may provide working capital. And some will tell you it can't be done. But it's *your* business, and *your* responsibility. It's you, and you alone, who'll have to make the decisions and take the risks. Do you have the motivation? The desire? The dream? If so, keep reading. I can't provide everything you need to succeed in paint contracting. The desire and motivation have to come from you.

Reasons for Going into Business

Most painters go into paint contracting after working for someone else for several years. They feel they know as much or more than the boss, and could do better working for themselves. That's the most common reason for getting into the paint contracting business. The second most common reason relates to the first. It's not liking working for others. That's nothing to be ashamed of. In fact, it's probably an advantage. Many people just have difficulty with rules and requirements laid down by others.

But there are other reasons:

- You want to build a future for yourself.
- You want to have something to leave your heirs.
- You want to say it belongs to you.
- You want to earn more than you could ever make as an employee.
- You're out of work, with nothing else to do.
- You want your fair share of the profit from every job.

Whatever your reason for going into business, you'd better learn how to do it right.

Why Paint Contracting Businesses Fail

The main reason is lack of motivation. Painters just get tired of doing the same thing day after day for what seems like too little pay. They forget why they went into business, or give up the dream, or find a new dream or goal in life. Lack of management skills, lack of money, lack of customers, lack of proper controls, and lack of goals also sink some paint contractors. These secondary reasons stem from the first — most business problems can be solved given enough time, effort and determination.

Key Factors for Business Success

What are the key factors for a successful business? Here they are:

1) Motivation
2) Knowing the business
3) Promotion
4) Giving it time to work
5) Staying with it

Dream a dream, of yourself and the business you're creating. Dream of the prestige of owning a business. Dream of the freedom of being your own boss. Picture what you want to be doing in 5, 10 or 20 years. Think of your business as a child you're raising. Give it loving care, attention and nourishment as it grows and matures. Many times it will disappoint you or become a burden. But tended properly, your business will become a source of pride and admiration — as well as a good living.

Types of Business Structures

If you've been in business for some time, you probably know that there are four forms of business organization: *sole proprietorship, partnership, limited partnership,* and a *corporation.* The difference is in the type of ownership.

Sole proprietorship — This is the fastest, and easiest, type of business to set up. It's you doing business in your own name. You may have many employees and lots of assets, but you and the company are one and the same. You're liable for all damages if you're sued. You get all the profits; and you make up all the losses. Income and expenses show up on your personal tax form. All profits are income to you and you have to pay taxes on them each year.

Partnership, or general partnership— This type of setup splits the liability and profits between two or more people. All partners are legally responsible for everything the company does and all have a claim on profits. All partnerships are based on a partnership agreement — like a contract — that should spell out the rights and obligations of each partner. Be very sure that you have this agreement in writing, and as detailed as possible. It will prevent many future problems and misunderstandings.

Partnerships pay no taxes. Instead, they file an "information return" that shows income, expense, profit or loss. All profits and losses are passed to each partner and the partners get the benefit of the loss or carry the burden of the gain on their personal tax return.

Limited partnership— This is much like a general partnership, with one big exception. The liability of limited partners can't exceed the investment made by limited partners. The general partner usually runs the business from day to day and has unlimited liability. Limited partners are usually just investors with no management role.

The advantage of any partnership is that it combines the talents of two or more painters — maybe someone who's good at selling and office work and someone else who's an expert at running crews in the field. The disadvantage is that the partners have to agree on everything. If they can't agree, the only alternative is a dissolution of the company. Maybe that's why partnerships have a history of failure in the paint contracting business. The usual cause is that one or more partners feels cheated, overworked, mistrusted, or slighted.

Still, some painting partnerships have lasted for many years. But if your company has three or more principals, I strongly suggest doing business as a corporation. Disputes are resolved by voting the shares. Anyone outvoted can sell his shares, either to other shareholders or an outsider.

Corporation— This is a taxable entity by itself. Corporations pay tax on net income at the end of their fiscal year, which may not be the same as a calendar year. Corporate taxes are lower than individual tax rates, so it's easier to accumulate money inside a corporation.

Corporations issue stock in exchange for money or property invested in the corporation. Shareholders elect the board of directors and the board hires officers to run the company. In a small company the chief stockholder usually sits on the board and probably serves as president also.

Shareholders are not liable for debts of the corporation. They can lose only the amount they've invested.

Each type of business has its own advantages and disadvantages. Some of them are summarized in Figure 2-1. You have to decide which is best for your situation. Most painters and paint contractors are one-person businesses. They generally choose the sole proprietorship because it's easiest and cheapest to form.

Your Business Name

Selecting a business name can be important. If you want to use a name other than your own, decide what you want to be known for. Quality? Fairness? Low prices? Cut-rate? As a small, personal company? As a large, corporation-sounding company? What kind of painting will you specialize in? Residential? Office? Institutional? Apartments? All kinds?

Try to create a name that projects the image. List several, then pick the three best. Why three? Often, you'll choose a name that's already in use. So you'll have to use your second or third choice instead.

A few other things to consider when choosing a name: Is it easy to remember? Will it fit well into advertising copy? Where will it be in the phone book's Yellow Pages? Browse through the Yellow Pages for name ideas — but don't use a name that's similar to one that's already in use.

And finally, remember that newspaper classified ads charge by the word — so keep it short. "Able Home Painters," instead of "Ready and Able Home Painter of Anytown U.S.A."

Setting It Up, Legal-Like

Okay, you've decided which type of business organization is right for you. Before going any further, answer a question: Is this business an experiment? Is it just a way to meet next month's food bill? Or is it going to be your career? If you're experimenting, or you're after a quick dollar, then forget the legal part. Why spend money on something you won't pursue? Just grab a brush and go to work. But if you're serious about a painting career, start it right.

The license and DBA— Your first stop is city hall for a business license. Most cities require that painters doing business in their city have a business license, even if you do only an occasional job there. If you're doing business in an unincorporated area, you may not need a business license, but check and verify this with your local city, town, or county government.

Take your three name choices to city hall. Submit them, with your business license application and an *Assumed Name Certificate,* or *DBA*. That stands for *Doing Business As.*

Next, publish the DBA in the local newspapers. Usually, it must run for at least three consecutive issues in a "newspaper of record." Your local paper will advise you what's required. The license, the DBA, and newspaper publication of the DBA cost from $30 to $100, depending on where you're doing business. This cost is your only cost for setting up a sole proprietorship.

When to hire a lawyer— If you've decided to set up as a partnership or a corporation, things are more complicated. And more expensive. You should have a lawyer draw up and file the necessary legal paperwork. Costs run from as little as $250 to as much as $1,500. If you're going to use a lawyer, ask how much he or she will charge before starting work. And don't think it's tacky to shop around.

But you *can* do it yourself in most states. And for under $250. Many bookstores carry self-help legal guides. These take you, step-by-step, through the proper procedures. Yes, it's legal — and it's not really that complicated. The books have all the information you'll need. And they usually include all the required forms as well.

If you can't find what you need at a bookstore, one source is:

Nolo Press
950 Parker Street
Berkeley, CA 94710

You'll also find that many large office supply companies carry a stock of legal forms. These forms cover a broad range of subjects, including contracts, mortgages, rental agreements, loan agreements and more. Look into it. You can buy for a dollar or two the same forms that a lawyer might charge you hundreds for. Most are self-explanatory and fairly easy to fill out.

Does Your Area Need Another Paint Contractor?

I have a few more questions to ask. You have a dream, but will it work? Does your community need another painter? Is there enough business to keep you going, day after day, year after year? Will your potential customers pay your price? Even if there are hundreds of homes in need of painting,

TYPES OF BUSINESS ORGANIZATION

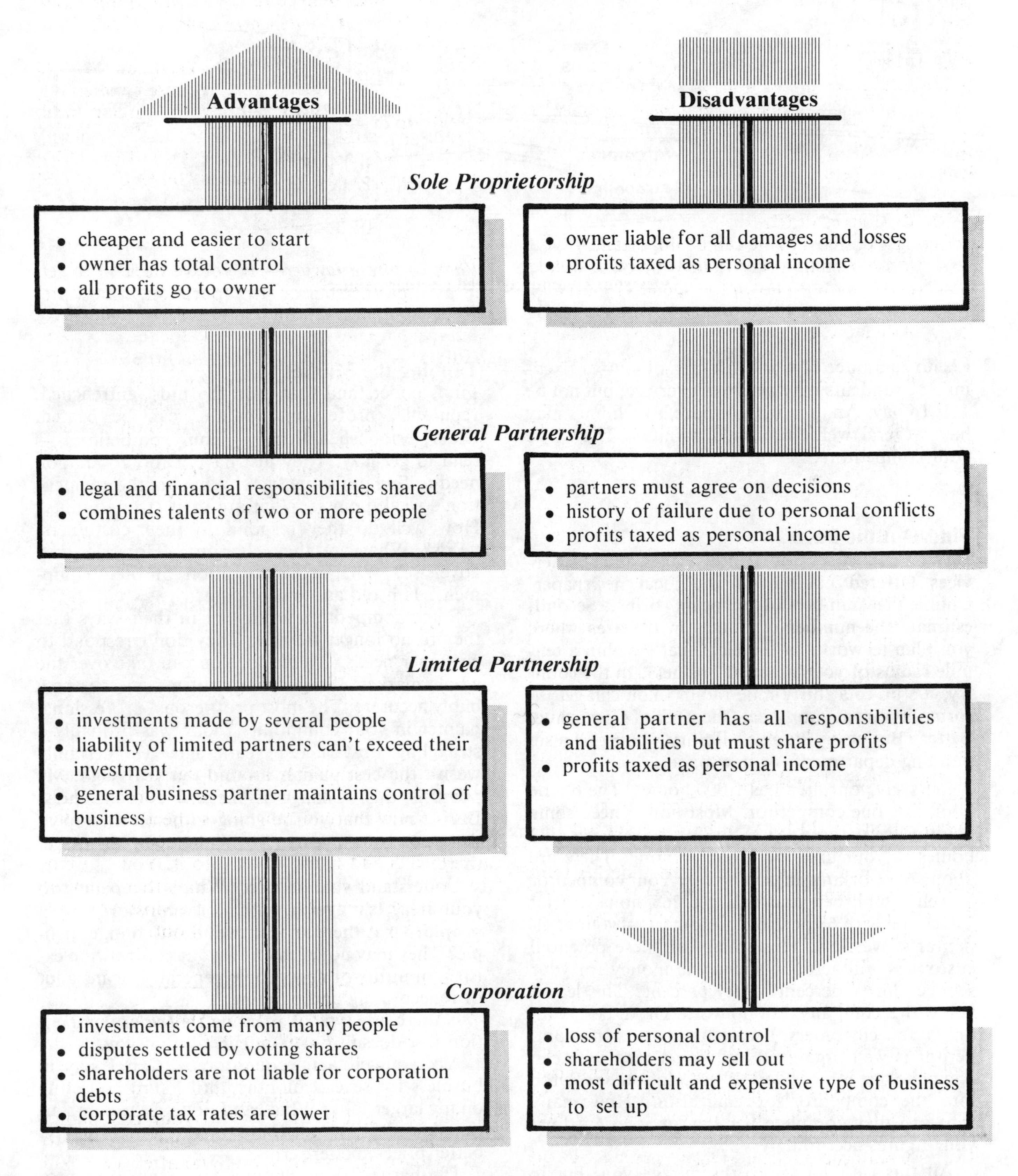

Types of business organization
Figure 2-1

Does your community need another painter?

it's no guarantee of work. An area of low- or fixed-income residents may need your service, but not be able to pay. An area with hundreds of homes may have several well-established painters. They'll be your competitors.

Find Out the Facts

How? First, check the Yellow Pages and the Services Offered column of the local newspaper. Count the number of painting firms. Second, estimate the number of homes in the area where you plan to work. In the city, that's within a ten-mile radius of your place of business. In the country, it's up to a thirty-mile radius. You can get information from your local Chamber of Commerce, Better Business Bureau, library, court house, building department, or newspaper.

Let's say your area has 1,000 homes. The phone book lists one competitor. Most homes need some painting every three years. So, for a given year, 333 homes in your area will need painting. There are about 265 working days in a year. Your competitor is well established and will get 265 homes. That leaves you with 68 possible customers — your competitor's overflow. How many of these potential customers will actually spend the money for your service? Forty percent? Fifty percent? This leaves you with a company, but no work. Or, at best, with only a few customers. I'm assuming that this competitor is well organized and has a crew who can complete one job per day. If not, do some research into the competitor's organization. How many jobs can they complete per day or week? Adjust your figures accordingly.

All this isn't meant to discourage you, but to make you think.

Tapping the Market

First, understand that even the most entrenched, reputable, professional painter can be beaten. Older, established firms — your competitors — tend to get lazy. They just have it too good. You need to find out some facts. What are the competition's weak points? What are their strong points? How well do they respond to their customers' needs? What are they charging? How do they advertise? What's the condition of their equipment? Is it old and out-of-date?

Some competitors are so set in their ways that they're no longer flexible. They don't respond to customer needs. If *you* do, you can take over the whole market. I'm guessing, but my guess is probably accurate: The most prosperous, professional painter in your community today was probably a struggling upstart ten years ago. He certainly wasn't the best painter around ten years ago. My point is this: Change is inevitable in business. Decide now that you're going to be on the top of the heap in ten years. Then make a plan to get there.

Understand your market. What's happening to your area? Is it ageing, with more and more retired people? Or is there an influx of new, younger people? They may not even know your competition exists. An influx of new people generally means a lot of redecorating and painting.

Is there any type of painting that your competition just doesn't do? You can fill that void.

Ask yourself a few questions. "Can I stay in business for several months until I start collecting on my larger jobs? What can I do that the competition doesn't?" Also, "Can I provide better quality, better prices, better terms?"

The hardest thing about your dream of a profitable painting business is facing reality. It's so easy to be overly optimistic, to change facts to suit your desired outcome. This leads to many business failures.

Be truthful with yourself about the facts. It hurts. But knowing the truth will better prepare you for the business at hand: your business, your success. You'll face many problems, and maybe some tough competition. But you know that any problem can be overcome with time, thought, and imagination. The first problem to solve is getting your company established.

Establish Your Business

I was once asked where I'd spend my money if I was starting a new business and had only a few hundred dollars to spare. Would it be a down payment on a truck? Or a supply of brushes, rollers, and paint? My answer was: Neither. I have a car I can use. I'll buy paint and brushes with the down payment on my first job. But I have to *get* that first job.

Customers are the key to any business. Customers are those nice people who are going to pay me for my services. That's where to spend those limited dollars: on getting customers. You don't want expensive equipment sitting idle in the driveway while you're drumming your fingers on the desk.

Most painters start out doing their own home, then that of a friend, neighbor, or relative. Do a good job, then ask these people to recommend you to others. Then word-of-mouth starts the ball rolling. Note that I said *ask* them to recommend you. Most people don't say much, unless asked to.

But word-of-mouth can't do it all. You need to tell as many people as possible about your business. That's where advertising comes in.

Advertise Your Business

What's an effective ad? That's easy: it's one that gets results. A good ad makes people take action — call you for estimates. It's not necessarily a large or expensive ad. A big ad helps sometimes, but it isn't necessary. My advice is to find a painter's ad that runs in every issue of your local paper. That's a dead giveaway. The ad works, or he wouldn't keep putting it in! Make up a similar ad for yourself and run it a few times. You should get the same results.

There's one thing to remember. All advertising must be repeated, over and over again. It must be there when the need is there. People read thousands of ads every day. There's no way they can — or even want to — remember all they read. Display your ad where people will find it when they need your service. This is why the "Services Offered" columns of the newspapers work. And why the Yellow Pages work. Find an effective ad and repeat it as often as you can afford to.

The following sections describe good advertising media for you to use. And you'll find some tips on how to use them most effectively.

Use the Yellow Pages

These are printed only once or twice a year. Get in the next printing and include all the correct information: address, phone number, contractor license number, hours of operation, what you do, what terms you extend.

Make your Yellow Pages ad as big as you can afford. Remember, you're charged each and every month, all year long. I recommend against saying you use a particular brand of paint. You might want to change brands during the year. You can use terms like, "Specializing in Interior House Painting," but these tend to limit your customer base. Better to just say "General Painting Contractor."

Know the cutoff date— There's something I want you to do, right now. Call the phone company business office for your area. Ask for their Yellow Pages Representative. Ask what the cutoff date is for inserting an ad in their next issue. Write the date here:______________. Do it now.

Okay, done? Good. Here's why. Phone books have long lead times: from three to six months. That's when they do their layout and printing. If your ad isn't in before the cutoff date, you'll have no Yellow Pages advertising — for a whole year!

No big deal, you say? You can advertise in the newspapers? I'll let you in on a little secret. The phone companies have spent hundreds of millions of dollars convincing people to use the Yellow Pages. It's the first place people look when they need something. In many cases, it's the only place. A lot of people believe that if your company's not listed in the Yellow Pages, it doesn't exist. Not nice, but it's a cold, hard fact. You'd better believe it. And, yes, it's probably the most effective advertising medium you can find.

To make your Yellow Pages ads — or any ads — more effective, use some illustrations along with the type. Consider using what printers call *clip art*. Clip art is available at many print shops and some bookstores.

Clip art is attention-getting. It's available in

thousands of different designs and themes. You can use it to create your flyers and business card, as well as your Yellow Pages and newspaper ads. The cost is low. You can get a complete book for under $20, separate sheets for about a dollar. Be creative.

Use Business Cards

Have a stock of business cards printed up, ready to hand out. Everyone you meet is a potential customer. So everyone you meet gets a card. This very effective advertising costs from $15 to $30 per 1000 cards. The printer will even help you with the design.

Use Neighborhood Flyers

The neighborhood flyer is a one-page, 8½ x 11 sales pitch. It tells who you are, what you do, and why you're best for the job. Don't forget to put your name and phone number on the flyer. (No joke. I've seen flyers without any phone number listed.)

Don't quote prices in your ad. You only talk price after the job is sold. *After* the job is sold? Yes: You're selling yourself, and what you can do. Convince your potential customer that you're best for the job. Convince them that they're not making a mistake by contracting you. Do that, and they're sold. Price becomes secondary.

Deliver between 500 and a few thousand flyers, door-to-door. Do it yourself, or have some kids do it for a few dollars. It'll cost you a day's time, and about $40 for printing. It should bring you a few customer calls. Only a few calls? Yes. Not many will be in the market for your services at that moment. The idea is to let people know you're in business.

Okay, what's next? Put up some signs and posters. And start your newspaper advertising.

Use Signs and Posters

Both of these are inexpensive. And they get your name noticed by lots of people. Make them right, and they can't be ignored.

Signs— Make a few portable signs. Make them big enough to be read from a moving vehicle, at a distance. About 20 inches square will do. Keep them readable and uncluttered. Use lettering 3, 6, 9, or even 12 inches high. For maximum contrast, use black letters on a white background.

Use these signs when you're on the job. Get the customer's permission to display the signs while you're working on their home.

Posters— I've found that an 11 x 17-inch poster works well. Use white, clay-coated card stock, with 2 to 4-inch high lettering of black, rubber-based ink. You'll need about 100 posters, which will cost you around $40.

Some people starting a new business staple or tape posters to every lamp post, power pole, store window, and tree in town. This may be illegal, so if you do it, be careful. Get permission from store owners. The power companies won't give you permission — their employees are frequently injured by nails and broken staples left by such advertising. Tape may not upset them quite so much. But they'll likely not bother to prosecute, either. Some will just ignore the posters for a few weeks, then remove them. Some will call or visit you. Others may have the local police call you. It's a judgment call, on your part, whether you want to try this. It is, however, very effective — for one reason: People are creatures of habit. They take the same routes to work or play each day. After looking at your sign for two or three weeks, they'll remember who you are.

Use Newspaper Ads

Small ads can be effective. Put your ads in the "Services Offered" column, for maximum drawing power. You could use display-type ads, but they're not very effective. Use them only with a picture and an offer on a sale item.

Writing ads— Try writing several small classified ads. What's the best way to write a good ad? That's no problem: Copy the ones you see frequently: they work. There are two approaches you can take:

1) Start the ad with what you're offering. "Painter," "House Painting," and so on. This ad will capture the people who need a painter immediately.

2) Sell a dream or an idea. "Dreaming of a beautiful home this summer?" or "Color your life with shades of nature." These will attract attention. Everyone who sees them will read them.

If you're running a sale or a special, start with proven key words: "Sale," "Special," "New," "Improved," "Deluxe." The idea is to draw attention to one ad out of hundreds — yours. Pick the best two or three ads you wrote, test each to see which is most effective. How do you test an ad? Run several different ads but each with a "key" difference. If you have only one phone line, you can end each ad with a name: "Ask for Bill (or Tom or Sally or Jim)." If the callers ask for Jim five times more than Bill or Tom or Sally, then that's the most effective ad. Stick with it.

Some tricks— One trick I sometimes use is to buy white space. A six- or seven-line ad, with only the following words, printed dead center, in it. "Don't Read Next Ad." The next ad is my sales pitch. Human nature being what it is, everyone will read it. They might not buy, but they'll read it.

Color is good. but only up to a point. When over 30% of the classified advertisers use color, the effect falls to near zero. Best to keep the ad small, to the point, and in constant view.

How long to advertise— Don't buy less than six weeks' worth. Any less than that, and your name disappears from view just when people were starting to remember it. They soon forget it entirely.

What are the best media to place it in? If your town has six newspapers, spend a couple of bucks and buy all six. Which one has the most ads for painters and contractors? That's the one you advertise in.

Good ads cost nothing: They generate business, success, and profits. Bad ads cost more than you can afford.

Want to run a special? *"Entire home exterior painted for $495.95. Week of July 5 to July 10 only. Up to 1,400 square feet of living area."*

The Contractor's License

In some states, painting contractors have to be licensed. In California, for example, it's illegal to advertise for anything over $200 — or, in effect, to even bid a job at over $200 — if you're not a licensed contractor. In states that require a contractor's license to paint, the types of jobs you can get without one are very limited. You'll also find that many newspapers won't even print your ads unless you include a license number.

Check with your contractor's license board to see if you need a contractor's license. It will cost from $150 to $500.

Many states require college, or four years' experience, or a combination of both, before you can apply. That's called protectionism. It's the way established painters limit competition. There are good and bad points. The good is that the consumer has some protection. The bad is that it handicaps otherwise good talent. The college education or four years' experience doesn't guarantee that painters know what they're doing. But that's the system you have to work with in many states. In many states, only licensed contractors can file a lien against a customer who balks at paying. If you're painting without a license in a state where one is required, you're at the mercy of your customer's goodwill and decency. Personally, I'll take the license. In California you can, if licensed, file a *"California Preliminary 20-Day Notice, Rediform 4S449 or 4P449."* This legally allows you to place a "lien" on the property. This lien prevents resale of the property until you're paid. Your state probably has similar requirements. Get the forms you need from an office supply store.

Timing Your Start-Up

When is the best time to start your company? Right now! There's no better time. You're already on your way. Reading this book is your first step to starting a successful paint contracting business.

Actually, you can start on any date. Some times, though, are better than others. If you were starting a retail business, I'd say: Start the research in April. Line up suppliers in May and June. Rent your space in August or late July. And be open the last week of August. This way, your first six months cover the Christmas season. That might help maximize sales in a retail business. But that's probably the *worst* time to start your paint contracting business. Here are my recommendations:

- Start research in November and December.
- Line up supplies and suppliers in January and February.
- Rent your place and start your ads in March.
- Open for business on April 2. This is the start of both the spring-cleanup season and the building season.

There's one factor we've already covered: Yellow Pages advertising. These usually come out in the spring, but not always. Be open for business on the day your ad appears, even if that day is in January.

We now have the dream, the legal set-up, the market, the advertising, and the right timing. What's next?

Dollars for Your Start-Up

Now, jump back to Day One. As you've just seen, it costs money to go into business. And we haven't even talked about supplies and tools.

"Tell you what, I'll paint your living room in exchange for your boat. That way, Uncle Sam doesn't get his fingers in the deal."

Where does your start-up money come from? Usually savings — yours. Do you have the cash available? These days, you need $8,000 to $10,000 for most business start-ups. The best way is to keep working for someone else while you start up your business. Work at your own business evenings and weekends, until you begin to get a feel for what's happening. This keeps expenses down. And you're still getting your regular paycheck. Don't forget that you need money to live on as well as for the business start-up. This may be true for up to eighteen months, until your business starts running in the black. Do a personal budget, including food, housing, insurance, vehicles, and anything else you *need* to live on. Make sure you have a way to meet these expenses until the business generates enough income.

There are other sources of money besides savings. Use your income tax refund. Sell something you no longer need. Use profits from small jobs you handle. And barter. Barter? Yep, it's still around and going strong. You trade your services for someone else's. Paint the lobby of the print shop. In return, get business cards, flyers, or posters. Paint your neighbor's house. In return, get the boat he no longer wants. Trade the boat for cash, or for an airless spray unit. Many people who barter and trade don't report this income on their tax return. That, of course, is illegal.

Loans of All Kinds

You can get loans from parents, friends, or relatives. But don't go overboard on this — it puts you in debt right from the start. If you have a Visa or MasterCard, it may be good for a grand or two, though the interest charges are high. If nothing else is available, sell a few non-essential items, like the stereo or TV.

My advice is not to borrow against your car or home. The bucks may be there, but it's a big risk. And it can cost you a lot. Besides, why put a huge investment on the line for small loan?

I'll caution you right off the bat: Most sole proprietorships and small companies fail in the first year. Don't put a lot of money into equipment and supplies until you're sure painting is what you want to do. Besides, say you do make it into a third or fourth year. You'll then need more money, for growth and expansion. That's when you may need to use your home as collateral for a loan.

Bank loans— Why not just go to a bank in the first place? Simple. They won't loan you a dime on a new business start-up, especially paint contracting. It's too unpredictable and seasonal. I've even tried for a loan with an established business. It was already into its fourth year — doing a six-figure volume with good profits. Banks make loans against assets like buildings, cars and equipment.

If you don't want to start a business from scratch, you can always buy one that's already running.

They also loan to large businesses and to smaller businesses when the loan is guaranteed by the Small Business Administration (SBA). Most banks won't make you an unsecured personal loan to start a business.

SBA loans— The SBA has done some direct lending in the past and guarantees loans made by commercial banks. Your local SBA office can explain what programs are active now and how you can qualify.

Silent partners— Another source of money is a silent partner — a relative or friend who has faith in you and wants you to succeed but doesn't want to be involved in day-to-day business affairs. They may even be willing to defer loan payments until you get on your feet. Consider offering them a percentage of the business in exchange for an investment now. But put the details in writing, especially if your silent partner is a relative. You've got enough problems running a paint contracting business. You don't need extra harassment from a relative.

Once your business starts to pay its way, it will finance itself. But only up to a point. A mistake made by many companies — including some of the world's biggest — is to forget that they'll need money in the future. As your business grows, you'll need to replace equipment and increase working capital. Get in the habit of setting aside a small percentage of gross each month. Put it in a separate bank account and try to forget it. That's your nestegg and guarantee of future prosperity.

Buying an Existing Business

So far, we've talked about starting a business from scratch. But there's another way to do it: Buy one that's already running. You'll need more money up front, but your chances of success are far greater. Someone else already established your market. Someone else solved many of the problems. Someone else spent the money experimenting. And someone else learned how to make the business grow.

A local business broker in your community has a listing of all kinds of businesses for sale. Some will be offered for far more than they're worth. But occasionally you'll find a motivated seller who only wants to cash out at the net asset value of the business. That's the value of equipment, inventory and work in progress, without any premium for goodwill. At least see what's for sale before starting your own business. Usually you can buy a business with minimum cash and payments over several years.

What to Look For

If you're buying a going business, buy from a painter who's changed his mind, is ready to retire, or is ill. There's not much demand for paint contracting companies. People who want to buy a business are usually looking for a retail store. They don't want a service-type business. Use this to your advantage: You may be the only customer this seller has. He has to either stay in business or sell out at a bargain basement price. Start by offering about what the equipment is worth on the used equipment market.

How to Buy It

Need financing? Then pay the asking price, but at your terms. Offer little or no cash with several years to pay, at a low interest rate.

The last business I bought was worth $120,000. I got it for $15,000, and the owner financed it at 2% for three years. Try to do as well. This particular business was already making a profit, so it worked out very well for me. But it doesn't matter if the business you buy is losing money. After all, your new start-up business would also be losing money at first. Think about it.

Start-Up Expenses

How much money you need depends on you, and on how ingenious and committed you are. I've seen companies grow into million-dollar concerns, with a start of only $5. I've watched others fail, even though they invested a million in start-up funds. How you do it is more important than the size of your bank account. The easiest way to pay your start-up expenses is with cash from your personal checking account. But don't do it. Open a business account at a commercial bank. You'll probably need a business license first, but not always.

Tax Deductions

The business account checkbook is your tax record, for the time being. Remember, all business start-up expenses are tax deductions *if* you can prove they're business expenses. This book is deductible. So are the business license, the DBA, and the newspaper DBA announcements. Add the mileage, gas, oil, and so on for your car while setting things up. Don't forget the business cards, legal fees, legal forms, supplies, etc.

Go to your nearest IRS office and ask for a copy of the *1040 Tax Guide for Individuals,* and the *Tax Guide for Small Business.* They're free, and have all the information you'll need. You can also get tax guides for partnerships and corporations, but you have to ask. Contact your state Franchise Tax Board for the state tax guides. Do yourself a favor. Spend the time and effort to get these guides, read them, and study them. Then apply what these guides suggest. It'll save you a lot of grief in the future.

Sales tax and the resale license— Do you want to buy your expendable supplies and paint without paying sales tax? Then you'll also need a resale license from your state's Board of Equalization. Most painters pay the sales tax on purchases and include the tax in the job. If you get a sales tax license, you have to collect the tax from your customers and forward the tax to the state on a monthly or quarterly basis. You become, in effect, the state's collection agency. That's usually unnecessary trouble.

Then why get a resale license? For one thing, many suppliers will only give you a discount if you have one. This is especially true of wholesale houses. They're not generally set up to collect the tax from you. When you invoice your customer, you must break out the materials and labor separately. You then add tax on the materials only.

Note that *all* money collected as tax must be reported and turned in, even if you overcharged. But if you undercollect, it comes out of your profits. That's just one reason to keep good records.

Many areas also have a county tax, a city tax, and sometimes a special tax. These are usually included on the state sales tax reporting form. The state collects from you, and then pays the county, the city, and so on.

Another tax— Some states take another tax bite. It's the end-of-year inventory tax. Better find out if your state does, and when it does. Then just see to it that you haven't any inventory to report at that time. Once again, your state and local governments can fill you in on the requirements.

Insurance

Do you need insurance? It's up to you. But I say yes. You have to understand why insurance came about in the first place. It was designed to help people in times of tragedy — a major illness or a home burning to the ground. It wasn't designed to cover every sneeze, bump, and scratch. You can usually cover the small claims as a cost of doing business. So buy insurance to cover only major problems or losses. Large claims can not only put you out of business but also cost you your home, bank accounts, and future earnings.

Are you now working for a company with insurance benefits? If so, then do yourself a favor before you leave. Get sick. Get a belly ache, a back ache, chest pains, whatever. Whatever will let you get a complete physical checkup, at their expense. If you're not employed, then I suggest you spend the money for the physical. Do it before you open up shop. Find out about any major ailments now. That can save you dollars later — and possibly your new business.

Health and Life Insurance

There's another reason for this checkup. Most insurance companies require a recent physical before they'll issue health insurance or a large life insurance policy. You should have both policies if you own your own business. Buy basic coverage only — these policies aren't cheap.

How much health insurance?— Even the most basic major medical and hospitalization insurance is expensive. But it's less if you accept a high deductible. If you can afford to pay the first $1000 of any doctor bills, your deductible should be $1000. If it's only $200, then have $200 deductible. It's much less expensive, in the long run, to self-insure for $1000 or $2000 than to pay the premiums on that first $1000 or $2000 of coverage.

How much life insurance?— Basic term life insurance runs from $15 to several hundred dollars per month, depending on your age. Whole life, and several other policies, include savings plans — expensive savings plans. And they include a big salesman's commission. Buy term life insurance. It's pure coverage at an affordable price.

Business Insurance

Business insurance packages cover fire and theft on equipment and records and loss of income after an accident. They cover customer damage liability, customer accident medical (caused by something you did). And they cover fire in your office or building. Terms and costs vary from company to company. Expect to pay about $300 for the most basic coverage during your first year.

As with other insurances, self-insure as much as you can. It'll cost you less in the long haul. One thing you'll discover about insurance salesmen: They work on commission. The more monthly premiums they load you up with, the more they earn. Many will try to sell you more than the minimums you need. Shop around. Find someone who can service your needs at a price you can afford.

More Insurance

Another insurance you need is on your vehicle. Get insurance on your business car or truck, even if it's your own personal vehicle. Most auto policies don't cover a vehicle used for business purposes. Check it out.

If you have employees, or plan to hire them, your insurance picture changes somewhat. Be sure your employees are covered on both your vehicle and business liability policies. To entice good people to work for you, you'll probably have to offer major medical and life insurance. But it may be enough to pay a portion of the premium. The employee pays the remainder. This depends on the practices of your competition and on the availability of good help.

Workers' Compensation Insurance— This is required by state law whenever you have an employee. In some business setups, it's required even if that other person is a partner. This insurance isn't voluntary, it's mandatory. The cost varies, but it may be up to 10% of wages. Talk this over with your insurance agent. Make sure you meet the legal requirements. Trouble with the government you don't need.

Insurance Deductions

Personal insurance on yourself is not deductible as a business expense. But whatever you're paying on insurance for employees is. So are the business insurance, and the extra premium for using your own vehicle. And so are any extra premiums you pay for an office in your home.

The home office— Be careful here. To be a tax deduction, a home office must be a dedicated office. That means it's used for no other purpose. Then you can deduct part of your rent or mortgage, phone, lights, heat, insurance, and so on. The amount you can deduct is based on the square footage of the office in relation to the rest of your home. The IRS is a little touchy about this, so the proof is up to you. You might consider having the company (corporation) lease the home office for you and from you. But you still must be able to prove that it's necessary for the running of the business. Be careful.

I suggest that you have a separate business phone installed. And get a separate insurance policy for that home office. But I wouldn't list the home office address on business cards or paperwork — even though this could provide proof of your use of the office-home. Use a P.O. box instead. Why? Because you really don't want customers and salesmen calling on you at home.

Finding More Information

The SBA publishes hundreds of books and booklets on all sorts of business-related subjects. Most are either free or cost very little. They cover most of the problems associated with setting up almost any business. Spend a few cents. Get their list of publications from the nearest SBA office or from:

Small Business Administration
Washington, D.C. 20416

Another source of information is the government printing office. You can get their list of publications from:

Superintendent of Documents
Government Printing Office
Washington, D.C. 20402

The Costs of Doing Business

I'm going to jump to another subject for a moment. Before getting on with the business of paperwork and control, let's talk about your salary and about the costs of doing business.

How much money do you expect to make a year? How much money would you like to make a year? What's the cost of living in your area? These are all variables that I have no control over. So I'm not about to tell you what you should be earning, or what you should charge your customers. I can, though, give you a formula for figuring it out.

Your Salary

First, if you're investing in a business, then I assume you plan to earn more than you could make working for someone else. Otherwise, why bother? How much you want to make is entirely up to you.

Here's the formula I promised you for figuring what to charge your customers to reach the income level you desire. There are 365 days in a year. There are 52 weekends, and 13 holidays. Set aside 10 days for vacation, and 5 days for sick time. Subtract the days off from days available, to find the number of days you're actually available to work. That's 365 minus 52 x 2 (weekends), minus 13 (holidays), minus 10 (vacations), minus 5 (sick days), which equals 233. You have 233 days available for work.

You'll spend about 10 hours per working day running your business. From these 10 hours, you must deduct non-earning time: Travel time to and from the job — say one hour. Buying supplies and materials, one hour. Paperwork, one hour. Breaks and lunch, one hour. Estimating and lost time, another hour. That's five hours. For those five hours, you're not earning any direct wages. But if you want to get paid for them, those hours must be included in your billable rate.

To summarize: 233 days, at 10 hours per day, gives you 2,330 total hours. You lose five hours per day, or one-half of 2,330. So you're actually billing for only 1,165 on-the-job hours. Say you want to earn $23,000 a year. This $23,000, divided by 1,165 hours is $19.74. So you must charge $19.74 for every hour you're actually on the job, painting, just to make your projected gross. But you've not yet included your expenses, or any materials.

Typical Business Expenses

So, let's look at some typical expenses for a one-person small business:

Rent or mortgage on office or shop	$ 3,600 per year
Utilities:	
Gas or heat	240
Electric or lights	360
Phone	360
Services:	
CPA	300
Bookkeeper	300
Insurance:	
Vehicle	300
Business	300
Health	840
Life	300
FICA, Social Security	1,980

Advertising:	
Yellow Pages	360
Other	360
Equipment:	
Vehicle	1,800
Other	600
Maintenance:	
Vehicle	300
Equipment	180
Office:	
Postage	60
Forms and paper, etc.	120
	$12,660

The total is $12,660. Gee! Divide this by your 1,165 hours. Now you have $10.87 per hour, added onto your $19.74 per hour. Total, $30.61 per hour. And it still doesn't include the cost of materials. You charge for those separately.

Now, let's figure your daily rate. Multiply this $30.61 per hour by the 5 hours a day you actually work. You have to charge $153.05 per day. Want to go on? Okay, let's look at a few other factors.

More Business Expenses

Expendable goods: brushes, roller covers, thinner, etc.	$ 300
Office equipment: calculator, desk, file cabinets	200
Equipment replacement fund	150
Profit, at 12% of gross earnings	4,260
Educational materials	100
Work clothes and uniforms	100
Repayment of loans on equipment purchases	600
Bank account and checkbook charges	100
Bank charges on extension of customer credit	350
Retirement fund for yourself (IRA)	2,000
Gifts to charity and others, such as police and fire department	100
Memberships in C of C, BBB, local organizations	100
Trade publications	30
Entertainment fund, making customer sales	150
Freight in/out charges on special supplies	50
Fund for accidental breakage of customers property	50
Self insurance reserve fund	500
Reserve for property taxes on inventory	200
	$9,340

This total is $9,340, or $8.02 per each hour of our five-hour day. This now means you have to charge $38.63 per hour, or $193.15 per day, plus materials.

Still more expenses— We still haven't set aside anything for the days you have no customers. Or the days you're rained out. Or the days when the customer gets mad and refuses to pay you. Or the days you do jury duty. Or the days you can't work, for any number of reasons. And we haven't included a fund for equipment losses, or accident losses, or losses from bad customer checks. Maybe you don't really want to go into the painting business after all!

The Formula

Okay, here's that formula I promised you. First, add what you want to earn to your total expenses. Next, divide that by 1,165 to find the hourly rate to charge. Finally, multiply the hourly rate by 5, for your daily rate. (To find the hourly rate for an eight-hour day, divide the daily rate by 8.) This is all for a one-person business.

By now I think you see why so many businesses fail each year. Many people haven't had the benefit of this little lesson. So they don't know how much it will cost. By now, also, you see why so many painters work out of the back seats of their cars, on a cash basis. They're called ''Fly-by-Nights,'' and they make it rough for those who want to operate a real business. The only good thing is that, at one time or another, most decide to operate as a business. And they either learn the rules, or fail quickly.

Hiring Employees

Let's imagine you're at the point where you need to hire a helper. What are you going to pay this person? What will you charge your customer for this person's labor? A business rule-of-thumb is that *for each dollar you pay out, you should collect three.* So, if you pay your helpers $5.00 an hour, charge your customer $15.00 an hour for their time. Figure they'll work an eight-hour day. This adds another $120.00 to your daily $193.15. This gives you a grand total of $313.15 per day, plus materials, that you have to charge your customer.

Frankly, you won't get it. The helper's salary will have to come out of your pocket. Out of your $23,000. But you can't afford that, either. Well, you say, what if I only charge my customer for the actual wages? That's $5.00 an hour, or $40.00 a day. This will work if that helper is a part-time employee. But if you pay him over $1,000 in wages in any three-month period, then by law he becomes a full-time employee, and you're subject to all the rules for employers. This means you have to pay taxes and other benefits set down by the government. It gets involved, but in practice a full-time employee who earns $5.00 an hour actually costs you $8.35 an hour to keep on the payroll. That's $66.80 a day.

You'd do better to hire a person who can work independently and professionally at $8.50 an hour. You can charge for their labor at the full $193.15 per day. This cuts your expenses by roughly 35%. So you can increase your earnings. Or you can decrease your daily rate, thus becoming more competitive.

Why do expenses only decrease 35%, and not 50%? That's easy. To keep this person busy, you have to increase sales. This means more advertising and more of your painting time lost to estimating and instructing them about the job. It also means higher expenses for taxes, insurances, and benefits.

If you hire painters as ''subcontractors'' *and* they have their own business license and contractor's license, you can bypass employee taxes. They have to pay them as ''self-employed'' persons. The problem here is that they'll probably end up going into competition with you.

More employees— Now both you and one employee are going strong. What's next? Add another employee? Not really. When you start adding employees, you start losing your own direct-labor time and billing. You become more of a supervisor, and more of an expense to the company.

In practice, a paint contracting company needs about five full-time employees before it can afford to lose the productive labor of the owner-operator. With three or four employees you'll be putting in 40-hour weeks painting and another 30 to 40 hours a week supervising and managing. Five, or, better yet, six employees will let you work a 40-hour week, keep control, do your estimating and supervising properly, and make money on top of it all. Employees can all work independently or form two-man teams, whichever works best for the people involved.

Your main problem is generating enough business to keep them all working eight hours a day, five days a week. You'll have to do three things: Expand your business area. Expand your advertising. And cut your prices — to somewhere near the level of the ''Fly-by-Nights.'' If it works, and you're so busy you're turning away work, you have two choices. Either raise your rates, or hire more people.

You can probably keep hiring, up to 10 or 11 people. That's when your overhead will once again jump. You'll need a full-time sales estimator and a supervisor. To pay their salaries, you'll need about 15 painters. And so it goes. For each 10 productive employees on the payroll, you can have one non-productive employee such as a supervisor, salesman, estimator or manager.

Following this formula, you can expand as fast as your working capital and customer base permit. In a large city or growing community, you can expand for several years without reaching growth limits.

The Costs of Growth

Now, let's run through a typical situation. Business got good, so you did add those five painters. At $8.50 per hour, per painter, the payroll is $340 per day. That's $1,700 per week. You've reduced the rate to customers to $150 per painter to generate new business. Your income is $150 times five painters, times five days, or $3,750 per week.

You're no longer painting, so part of this pays your salary. At $23,000 a year, you get $442.30 a week. Add this to the weekly payroll of $1,700, for total wages of $2,142.30. The weekly income of $3,750, less weekly payroll of $2,142.30, leaves you with $1,607.70. So far, not bad.

More Costs

Now, when you worked alone, your expenses were $423.07 a week. So, $1,607.70 minus $423.07 leaves you with $1,184.63 profit weekly, right? Well, almost. You forgot a few minor items. Like the fact that you've got to pay for the increased advertising. It was about 8% of gross sales. It should stay at that percentage. Advertising, then, comes to $300 a week. You have to add a new vehicle, new tools, sprayers, and so on.

You must match what your employees pay Uncle Sam for FICA (Social Security). That's about $160. You must pay FUI (unemployment insurance) and DI (disability insurance); about 2% of their salaries. Another $40. You're paying half their medical insurance premiums. That's another $45. Your bookkeepers have a lot more work, and double their rate to you. The insurance agent sees all the extra liability and doubles your premiums. Add another $70. Your phone now has more people using it. Up it goes — add $30. Your expendable items are increased to $40. Since you're a nice person, you've set up a benefit package for your employees. It's five days a year paid vacation, five days a year paid sick leave, and 13 paid holidays. This comes to $30 a week.

Since you're doing more supervision, you should get a 20% increase in salary. (Most managers make 20% more than their highest-paid employee.) That's a raise of $88.46 a week. Now, you're up to $530.76.

You should also increase all of your standby funds. And, most certainly, increase your profits. So add another $100. The weekly total is now $903.46, if I haven't forgotten anything. That $1,184.63 weekly income is now only $281.17. Not much of a margin.

Add to all this the lost time, say 10 days' worth of work per painter, at $150 a day per person. That's $150 times 5, times 10 days, or $7,500. Divide it by 52 weeks. Another $144.23 a week for expenses. Now, your $281.17, less $144.23, is $136.94, about enough for a set of work clothes and a fast-food meal. One more day of work lost and your profits evaporate entirely.

Stay in Control

Those big bucks rolling in the door look great. But unless you have tight control, excellent paperwork and accounting systems, and a good feel for what's going on, you can be in big trouble. And you can be there faster than you'd ever imagine.

Think it can't happen? It happens every day, to large companies and small ones. They grow so fast that management and systems just can't keep up. The last company I worked for lost $1.8 million on sales of $5.5 million. They didn't understand their cost of doing business.

I recommended that you use a CPA and a bookkeeper. But don't expect that they'll keep your business profitable. They'll keep you out of legal trouble with the IRS. They'll give you a summary report of what your business is doing. But you only get this summary monthly or quarterly. You can get in deep trouble between reports. Keep your own records. Learn to analyze them on a daily basis. Learn how to spot potential problems. And learn how to make decisions when they're needed. It's your company, your success, and your future at stake.

chapter three

An Accounting System for Your Business

Before we had income taxes in this country, it was perfectly acceptable for paint contractors to pocket all their receipts — without keeping any record of where the money came from or where it was spent. Unfortunately, that's illegal today. Under both state and federal law, you have to record income as it's received and expenses as they're paid. Any painter who isn't keeping adequate records is skating on thin ice with the Internal Revenue Service and is heading for financial disaster, even if his tax return is never audited.

Without adequate business records, the IRS has the right to estimate your income and assess whatever tax they feel is due. And if that happens, you'll have no way to challenge their estimate! Running a business without records may be appropriate for a drug dealer or an organized crime boss. But don't risk having your home or possessions seized to pay a tax lien. Do enough accounting to meet tax requirements. Along the way, you'll discover all sorts of interesting (and valuable) information these records reveal about your business. And here's another advantage: Down at the bank they read and believe in financial statements. They don't trust loan applicants who don't have them. Do what's suggested in this chapter to keep yourself out of trouble.

This chapter isn't intended to make you a tax and accounting whiz. Leave that to the experts. It's much easier to hire a good accountant than it is to become one. But it's foolish to pay an accountant to handle routine record keeping. That's something you or your clerical help should be doing. This chapter shows how. It also outlines a few of the key tax and accounting principles every painter should understand, and identifies the information your accountant needs to prepare tax returns and financial statements for your company.

Accounting for Time and Dollars

Accounting systems come in many varieties, from simple to complex. The only inflexible rule is that your system must fit your business: It should be as simple and convenient as possible, require a minimum of time, adapt itself to the business rather than making the business conform to the accounting system, and produce records good enough for the IRS and your personal use as manager of the business.

It's nearly impossible to make sound decisions if you don't have information about where your business has been and where it's going. The history of your company is being written every day in numbers — dollars of income and expense, money spent wisely and money spent foolishly. If you don't collect these numbers and understand what's happening, your future in the painting business is pretty unpredictable, but most likely short.

To explain what I mean, let's walk through a beginning painter's record-keeping system for a small job. I hope you don't see any similarities here with the record-keeping system you're using now.

Tuesday morning Jack Slapdash at Slapdash Painting gets a request for an estimate. Mr. Smith wants a figure for painting two bedrooms. Later that afternoon Jack hops in his truck and drives 10 miles to Mr. Smith's home. He introduces himself and starts writing up an estimate. One hour later, after some haggling, he leaves with a handshake. He got the job and will return later in the week to do the work.

Come Friday, Jack drives five miles to the paint store. He buys a roller cover, a plastic drop cloth, and two gallons of latex paint. Then he drives 15 miles to Mr. Smith's. (The paint store was in other direction.) Jack spends 15 minutes talking, then goes to work. Eight hours later, he's finished the job and hands Mr. Smith a bill for $105.

Mr. Smith can't pay because his wife has the checkbook. She won't be home for several hours. Jack leaves, returns the next day, and actually does get paid.

In his income record (a notebook), Jack enters the following:

Smith job:
Labor - 8 hours times $10 per hour - $80
Materials - $25
Total - $105 paid in full
Profit - $80

That note isn't very descriptive, but it's a lot better than nothing. If Jack doesn't staple the receipt for supplies and paint to the back, he may have trouble deducting the $25 from his taxable income. But that isn't the biggest problem. Look at Jack's estimated profit. It's pure fiction! I'll explain why.

Jack's Problems

What's wrong with Jack's evaluation of the Smith job? First, Jack didn't spend eight hours doing the work. True, he painted for eight hours. But he also spent one hour estimating, two hours picking up materials and driving to and from the job, and one hour going back to get paid. That's an additional four hours.

Also, he put 60 miles on his truck. He didn't add any markup on his materials. He didn't separate sales tax on the materials from other costs. He had no contract that set down terms, conditions and the payment date for the job. He didn't even have proof of the price he quoted to Mr. Smith. But Jack figures he made a profit, and he's happy.

Did he really make money from this job? Let's do a little comparing.

The accounting comparison— Here's a comparison of Jack's method of accounting with the way it should have been done:

Jack's accounting		**Our accounting**	
Job total	$105.00	Job total	$105.00
Less	00	Less 60 miles @ 20¢ per mile	- 12.00
Less	00	Less cost of the materials	- 23.50
Less	00	Less sales tax paid	- 1.50
Total earnings	$105.00	Total earnings	$68.00
Income tax @ 14%	(14.70)	Total tax @ 14%	- 9.52
FICA @ 7.8%	(8.19)	FICA @ 7.8%	- 5.30
Total net earnings	$ 82.11	Total net earnings	$ 53.18
Net/8 hours = $10.26 per hour		Net/12 hours = $4.43 per hour	

And we haven't even considered any other business expenses. Or any overhead. They'll reduce the "profit" even more.

Jack's other expenses— But, you say, Jack's system works! It shows a fair hourly wage. Yes, it does. But that figure, as you can see, is false. His actual earnings are the $4.43 an hour we came up with. Jack also paid Uncle Sam some $8.07 too much in taxes. If he does this on every job, he'll pay about $900 extra for the year.

Jack has fooled himself into thinking he's got a successful business going. In Chapter 2, you saw that expenses can run as much as $8 to $10 an hour above salary. Jack's accounting method shows he made $10.26 an hour. If we deduct the $8 to $10 figure, he earned somewhere between 26 cents and $2.26 an hour. But our figures show that Jack's actual earnings were $4.43 an hour. Deducting the $8 to $10 amount now leaves him somewhere between $3.57 and $5.57 in the hole for every hour he worked!

With Jack's accounting and record-keeping system, he'll only be in business a few months. Now, what if he does away with all the insurance, advertising, and other expenses that increased hourly labor costs between $8 and $10? Well, sooner or later someone's not going to like the color he used and insist on the job being redone — for free, of course. And eventually, someone's going to sue him. He'll lose not only his business, but probably also his car, home, and credit. Even if he doesn't end up in court, without advertising he'll be out of work for days at a time. This will lower his earnings — which are already close to minimum wage. Then again, he could fall off a ladder. With no insurance, he'll have no income.

The Fly-by-Night Painter

At this point, Jack's going to pop into our conversation with, "I don't care what I earn per hour. It's money and it's paying the bills. My truck is old, and, after all, I'd have to drive it to work anyway. As for taxes, well, if I paid taxes it would cost me more than that $900 extra per year for a bookkeeper. But I don't. I'm careful and I do all my business with cash." Sounds good to me, Jack. But the IRS may not be so enthusiastic about it. To put it bluntly, you're breaking the law. That's the long and short of it.

This example is typical of the "Fly-by-Nights," those bargain painters who always underbid the pros. And here's the rest of it: if Jack's like most of the other scofflaws, his painting skills are minimal. Mr. Smith thinks he got a good deal until next month when the paint starts peeling off the walls. This is when our Mr. Smith becomes a smart consumer. Jack's now out of business, and you and I have to tell Mr. Smith that he has to pay for more than a new paint job. He also has to pay for extensive prep work to correct Jack's job. All dear old Jack succeeded at was causing problems for Mr. Smith, other reputable professional painters, and himself. Such a success you don't need.

Accounting to the IRS

If you keep no other records, at least keep records you can use at tax time. If you're in business, the IRS wants to see records. And you can use these same records to — legitimately — reduce Uncle Sam's tax bite. The next section has information to help you keep it all straight.

But first, note that this book was written as 1985 tax returns were being filed. Tax laws change every day. What you read here may have changed. It's *your* responsibility to find out exactly which deductions are legitimate. Also, state income tax deductions vary from the federal deductions. The deductions on your federal return aren't necessarily the ones you can use on your state return.

Vehicle Costs

Let's look first at the cost of a critical part of your business: your vehicle. It can be a car, a van, or a truck. It can be bank-owned, paid for, or rented. It can be old or new, personal or company-registered. At some time in the future, you'll have to replace it. It doesn't much matter, it's still an expense. And that makes it a deductible item.

You need to determine both the true cost and the cost per mile for this vehicle. We'll work through an example so you see how to do it. Here's the information you need:

- Interest charges on the loan
- The total monthly payments
- The operating cost
- Actual cost per mile to operate
- The IRS depreciation allowance

Figuring It Out

Say you buy a new van. The purchase price is $9,500. You put $2,500 down and pay off the balance at 14% over the next 36 months. The dealer estimates that the resale value after 36 months will be $5,500. You'll use the van five days a week, 52 weeks a year, with average daily mileage of 50 miles. Gas is currently selling at $1.40 a gallon. The van's estimated MPG (Miles Per Gallon) is 15. Insurance costs $360 a year.

Interest charges— From the lender's amortization table, you find that each $100 borrowed at 14% costs $23.04 per month. Then, figure out the total interest this way:

Interest Charges	
Sale price	$9,500
Less down payment	- 2,500
Loan	$7,000

So $7,000/100 = 70 and 70 x $23.04 = $1,612.80

So, the total interest you'll pay on the loan is $1,612. If you divide this by three (years), you'll have the interest charge per year ($537.33).

The total interest in the example is correct for the three-year period. This method, however, of taking the total interest and dividing by three years, is not technically accurate. You actually pay interest on the declining balance of the loan. A lot of interest the first year and less interest the following years. I suggest you get an amortization chart from your dealer or bank when you take out the loan.

Total monthly payment— The loan principal amount is $7,000. Add to it the interest of $1,612. The total you'll pay over 36 months is $8,612. Divide by 36 (months) for a monthly payment of $239.22.

Insurance cost— We said that insurance for one year is $360. For three years it'll be $360 times 3, or $1,080.

Operating cost— How many total miles will you put on this van? You'll use it 52 weeks a year, times 5 days a week, times 50 miles per day, times 3 years. You'll probably put 39,000 miles on it.

How much gas will you use? If the van gets 15 MPG, divide 39,000 miles by 15. That's 2,600 gallons of gas. How much will it cost? Multiply 2,600 gallons by $1.40 per gallon. That's $3,640 for gasoline over the three years.

The van will need preventive maintenance over the years. An oil change and lube should be done every 5,000 miles. So, 39,000 miles, divided by 5,000 — that's 7.8. Eight oil changes, at an average cost of $15 each, will come to $120. It should have a minor tune-up every 20,000 miles, with new plugs, air filter, and so on. One tune-up in the 39,000 miles, at a cost of about $55. And you'd better figure on needing about $5 worth of parts each month. So, 36 months, times $5. That's $180.

And tires do wear out. The factory ones should last 30,000 miles, so you'll need one set in the three years. At $75 each, that's $300.

It's a business vehicle, so you'd better keep it looking clean, with a wash and cleaning every two weeks. So figuring four weeks to a month: 36 months times 2, equals 72 wash-and-wax jobs. At $5 each, that's $360.

Here's how much it will cost to operate this van for three years:

Cost to Operate	
Gasoline	$3,640
Maintenance	120
Tune-up	55
Miscellaneous parts	180
Tires	300
Wash and wax	360
Comes to a total of:	$4,655

Total vehicle cost— And here's the total cost of the van:

Total Cost of Vehicle	
Purchase price	$9,500
Interest paid	1,612
Insurance	1,080
Cost to operate	4,655
	$16,847

To find the cost per mile, divide this $16,847 by 39,000 miles (what you figured to put on it in three years). It comes to 43 cents per mile. Forty-three cents per mile! Yep, expensive, isn't it? If you add in the depreciation reserve account (explained in the next section,) you bring that cost up even more. Now you see why you must account for mileage, and bill your customer accordingly.

Depreciation

Buying a van isn't like buying sandpaper or brushes. You deduct the cost of most supplies and materials right away — in the same year the supplies or materials are bought. Deductions from income for expensive equipment like a truck have to be spread over several years. This spreading of deductions over several years is called *depreciation*.

Depreciation is a complex subject. Some accountants make a good living handling depreciation questions and little else. I'll only outline the basics here. If you need more information, buy one of the popular federal income tax preparation manuals you see sold in drug stores and supermarkets in February and March each year. These manuals cover the subject in more depth and will include recent changes in the law.

But unless you enjoy tax return preparation and take pride in doing your own return, it's probably better to leave depreciation questions to the expert who prepares your taxes. Here, though, are the basics of depreciation, if you really want to know.

First, you can expense (deduct from taxable income this year) up to $5,000 of the cost of any business equipment, even if you bought the equipment on the last day of the tax year. Any cost over $5,000 must be depreciated over several years. This is a big advantage and simplifies preparing most business returns. But note that you can't claim an investment tax credit on the portion of the cost that's treated as expense. Your accountant may advise you to depreciate rather than expense because a tax credit may be more valuable than the deduction.

Once you've decided to depreciate, the next choice is the method to use. Generally, you want to depreciate most equipment and buildings as quickly as possible. That cuts your taxes as much as possible as soon as possible.

In 1981 Congress adopted the Accelerated Cost Recovery System (ACRS). Under ACRS nearly all business equipment is depreciated over either three or five years. Cars and trucks are three-year property and nearly everything else is five-year property. For three-year property, deduct 25% of the cost of a truck the first year, 38% the next year and 37% the last year. These rules apply whether the equipment was bought new or used and no matter what day of the year the equipment was purchased. For five-year property, deduct 15% the first year, 22% the next year and 21% in each of the following three years.

ACRS ignores the salvage value of equipment. You can deduct the entire purchase price, regardless of what the equipment may be worth when it's sold or junked. But when the equipment is finally sold, you may have to "recapture" part of the depreciation as ordinary income if the depreciated value is less than the sales price. And note that no depreciation is allowed in the year equipment is sold. Also, special rules apply to equipment that's used partly in business and partly for non-business purposes. So if your company owns a car or computer and if you take either home occasionally, your accountant will have to figure the depreciation allowed.

ACRS also applies to real property (buildings). The recovery periods are either 15 or 18 years. Your accountant will set up a depreciation schedule on any buildings you or your company own and use in business.

The ACRS system permits depreciation of equipment and buildings much faster than under depreciation schedules that were used before 1981. But sometimes that isn't desirable. Suppose you set up a new painting business this year, buying a truck and some airless spray equipment. Profits this year may be slim or none. There's no advantage to faster depreciation. In fact, depreciation may be wasted for a year or two until profits start to roll in. It would be better to spread out depreciation over a longer term. Fortunately, that's exactly what the law allows. You can still use straight-line depreciation and select a term much longer than three or five years: up to 12 years for three-year property and 25 years for five-year property.

Suppose you elect to depreciate a $1,500 airless spray rig over 12 years, using straight-line depreciation. What's your annual depreciation deduction? Easy. $1,500 divided by 12 equals $125 per year. Straight-line depreciation deducts the same percentage of cost each year for the depreciable term.

Which Method Should You Use?

I have a suggestion. The first few years you're in business are usually the toughest. You'll have low earnings and high expenses. High depreciation deductions may be wasted. Use the straight-line method.

The only real caution I have is this: Once you begin to depreciate a piece of equipment, you must use the same method each year for that equipment. You can't use straight-line one year, ACRS next year, and so on. But you *can* use different methods on different pieces of equipment during the same year, providing, of course, you keep separate records on each one.

What Items to Depreciate

Depreciation can only be taken on equipment, vehicles and buildings used in your business or other income-producing activity. Don't try to depreciate anything used for other than business purposes or anything that's like inventory being held for sale.

You can't claim depreciation on low-value items, or items with short life-spans. This includes brushes, rollers, and other items with an initial cost of under $100. Deduct these in the year you buy them, as "Business Expense," or "Cost of Materials."

What are these deductions? That's what we'll cover in the next few pages. We'll start with Business Expenses, then cover Cost of Materials.

Business Expenses

The best way to explain these expenses is with an example. I'll show you a year's records for a typical small painting company, starting with interest expense.

Interest Expense

When we were discussing depreciation, I calculated the interest paid on the loan. This is deductible. The interest paid on every loan, and on every charge account that's company-owned, is deductible. You should keep records like this:

Interest Paid Out

Item	Purchase date	Cost	Interest paid
Vehicle No. 1 S/N 42643	2- 1-85	$9,500.00	$537.33 (1/3 of $1,612)
Airless No. 1 S/N 7645	4-15-85	$1,500.00	$ 86.40
Bank loan no. 1056004	3- 4-85	$1,000.00	$ 75.00
MasterCard no. 4792	Revolving acct	Varies	$150.00
	Total interest paid		$848.73

Sales Tax Paid

Another deductible expense is the sales tax you pay on purchases. The van, for instance, was purchased in a state with a 6.5% state sales tax. That's $9,500 times 6.5% (0.065) or $617.50. That's a nice tax deduction.

Federal Excise Tax (FET), however, is not deductible. FET is charged on vehicles and tires. A portion of your registration fee is deductible in some states. Check with your Department of Motor Vehicles.

Record the sales tax on every purchase. Here's the way to do it:

Sales Tax Paid Out

Item	Purchase date	Cost	Sales tax paid
Vehicle no. 1	2- 1-85	$9,500.00	$617.50
Airless no. 1	4-15-85	$1,500.00	$ 95.00
	Total tax paid		$712.50

If, however, you have a resale license and buy for resale, you pay no sales tax. Obviously, you can't deduct sales tax you didn't pay. If you buy an item through the mail from a company that doesn't have an office or salesmen in your state, no sales tax will be charged. There may be a "use tax" due, but no sales tax is payable.

Freight Charges

Shipping costs may be added to the cost of materials and equipment you buy. Any time a shipping expense is shown separately on a bill, you have a charge called *Freight In. FRT/IN* is a deductible item. Any handling charge is not.

And if you're shipping materials, you'll pay what's called *Freight Out*, or *FRT/OUT*. This, too, is a deductible expense. United Parcel Service (UPS), air freight, private trucking, all fall into this freight in, freight out classification. But the cost of your carrying a bucket of paint to your job site doesn't qualify. It might be different if you're delivering paint to another contractor and he's being charged for the paint.

Here's the record:

Freight In/Out

Item	Date	Freight/In	Freight/out
Vehicle no. 1	2- 1-85	$187.50	--
Airless no. 1	4-15-85	$ 22.50	--
	Total	$210.00	--

Credit Company Charges

I've got a question: Do you now, or do you plan to, extend credit to your customers? You can do so via MasterCard, Visa, or a credit company like General Electric Credit Corporation (GECC). On a large job, offering credit can mean the difference between getting the work and losing it to a competitor.

Credit companies, of course, charge your customers monthly interest on the unpaid balance. What many people don't know is that credit companies also charge the assignor of the note a flat-rate charge. The assignor is you, if you sell the note to your customers. The rate is negotiable, between you and the credit company. It runs from as low as 1.5% to as high as 7%, depending on your volume and the average account size. The larger the average billing, the smaller the rate. This charge is a cost of doing business and is deductible.

Keep records like this:

Credit Extended

Customer	Job number	Total $	Credit company	Rate	Amount
Smith	25402	$400.00	MC	2%	$ 8.00
Jones	25403	$550.00	Visa	4%	22.00
Jacks	25404	$180.00	GECC	5%	9.00
			Total credit charges		$39.00

Bank Charges

The bank where you have your company checking account will generally have a monthly service charge. Usually it's either several dollars a month or from 10 to 25 cents per check. That can add up fast. Either record this expense monthly or find the accumulated total at year end. The bank should give you a year-end statement that shows all service charges. These service charges are also a cost of doing business and are therefore deductible.

Interest on accounts— Speaking of banks, some checking accounts and almost all savings accounts pay interest. This interest must be recorded as "Interest Income." This category also includes interest earned on any loans or credit that you extend to your customers, if you carry the loan yourself and charge interest.

In your records, note the bank, the account number, the kind of account, and the interest you earned that year.

Advertising Cost

I've mentioned advertising several times. Advertising will be a major cost of doing business. Budget at least 4% of your gross receipts (that's income, before taking out any deductions). A growing paint contractor will spend closer to 8% of gross on advertising. No matter how much you spend on advertising, record the purchase date, date paid, check number, and amount. An advertising record might look like this:

Advertising Cost

Item	Purchase date	Date paid	Check no.	Amount
Truck sign	4-12-85	4-12-85	0121	$148.00
Daily News Classified	4- 1-85	4-15-85	0135	25.00
Business cards	4- 3-85	4-10-85	0118	25.00
		Total advertising		$198.00

Advertising includes anything that promotes your company's name, image, and business. It includes business cards, signs, posters, flyers, TV spots, radio spots, newspaper ads, calendars, pens, pencils, or caps or T-shirts imprinted with your company name.

Advertising also includes the cost of materials used to create advertising, the labor involved in creating the advertising, and any travel involved in creating or delivering the advertising. If you hire an agency, their fee is deductible as advertising. Direct customer mailings of specials, notices, flyers, are considered advertising.

Separating items for budget control— You can lump all these items together, or separate them out. Separating them helps you to analyze which advertising method brought in the most jobs. Then you'll know which method of advertising to do again. You want to get the most for your advertising dollar. We'll talk more about this kind of analysis in the next chapter. As an example, direct mailings can be split like this:

Mailings

Type	Quantity	Office supply account	Postage account	Advertising account
Advertising	500	--	$125.00	$25.00
Billing	50	$5.00	12.50	--
Special	300	--	75.00	20.00
Totals	850	$5.00	$212.50	$45.00

This gives you some control over where each dollar is spent. If you lump them together, you've recorded enough for tax purposes, but can't evaluate what money was well-spent and which promotions were money wasted.

Office Supplies

Your office supplies include pens, pencils, paper clips, forms, typewriter ribbon, and so on.

Office Supplies

Type	Quantity	Date purchased	From	Cost	Sales tax paid
Invoices	500 bulk	4-1-85	ABC Office	$45.00	$2.70
Pencils	1 box of 12	4-2-85	ABC Office	1.50	.09
Blank forms	500 bulk	4-4-85	Able Printing	30.00	1.80
			Totals	$76.50	$4.59

Office Equipment

Office supplies are expendable items, necessary to

all business, and therefore deductible. But remember, this doesn't include office *equipment.* Office equipment is an asset, and if it costs over $100, it must be depreciated using one of the methods we covered. Deduct any equipment that costs under $100 as a cost of doing business in the year it was purchased. *Caution!* The IRS can get sensitive about this. They consider low-cost office equipment an asset which must be depreciated over years, not expensed in the year purchased.

In the example below, I show depreciation on two items. They're expensive enough to be considered capital assets and to have depreciation schedules. The stapler and calculator would normally be expensed. But this is a new painting company and the accountant recommended that these start-up costs be capitalized and depreciated over several years.

All these items should be shown on your equipment inventory list.

Office Equipment

Item	Purchased from	Date	Cost	Deduct	Sales tax
Desk	ABC Office	4- 1-85	$ 450.00	Depreciation	$ 27.00
Calculator	ABC Office	4- 1-85	25.00	--	1.50
Computer	Info Electric	4-15-85	3,000.00	Depreciation	180.00
Stapler	ABC Office	4-18-85	12.00	--	.72
	Office equipment totals		$3,487.00		$209.22

Equipment Repairs

The IRS *does* allow an immediate deduction for equipment repair or service. If the calculator you bought for $25 breaks, the repair cost is probably deductible, even if the repair cost more than the calculator itself. This goes for most of your non-disposable tools and equipment — calculators, wall clocks, sanders, drills, paper punches, staplers, power saws, and so on. But don't try this on your paint brushes and roller frames. Those are disposables.

Equipment Repairs

Item	Repair date	From	Cost/Deduction	Sales tax
Stapler	8-20-85	ABC Office	$12.00	$ .72
Sander, disc	9- 1-85	Jay's shop	68.00	4.08
		Totals	$80.00	$4.80

Educational Materials

Schooling — yours and your employees' — is deductible, if it's required for the job and if you pay the tuition fees. Classes in auto repair and butterfly collecting don't qualify, of course. But classes on accounting, business management, and paint contracting will be deductible.

The cost of educational material, trade magazines and trade memberships are deductible items if they're required for your business success. The price of this book is deductible. So are union dues. Deduct for yourself and anything you contribute to your employees' dues.

Educational and Memberships

Item	Date	Cost	Sales tax
BBB (Better Business Bureau)	4- 1-85	$100.00	--
CC (Chamber of Commerce)	4- 1-85	100.00	--
Painter Weekly (magazine)	4-15-85	25.00 yr	$1.50
Contractor Guide Book	4- 1-85	29.95	1.79
	Totals	$254.95	$3.29

Equipment Operating Expenses

As I mentioned earlier, operating expenses — or better yet, "Operating and Maintenance" expenses — on vehicles and equipment, are an ordinary and necessary cost of doing business and are therefore deductible.

Operating and Maintenance Expense

Item	Date	Invoice no.	From	Cost	Sales tax
Truck transmission	9-20-85	2304	AJAX Trans.	$455.00	$17.30
Airless, overhaul	11-11-85	10102	Bills Mach.	155.00	9.30
Car wash and wax	Semi-weekly	--	--	60.00	--
Oil and lube	6-15-85	A-452	Johns Serv.	15.00	.40
		Totals		$685.00	$27.00

Expendable Expenses

Shop supplies (expendables like soap and paper towels) are deductible as a business expense.

Shop Expendables Expense

Item	Date	From	Quantity	Cost	Sales tax
Hand soap	4-1-85	Joe's Market	2	$1.25	$ .07
Paper towels	4-1-85	Joe's Market	2	2.50	.15
Brush, paint 1"	4-1-85	Bill's Paint	10	10.00	.60
Tarp, plastic 9 x 12	4-1-85	Bill's Paint	10	6.50	.39
		Totals		$20.25	$1.21

Note that we paid sales tax on these items. If we planned to list the brushes and tarps on the customer's invoice as part of the job and then collect sales tax from the customer, we wouldn't have paid this tax. The soap and towels, though, are strictly for in-house use. Therefore we must pay the sales tax on them.

Protective Clothing

Uniforms and special protective clothing, including hard hats and work shoes, are deductible only if they're used exclusively at work. Jeans and a T-shirt aren't deductible because they're also appropriate for casual use off the job. Painters' whites are a uniform of the trade and should be deductible. Have your company name embroidered or silkscreened on the whites or smocks. If disallowed as a clothing expense, you may be able to deduct them as an advertising expense. Work shoes should be just that — not an old pair of sneakers.

If you supply your employees with work clothing, that's deductible too. Dust masks, respirators, rubber gloves, safety glasses, smocks, and ear protectors will be deductible items.

Uniforms and Protective Clothing Expense

Item	Date	From	Quantity	Each	Extension	Sales tax
Uniform, white	4-1-85	Painter Sup.	5	$22.00	$110.00	$6.60
Respirator	4-1-85	Painter Sup.	2	35.00	70.00	4.20
Dust mask	4-1-85	Painter Sup.	50	1.00	50.00	3.00
			Totals		$230.00	$13.80

Equipment Rentals

If you rent some or all of your major equipment, the rental fee is a deductible cost of doing business. Be sure to separate out the sales tax for end-of-year accounting.

Equipment Rental Expense

Item	Date	From	Check no.	Amount	Sales tax
Road striper	4-18-85	J's Rents	0128	$ 55.00	$3.30
Hopper unit	5-25-85	J's Rents	0142	110.00	6.60
			Totals	$165.00	$9.90

Renting equipment for a job is known as a "Pass Through" expense. List the rental charge on the customer's job invoice. The customer pays for the rental.

Adding markup— If you add a markup (M/U) to the rental cost, this charge is income. You must report it as such. Markup should cover your cost of renting the equipment: mileage and travel time. Be sure to keep documentation of each rental expense so they can all be deducted from your income.

Long-term rentals — equipment, vehicles, and buildings — are also deductible. Equipment leased for several months, like an airless unit, will be used on several jobs. Don't waste time breaking out the rental cost for each job. Instead, the rent should become part of your overhead and be included in your hourly billing rate.

Contributions

Most contributions and gifts will be deductible. But be careful on gifts. If you want to give a bottle of whiskey to a good, steady customer, be sure to put your company name on the bottle. That helps identify it as an advertising expense.

You can make gifts to most legitimate charities and deduct the amount of the gift in the year made. But if you make a gift to a son or daughter, there may be a gift tax due. Check with your accountant or CPA before making any substantial gifts to relatives or friends.

You can give away old equipment and supplies no longer needed by your company if they have no salvage value. And of course, you can give it away to a charity even if it *has* a salvage value. For example, if a spray outfit cost $250 new and has a salvage value after depreciation of $25, the $25 is deductible as a gift.

But, as with all your business transactions, be sure to get a receipt so there's no question that the gift was made.

Cost of Materials

The cost of paint, colorants, roller covers, sandpaper, and so on, is deductible as a cost of materials when used on a job. But any materials held in stock awaiting use will be inventory. This inventory is considered an asset and is not deductible until actually used.

Most small painting contractors should keep as little inventory as possible. Buy for each job individually and return anything unopened and unused when the job is complete. As you grow and become more competitive, buying in bulk quantities may be justified. But that means you'll need some sort of inventory control system. We'll talk about that in the next chapter. For now, this is the kind of record you should keep for your cost of materials:

Cost of Materials Used

Job no.	Item- description	Quantity	Cost	Extension	Sold	M/U
10001	Paint, latex, special Y2D1	1 gal.	$12.50	$12.50	$ 16.50	$ 4.00
10002	Paint, poly	1 gal.	24.00	24.00	32.00	8.00
10002	Brush, 2"	1 each	3.00	3.00	5.00	2.00
10002	Tarp, plastic 9 x 12	1 each	2.50	2.50	4.00	1.50
10003	Paint, latex, eggshell white	5 gal.	8.50	42.50	65.00	22.50
		Totals		$84.50	$122.50	$38.00

If you paid sales tax on any of these items, have a column marked "Sales Tax Paid." Your invoice to your customer need not show cost of materials or sales tax, but only the total of all materials, tax, and labor.

If you used your resale license to buy materials for resale, have a column marked "Sales Tax Collected." In this case, your invoice to your customer must itemize the cost of materials and sales tax. My invoices usually show the cost of materials as a separate item, unless materials are a very small part of the job. Showing materials separately makes customers feel you're being completely honest with them.

Insurance

Insurance is a necessary evil — until a tragedy happens. Then it's a blessing. Doing business without adequate insurance is both foolish and irresponsible. It only takes one accident by you, a customer, an employee, or even some stranger, to put you out of business. Don't let one break-in, one pilot light you forgot to turn off, or one second of carelessness wipe out years of hard work. Carry enough insurance to cover the risks most painters face.

In the last chapter I mentioned a company I bought worth $120,000. Remember? Well, I lost the company because of a robbery. Nearly all my equipment and supplies were stolen over a long holiday weekend. Sure, I was insured, and all customers and creditors were paid off by my insurance carrier. But I didn't have loss-of-income coverage. I had a company with no inventory and no equipment! I couldn't do business on that basis. The insurance company was in no hurry to pay off the claim. So I sat in an empty building with nothing to sell.

Loss-of-Income Insurance

I strongly suggest that your insurance coverage include a loss-of-income clause. Fall off a ladder and break a leg, and you'll be out of work for many weeks: weeks you can't spend painting; weeks the bills keep rolling in. Medical insurance will pay most medical bills. But if your company folds because no one is there to run it, the economic loss will be bigger still.

Insurance Records

A sample record is shown below. Check with your insurance agent or accountant about IRS deductions for health and life premiums. There's a fine line between what's deductible as a business expense and what's considered a personal expense. This is especially true if you're a sole-proprietorship or a general partnership. In a limited partnership or a corporation, insurance will usually be fully deductible. Vehicle and business insurance premiums, though, are usually deductible for all types of business set-ups.

Insurance File

Policy no.	Company	Type coverage	Cost	Deductible
1354615	Allied	Life, term	$276.00	No
5432165	Allied	Vehicle no. 1	360.00	Yes
846392	Insurco	Business	360.00	Yes
9354616	Allied	Health	840.00	Part
	Total premium		$1,836.00	

Deductible portions of premiums $1,320.00

Other Deductions

It's common for the IRS to question the use of your personal vehicle for business. And the IRS will nearly always scrutinize any deduction for an office in a home — even if it's your only office. If you keep a log of business use of the car and home and pay for all personal use, there won't be any problem. But if you have no records, the IRS will probably disallow any deduction.

Personal Vehicle Use

A personal vehicle, used to go to and from work, is not a deductible item. When used to go from office to job site and back to the office, it becomes a deductible business expense. If your office is your home, then going to the job site from home and back may be business use. So it becomes deductible — but only during normal working hours. Going to the beach or to the movies returns the vehicle to personal status. If you're audited, use of the car will be one of the first items challenged. Record every business trip by date, reason and mileage.

Your Home Office

As I said earlier, your office at home is deductible if it's the place where you usually meet clients and customers, or if it's your principal place of business. If your home meets either of these tests, both the cost of owning and maintaining the part of the home used as an office will be deductible. If your home has 1,400 square feet of living space, and your office is 8 x 10, or 80 square feet, you'd divide 80 by 1,400, to get 0.057. That means 5.7% of your home expenses can be deducted. Therefore:

Home Office Expense (Monthly)

Item	Total cost	x 5.7%	=	Deduction
Electric	$ 58.00	x 5.7%	=	$ 3.31
Gas heat	125.00	x 5.7%	=	7.13
Property taxes	100.00	x 5.7%	=	5.70
Cleaning	25.00	x 5.7%	=	1.43
		Total		$17.57

If your home is rented, deduct 5.7% of your rent. If it's owned, depreciate the portion that you use. If you repaint or make repairs to the office, these are fully-deductible expenses. Note that I said repairs, not improvements. Improvements are not deductible as a business expense — only repair and maintenance of what already exists. That's another IRS fine point.

Telephone Expenses

Notice that I haven't discussed the telephone yet. As I said in the previous chapter, get a separate business line for your business. Using your personal phone for business makes it hard to deduct the business portion of phone bills and makes it harder to claim an office-in-the-home deduction.

Here again, you'll find an FET on the billing. It's not deductible; only the state sales tax is. And put the Yellow Pages part of the billing in your advertising budget, not your phone budget.

When you keep all these records, you'll simplify your life if you have columns for date paid and check number. It makes checking back on your records a lot easier.

Phone Expense

Monthly billing date	Monthly billing	Service cost	Yellow pages	Sales tax
1-21-85	$ 38.73	$13.00	$25.00	$ .73
2-21-85	55.04	29.00	25.00	1.04
3-21-85	42.20	16.50	25.00	.70
Totals	$135.97	$58.50	$75.00	$2.47

We've covered most of the expenses, and you know how to record them. Now, we can consolidate the sales tax:

Consolidated Sales Tax Paid

Sales tax paid out	$712.50
Office supplies account	4.59
Office equipment account	209.22
Equipment repairs account	4.80
Education and membership account	3.29
Operating and maintenance account	27.00
Shop expendables	1.21
Uniform and protective clothing	13.80
Equipment rental account	9.90
Cost of materials account: Include, if you paid the tax	
Phone cost account	2.47
Total sales tax paid	$988.78

Taxable Income

What do all these deductions do for you? Let's say your gross receipts for the year are $46,000. With no deductions, you'll pay about 25% of this amount as Federal Income Tax. That's $10,750 in taxes.

Take Business Deductions

Business deductions will decrease gross earnings and reduce the tax. If your deductions total $16,000, taxable income is $46,000 less $16,000, or $30,000. Your tax will be about 20% of $30,000 or $6,000. That's $4,700 in *your* pocket!

Open a Retirement Account

Since you're not working for someone else, there's no pension plan for your retirement. You have to take care of that yourself. So, if you can possibly afford it, put $2,000 in a tax-deferred savings or retirement account (IRA) at your local bank. This lowers your taxable income by $2,000 and may lower your tax bracket, too. But once again, these laws are changing. Check with the IRS or your accountant before you take any action. Only they know the *current* regulations.

As a one-person business, you'll spend about 250 hours a year on paperwork and accounting. By keeping good records, you just dropped your income tax by $4,700. Divide this by the 250 hours you spent on paperwork: You "earned" $18.80 an hour for doing paperwork! Not bad, considering you already got paid for this time in your hourly billing rate charged to your customer. You see, good paperwork *does* pay.

The IRS Audit

All this paperwork pays off even more when you're audited by the IRS. The audit will usually come in the second year of business. They usually leave new businesses alone until year two. Then they drop the boom. If the audit turns up a deficiency, you're subject to back taxes, fines, and future audits. If you have correct and accurate records, you probably won't see the IRS again for several years.

The examples I've shown in this chapter are simple. They're easy to follow, will keep you out of trouble and will help you make decisions. The next step is yours. You have to do what I've outlined.

Paying Taxes and Getting Bank Loans

Everyone in business gets audited occasionally. Auditing keeps small business honest. But even if there were no income taxes, it would be wise to keep good records and know where the money is coming from and going to.

It's tempting to simply pocket the cash in this type of business. It's also easy to declare almost everything you buy as a business expense. But that's no way to run a legitimate business. You want to show profits and pay taxes. That gives credibility to your company and makes you a good prospect for an expansion loan. Banks would rather dance with the devil than lend money to a tax cheater.

Banks usually require two to three years of past IRS 1040's. They use these to verify your income. Show up with all deductions and no earnings and you can forget about the loan. The bank doesn't care that you *really* made $30,000 last year. Companies with no assets and microscopic taxable income don't get much notice from loan officers.

Drawbacks to Self-Employment

Running your own company has lots of advantages. But if you're self-employed, there are some major drawbacks. What happens if you quit the business? The self-employed usually don't have state unemployment or disability coverage. This can hurt if you haven't planned for it. Also, there's no retirement income unless you plan for it. Remember that retirement account I mentioned? I strongly suggest that you sit down with your CPA and make some long range plans to overcome these problems.

Accounting, Inventory, & Management Control

In this chapter we'll go into some of the procedures that I've found to be important in running my paint contracting business. We'll cover budgeting — how to do it and what a good budget can do for you. We'll look at some accounting details — forms, logs, and files. I'll explain what you should know about inventory management. We'll examine some simple but effective ways to control what's spent on materials and labor. I'll describe how charts can help you follow your company's progress. And finally, we'll discuss computerized accounting.

A Typical Company Budget

Do you know what percent of your total income is spent on advertising? Rent? Payroll? Maintenance? Is the percentage spent on advertising today more or less than the percentage spent last year or the year before? Making comparisons like this can be very revealing.

Let's say your company is operating profitably and business volume is growing. You're spending 5% of gross on advertising, 35% on payroll and 10% on rent. Next year revenue should grow by $100,000. How much more can you plan to spend for advertising, payroll and rent? Easy! If all other expense percentages remain the same, you can spend $5,000 more on advertising, $35,000 more on payroll and $10,000 more on rent.

As your sales rise, the dollars spent on advertising, rent, payroll, etc. can rise proportionally without changing the percentage spent on each category. Conversely, when sales are dropping, cutting back proportionately in each category will keep you profitable. That's not an easy trick in a down market. But some painters do it.

My point is that once you know the correct percentage to allocate to each category, it's easy to draw up a budget for the next week, month or year. These percentages become a powerful tool — helping you plan for expansion as sales rise and keeping you out of financial trouble as sales decline.

Let's draw up a proposed budget so you can see how it works. If your company is just starting out, you don't have much history to rely on. Your best guess will be the only guidance available. Use the percentages shown below unless you know that they're going to be wrong for you.

If your company has been in business for several years, go back over your check stubs (remember that business checking account?) to see what percentage of gross sales has been spent on each expense category.

Now, let's make the budget. Suppose you come up with the following percentages.

Proposed Budget	
Salary	34 %*
Materials	22 %*

Advertising	8 %	
Rent	12 %	
Postage	.5%	
Utilities	1.5%	
Taxes	12 %	*Labor to material ratio is approximately 60:40, which is close to standard.
Vehicles	7 %	
Office	2 %	
Insurance	1 %	
Total	100 %	

Okay, now what? Now we have to find the actual dollar amounts. Earlier, we said that you want to earn $23,000 per year.

How Much Gross Income?

First, we have to know how much money you need to bring in. It must cover all expenses and still give you a minimum salary of $23,000 a year. That salary is 34% of your gross revenue. What's your company's gross? Here comes the math: $23,000 is 34% of X. To find X, divide 23000 by 0.34, for an answer of 67647.50. Let's round it to $68,000. That's what you must bring in per year to clear $23,000 for yourself.

The Operating Expense Budget

Next we can figure out the dollar amounts that go with those percentages:

Dollar Amounts, Proposed Budget

Materials	Advertising	Rent	Postage	Utilities
68,000	68,000	68,000	68,000	68,000
x .22	x .08	x .12	x .005	x. 015
$14,960	$5,440	$8,160	$340	$1,020

Taxes	Vehicles	Office	Insurance
68,000	68,000	68,000	68,000
x .12	x .07	x .02	x. 01
$8,160	$4,760	$1,320	$680

That's your operating expense budget. But will it work? Did you spot the problems? First, the insurance budget is much too low. In the last chapter, we said you needed about $1,836. That's three times the $680 you have here. So let's change that 1% to 3%. Now, it's $68,000 times 0.03, or $2,040. That's more than enough to cover you.

But now our figures come to 102% of the gross! We can't have that. The total must be 100%. So we have to borrow 2% from somewhere else. Let's look further.

The biggest mistake is in the taxes. That 12% tax is on your gross income of $68,000, not on your net income. (You're taxed on net income: what's left over after you deduct expenses.) Net income is your salary, or $23,000. Let's re-figure the tax: $23,000 times 0.12 is $2,760. That's a difference of $5,400.

Now what percent are you paying for taxes? Divide 2,760 by 68,000, for 0.04, or 4%. This lets us change our tax budget from 12% to 4%, leaving 8% to use elsewhere. We already used 2% of this 8% for the increase in insurance. We have 6% left. All the rest of the figures look pretty good. You should be able to live with them. That 6%? Well there's one item we didn't budget. And it's an important item for success:

Retained earnings and profit— You're in business to make a profit. Retained earnings are used to buy replacement equipment, or for expansion, or to get through hard times. For a revised budget, we have:

Revised Dollar Amounts

Taxes	Insurance	Retained Earnings and Profit
68,000	68,000	68,000
x .04	x .03	x .06
$2,720	$2,040	$4,080

Accounting Errors

All these budget figures should add up to a total of $68,000. They don't. Let's recheck all the figures. Okay, we found two more mistakes. First, the office budget should be $1,360, not $1,320. Second, we used $23,000 as salary and figured it would be 34% of the gross. Remember when we rounded up the $67,647.50 to $68,000? Well, $68,000 times 0.34 is $23,120. You just got a raise. Now the total is correct.

These mistakes illustrate a point. *Double-check your calculations.* A simple error can cause big problems when it comes to dishing out the dollars.

The Realistic Budget

The next thing to check is this: Is that $68,000 realistic? In an earlier chapter we said you'd work

233 days out of 365. Back then, we deducted 13 holidays, 5 sick days, and 2 weeks of vacation. And you didn't work weekends, either. So $68,000 divided by 233 days is $291.84. You have to take in $291.84 each working day! Unfortunately, this is about $100 a day more than you're likely to earn.

Let's see, then, just how much the difference is. That's 233 working days, times $100 a day; then subtracted from $68,000. This gives a true expected gross income of $44,700.

What to do now? You have several choices. You can cut out some of the time off. Take 9 holidays instead of 13. Take one week of vacation instead of two. This gives you 9 more working days, for a total of 241 days. If you can take in $191.84 a day, this 9 days equals another $1,722.56.

You can work out of the house instead of renting an office. This cuts expenses by $8,160. And, by working out of the home, you decrease the business portion of the utility budget — maybe by as much as $700.

You can buy a two-year-old vehicle instead of a new one. It will still have 36 months of good life on it. And it's already been depreciated by its prior owner, so it costs less. Resale value drops most in the first two years of ownership. Let's say your $9,500 is now only $7,000. That's a $2,500 savings. You also reduce the taxes, interest payments, and license and insurance fees by buying a used vehicle. Another $1,000 saved.

Let's add it all up. $1,722.56 plus $8,160, plus $700, plus $2,500, plus $1,000, gives a total of $14,082.56. Now let's distribute the rent budget of 12%. Move it to labor, materials, and advertising by adding 4% to each. Now we can make up a new budget — one that reflects your true potential earnings of $44,700. Grab a pencil and try it. Do the same exercise I did, but use the new figures.

A One-Man Contracting Business Budget

You've created a pretty good working budget for your new one-man paint contracting business. Percentages are a powerful tool. They help you see what you have to earn. They help you see what you can spend. And they help you see where spending more money will get the best results. There are only two rules to follow:

Rule 1. The total of all percentages used must equal 100.

Rule 2. The total of all dollar figures must equal the estimated gross income.

As I said, grab a pencil and try it.

Math and Formulas

Check your library or bookstore for *Fundamentals of Business Mathematics* by Nelda W. Roueche (Prentice-Hall, $11.95). It's one of the better guides and has nearly all the math you'll need to run any painting business.

Percentage Formulas

Here are some useful "quicky" formulas for percentage:

Quicky Formulas For Percentage

Formula No. 1 A is B % of X

$$X = \frac{A}{B/100}$$

Example No. 1:

50 is 10% of what?

$$X = \frac{50}{10/100}$$

$$X = \frac{50}{.10}$$

$$X = 500$$

Formula No. 2:

X is A% of B

$$X = B \times \left(\frac{A}{100}\right)$$

Example No. 2:

What is 5% of 200?

$$X = 200 \times \left(\frac{5}{100}\right)$$

$$X = 200 \times .05$$

$$X = 10$$

Markup (M/U) Percentages

And here are some more "quicky" formulas. These show two ways to figure how much markup to add to your customer's invoice. As we go through this chapter, you'll see how important markup is. The second method, "M/U Based on Selling Price," is the industry standard. Look at the figures, and you'll see why:

Quicky Formulas For M/U

M/U Based on Cost

Cost + (overhead + profit) = selling price

C + (OH + P) = S

Example:

A gallon of paint costs you $8.00. Your OH is 38% and you want a 6% profit.

C + [(.38 x C) + (.06 x C)] = S

8 + [(.38 x 8) + (.06 x 8)] = S

8 + (3.04 + .48) = S

8 + (3.52) = S

8 + 3.52 = S

Selling price = $11.52 per gallon

M/U Based on Selling Price

SPF x C = S

SPF (selling price factor) is 1/ (100% - M/U%)

1/ (100% - 42%) = SPF

We'll use same % and cost as we did for M/U based on cost for this example.

So 1/ (1 - .42) = SPF

SPF = 1/ (.58)

SPF = 1.72

SPF x C = S

1.72 x 8 = S

Selling price = $13.76 per gallon

Learn to use markup formulas— Notice that we made $2.24 more per gallon by using the industry-standard method. The markup percentage I used is also very close to the markup on most retail items you'll buy.

As an added note, if you want to figure out what something cost, divide the selling price by the SPF. We added 38% and 6% to give us 42% overall gross profit. Therefore 100% (cost) minus 42% equals 58% (.58) and 1/.58 equals 1.72. So 1.72 is the Selling Price Factor. Once you've figured out the SPF, it's easier to compute selling prices that way than by using the formula for markup based on cost.

The SPF for a 33% markup is 1.49. I suggest not using an SPF lower than that. Most newcomers to the business world think that markup equals cost times some percent. It's not, as our example shows. But let's try it: A 42% markup on $8.00 is $3.36. And $8.00 plus $3.36 equals $11.36. Not the same at all.

How much difference can it make? Let's see: You use about 4 gallons of paint each of the 233 working days in a year. That's 932 gallons. Say you're paying an average of $12 a gallon. That's 932 times $12, or $11,184 at your cost. Now, in the "cost times some percent" method, that's $11,184 times 42%, or $4,697.28. That's your gross profit on the paint. With the "M/U based on cost" method, we have $4,920.96 gross profit. With the "M/U based on selling price" method, we have $8,052.48 gross profit. Just a little difference — but enough of a difference to keep you in business another year.

There's a lesson to be learned in this chapter. *You must understand your cost of doing business. And you must keep tabs on those costs.* Paperwork is necessary. Keep accurate records. Study them. Chart them out. Set up budgets and markups, and stick to them.

Handling Paperwork

There was once a time when paperwork could be ignored by many of the smallest paint contracting companies. No more. Try ignoring your tax obligations today and you'll be out of business pretty quick, especially if you have employees.

We're living in a complex commercial society where every business is expected to file monthly, quarterly and annual tax forms with local, state and federal governments. Tax law requires that you keep track of income and expense — receipts, check stubs and invoices — and that you withhold taxes from employees and submit those taxes by the dates due. State law requires that you buy workers' compensation insurance, and good sense dictates that you have liability insurance. Your customers will expect nicely prepared written estimates and your attorney will advise you to use legally binding contract forms. Suppliers will expect that you honor their invoices when due, and your customers will be quicker to honor their obligations to you when billed promptly, efficiently and accurately.

Here's my point: There's no way to escape paperwork. You might as well accept that as a fact of life in the painting business. Naturally, you want to do as little paper-shuffling as possible. But there is some irreducible minimum below which you cannot go. And keep this in mind. The most profitable, the most professional painting companies I know seem to be the best at paperwork. I don't know whether they grew and prospered because they followed good office procedures or whether they learned to manage paperwork effectively because they became profitable. But it's true. Good procedures and good profits seem to go hand-in-hand.

Let me sketch the outline of a low-cost but effective paperwork system that should be adequate to meet your needs.

Forms and Ledgers

Most larger office supply stores sell many types of blank forms. You can buy invoices, receipt books, ledgers, and so on. Most are already numbered. Most have space for your company name. Most come with NCR or carbon for making multiple copies. And most are expensive on a per-sheet basis.

Don't buy forms in booklets. The invoices are pre-numbered, and the numbers often duplicate themselves from book to book. This can lead to accounting problems. And since they don't include your company name, you'll have to buy a rubber stamp and stamp each copy. That doesn't look particularly professional.

Most standard ledgers are general-purpose and won't fit your business very well. Also, it's easy to fill up a ledger. You'll soon find that you have to go searching for information in several different places. So what's the solution?

Quote/invoice forms— Most office supply stores sell blank, unnumbered quote/invoice forms. (I use a Rediform #7S729, but many others are available.) They come in bulk packages of 25, 100, 250, and 500. As you'll see later on, this one form can serve several purposes.

These are loose forms and can be "crash imprinted" by most print shops. Crash imprinting puts your company name, phone, address, advertising, and numbers on all copies of each invoice. Yes, it costs a little more. But buying the forms in bulk helps offset the printing cost. And you now have a high-quality, professional-looking form. You'll be giving this form out to your customers. You want it to reflect your professionalism.

In-house forms— You don't need fancy printed forms for documents that never leave your office. But you *do* need neat, well thought-out office forms that have all the information you need to collect. They say a picture is worth a thousand words. In the next few pages I'll show you why.

Figure 4-1 is a sample of the blank form I use. You can copy or re-draw it for yourself. It serves several purposes:

It's my job sheet. I record each day's jobs on one sheet. At the end of each day, I transfer the daily totals to a monthly sheet. For yearly totals, I just total the monthly sheets.

It also becomes my expense sheet. I use the same sheet for the whole month. Then I transfer the totals to a monthly sheet. Again, I just add up the monthly sheets for the yearly expense totals.

I then have all the information I need to do my P & L (Profit and Loss) statement and my income taxes. It only takes a short time each day to fill in the blanks. But it sure saves a lot of headaches at tax time.

I'll describe all these forms in a minute. But, first, I want to talk about something you may not have considered. Where are you going to put these records?

Where to Keep Your Files

Keeping files in some sort of order can be complex. Or simple. You could buy a file cabinet for $150, file folders for $50, file folder hangers for $15, and put together a nice system. Or, you can use standard 4⅛" by 9½" all-purpose envelopes from most any stationery store.

Use one envelope for each job. Use one for each expense account. A simple plastic, wood or metal tray (about $5) can keep them handy. Remember, even though you transfer information to your log sheet or a computer disk, you need backup paperwork to support each record.

About your log sheets: Make two copies of each one, each day. One copy is for your file folder. Keep the other copy at another location. This is in case of fire. If you lose your records, you lose your business. If you don't want to start keeping copies elsewhere, buy a small fireproof safe. A new one will cost you about $250.

Graphic Arts for Effect

One last note on paperwork: From time to time, you'll need graphic arts materials. You can hire others to lay out your forms, flyers, and other advertising, but it gets very expensive. Learn to do it yourself — you can.

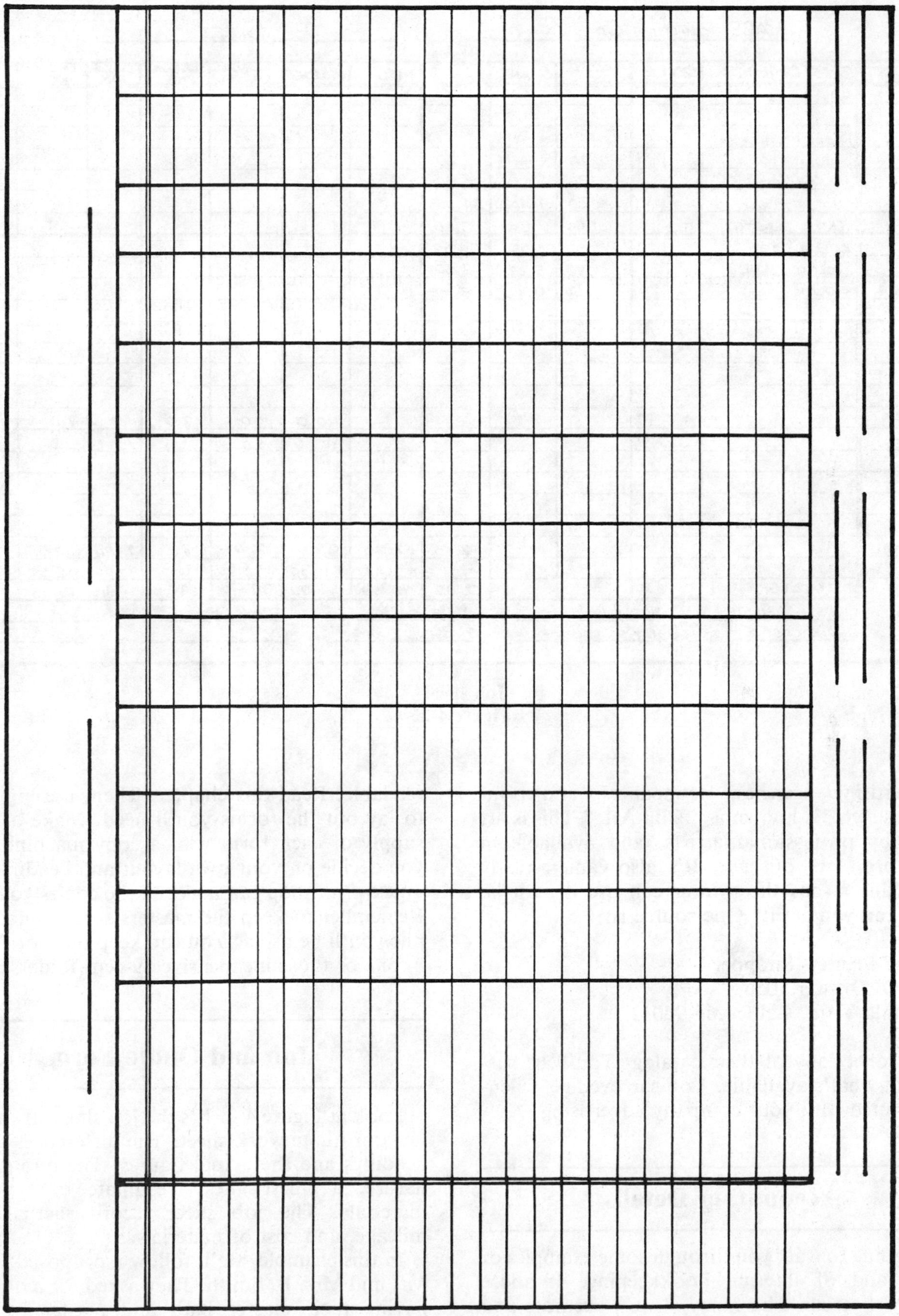

Blank log form
Figure 4-1

ACE PAINTING CO. — JOB SHEET

JOB #	CUSTOMER	DATE	MATER. COST	QUAN	EXT. COST	M/U	MILEAGE	LABOR TRAVEL	LABOR JOB	SALES TAX	LABOR COST	
10001	SMITH, B	4-1-87										
P/N	DESCRIPTION											
10354	BRUSH 1"		2.20	1 EA	2.20							
10355	ROLLER COVER 9"		4.80	1 EA	4.80							
10103	LATEX, BASE		8.50	4 GAL	34.00							
10632	TARP, PLASTIC		1.50	2 EA	3.00							
20101	COLOR, RED		.50	1 OZ	.50							
20102	COLOR, BLUE		.50	5 OZ	2.50							
20103	COLOR, TAN		.50	10 OZ	5.00							
JOB		4-1-87					40.0	1.0	7.0		8.0	
QUOTE		3-25-87					40.0	1.0	1.5		2.5	
SUB TOTAL					52.00	52.00	80.0	2.0	8.5	73.84	10.5	
RATE					—	x 1.42	x .30	x19.95	x19.95	x .06	x 8.50	
	TOTALS				52.00	73.84	24.00	39.90	169.58	4.43	89.25	

GROSS $311.75 NET $142.07 % = 46% TRAV. SELL 63.90

QUOTE COST $33.25 MAT VS LAB 24:76 GROSS VS QUOTE 11%

Job sheet
Figure 4-2

Advertising agencies, newspapers, and print shops use what's known as "Clip Art." This is art drawn by professional artists, and available in book form for your use. It's also camera-ready copy. This means the printer can use it as it is. Where can you get it? One source is:

The Printers Shopper
P.O. Drawer 1056
Chula Vista, California 92012

Ask for a copy of their catalog. You'll be surprised at what's available. You can even get calendars printed with your company advertising.

Accounting Details

I'm going to walk you through some examples of forms, and of internal bookkeeping. In-house paperwork needn't be fancy — just accurate and readable. Read this chapter. Then, use Figure 4-1 to lay out the forms you'll need. Make a 30-day supply of each form with a copy machine. Once you decide on your own layout and headings, any quick-print shop can make all the copies you want. Remember to keep the masters separate from the pile you'll be using. You can keep the work copies in one of those inexpensive 59-cent folders.

Job and Quote Forms

Look at Figure 4-2. It's the job sheet. It can also be your quote work sheet. The differences are the headings and the actual figures. The quote is, by nature, a "best guess" estimate of time and materials. The job sheet records actual time, mileage, and cost of materials.

In this example, we'll follow a proposed job for Mr. and Mrs. B. Smith. They asked for a quote on having their master bath and master bedroom painted. The bath is to be tan; the bedroom is to match sample #302.

The Job Sheet

Use this sheet to figure what the job will cost you and how much to charge the customer. First, list all the materials you'll use for the job. Include the part numbers that you've assigned to them. (More on part numbers in the section on inventory control later in this chapter.) Next, list the mileage and time needed for travel and labor for the actual job. Add up the subtotals. Finally, multiply by the rates to get totals.

I've used $8.50 as the hourly labor cost, $19.95 as the labor selling price, $0.30 for mileage cost; and 142% as the materials markup. Your figures will almost certainly be different. Across the bottom of the sheet are several figures derived from this information:

- "Gross $" is the price to the customer. It's the total of materials at M/U (markup), mileage at M/U, travel and standard labor at M/U, and the sales tax.

- "Net $" is "Gross $" less extended cost of materials, cost of total labor used, and sales tax.

- "%" is the percentage your net amount is of the gross amount. Find it by dividing "Net $" by "Gross $."

- "Trav Sell" means travel costs sold to the customer. It's the total mileage at M/U, plus the total travel hours at M/U.

- "Quote Cost" is the total of quote mileage at M/U, and labor used during the quote at cost ($8.50).

- "Mat vs. Lab" means material versus labor. It tells you what percentage of the gross came from materials, and what percentage from labor. First, find what percentage of the gross materials are. Do this by dividing total materials sold ($73.84) by "$ Gross." The answer is 24%. Since material sold is 24% of gross, labor sold must be 76% of gross.

ACE PAINTING CO.
2468 INDUSTRIAL RD.
GLIDDEN, WA 20070

(206) 555-1234

No. Q 12500 Customer's Order

Date 3-25-87

SOLD TO: MR. & MRS. B. SMITH
S 26 ELM ST.
ANYTOWN, USA
PHONE XXX-XXXX

SHIPPED TO: ACCEPTED BY:
B. Smith 3-25-87

DATE SHIPPED	SHIPPED VIA	TERMS	F.O.B.	SALESMAN	
		50% DOWN BAL ON COMP	SITE	BILL (OK)	WORK TO START 8AM 4-1-87

ESTIMATE TO PAINT MASTER BATH, TAN SEMI-GLOSS ENAMEL;
MASTER BEDROOM, PER SAMPLE. #302, LATEX FLAT;
TO INCLUDE ALL MATERIALS, LABOR, AND SALES TAX.

RECEIVED $157.50 CK# 426
ON 3-25-87 Bill Jones

ESTIMATE TOTAL $315.00
+/- 5%

Sample customer quote
Figure 4-3

• "Gross vs. Quote" tells you how much the quote cost in relation to the gross income from the job. It's written as a percentage. To find it, divide "Quote Cost" by "Gross $." In this case, the quote cost 11% of the gross. Not bad.

The Customer Quote

Since the job sheet has most of the job information, it becomes the quote work sheet. Use it for filling out the customer quote, shown in Figure 4-3. I've done the heading on the quote in ink, but you'd do better to have your company name, logo, and phone number printed on the form. I also suggest that you include your advertising and professional memberships.

Avoid quoting exact amounts, if possible— When you quote a job to your customer, it's perfectly acceptable to quote an exact cost per hour and only the estimated hours with a plus and minus percentage. That way, you're covered if you can't do the job at exactly the stated time. You'll collect for all time spent on the job so long as it doesn't exceed the maximum. I usually quote plus or minus 5%, but in practice, most customers will accept a plus or minus 10%. Just be sure that's in your quote and your customers understand that their cost depends on how long the job takes.

Of course, on smaller jobs or if your competition is expected to quote exact costs, you'll probably want to provide an exact price for the work.

Note in Figure 4-3 that I didn't give a complete job breakdown. I just wrote the minimum necessary to tell the customer the price, and the general work to be done. For most customers, this is enough. For professional and government customers, much more detail will be needed — and a formal contract.

Inventory Pull List

Another nice thing about the job sheet is that it's more than just the quote work sheet. It's also the inventory pull list, since it lists all the materials needed for the job.

The Customer Invoice

Figure 4-4 is the actual customer invoice. It should

ACE PAINTING CO.
2468 INDUSTRIAL RD.
GLIDDEN, WA 20030
(206) 555-1234

No. 10001 Customer's Order Date 4-1-87

SOLD TO: MR. & MRS. B. SMITH
526 ELM ST.
ANYTOWN, USA
PHONE XXX-XXXX

SHIPPED TO: COMPLETION ACCEPTED BY: B Smith 4-1-87

DATE SHIPPED	SHIPPED VIA	TERMS	F.O.B.	SALESMAN
4-1-87	—	—	—	BILL

TOTAL MATERIALS		73 84
TOTAL TAX		4 43
TOTAL LABOR		233 48
PAINTED MASTERBATH, TAN SEMI-GLOSS	TOTAL	311 75
PAINTED MASTER BEDROOM, FLAT #302	LESS DOWN	157 50
	BAL DUE	154 25

PAID IN FULL. CK #435 FOR $154.25
ON 4-1-87. THANK YOU Bill Jones

Sample customer invoice
Figure 4-4

balance with all the figures on the job sheet. Instead of listing a separate mileage charge, we've just lumped it in with the labor charge. That's fine — as long as the figures balance.

Again, make the invoice as general or as detailed as you want. It's really only a summary that identifies the work. The details are all on the job sheet which doesn't leave your office.

The Quote Log

From all the quotes, we generate a quote log, Figure 4-5. It tells you how you're doing in the sales department. Here's what some of those column headings mean. In the "Sold/Age" column, the "Age" figure is the number of days old the quote is. Quote 12503 was converted to a job in four days. Quote 12507 is now two days old. It's time to follow up and see what the problem is.

The "Act $" column is the actual amount the job was sold at. It's what the customer paid.

The "delta" (small triangle) column is the total of each quote, less the actual dollars the job was finally sold at. Delta is the math symbol for "difference." Brackets around any amount mean that the job sold for less than the quote.

The "#" in the last column shows the account or job number the money is actually charged-off to. As you can see, the quotes that didn't turn into jobs get charged to 3500, the sales account.

Now for an explanation of those figures at the bottom of the quote log. Here's what they all mean:

- "% Converted" tells you how many of your quotes turned into jobs. Find it by dividing the number of quotes that became jobs by the number of jobs bid. Here, 42% of your bids turned into jobs.

- "$% Conv" compares the dollars earned by the jobs you did to the total dollars possible, if all bids had turned into jobs. Find it by dividing the total dollars of accepted bids by the total dollars of all bids. Round numbers off to make it easier. The

ACE PAINTING CO. — QUOTE LOG

QUOTE #	CUSTOMER	TOTAL	DATE	DATE ACCEPTED	SOLD	AGE	JOB DATE	ACT $	Δ	MILEAGE COST	LAB HRS.	CHARGE TO	#
12500	SMITH, B	315.00	3-25-87	3-25-87	Y	0	4-1-87	311.74	<3.26>	12.00	2.5	JOB	10001
12501	WEEKS, J	135.00	3-25-87	—	N		—	—	—	3.60	.5	SALES	3500
12502	MENCO	487.00	3-26-87	—	N		—	—	—	4.80	1.0	SALES	3500
12503	JONES, B	90.00	3-26-87	4-1-87	Y	4	4-2-87	101.03	+11.03	9.60	.75	JOB	10002
12504	MARKS, K	75.00	3-27-87	—	N		—	—	—	6.00	.75	SALES	3500
12505	HOWARD, S	110.00	3-28-87	3-30-87	Y	2	4-2-87	106.13	<3.87>	10.80	1.0	JOB	10003
12506	PETERS, M	550.00	3-28-87	—	N		—	—	—	3.60	1.5	SALES	3500
12507	JAMES, T	895.00	4-3-87			''				5.40	1.5	*	
* NOTE:	THIS IS OPEN SO NO FIGURES ARE INCLUDED IN TOTALS.												
											8.0		
											x 8.5		
	TOTALS	1,762	—	—	—		—	518.90	+3.90	50.40	68.00	—	—

% CONVERTED = 42% — $ % CONV = 29 % — QUOTE VS ACT Δ = 101% — QUOTE COST = 118.40 — QUOTE VS SALES = 23%

TO JOBS = $68.52 — TO SALES = $49.88

Quote log
Figure 4-5

jobs brought in $515. That's $515 divided by $1762, or 0.29 — 29%. I didn't include the last bid, since it's still open. Note the difference between "% Converted" and "$% Conv." You're selling 42% of the jobs, but only getting 29% of the available dollars.

• The "Quote vs. Act" delta tells you just how close your quote was to the actual selling price of the job. Divide "Act $" by the total of the accepted quotes. In this case, $518.90 divided by $515, or 1.01 — 101%. Your actual dollars were within one percent of your quoted dollars. If you're within plus or minus 5%, you're usually doing fine. More or less than 5%, and you're doing something wrong.

• "Quote Cost" is the total mileage cost for all quotes, plus the total labor used to do all the quotes. As you can see, it's a major expense.

• "Quote vs. Sales" shows what percentage quote cost is of the actual dollars brought in. Find it by dividing the quote cost by the actual dollars of jobs sold. Here, it's $118.40 divided by $518.90, or 0.23 — 23%. Our example is very high. This percentage should be down around 6%, but certainly not more than 10%.

• "To Jobs" and "To Sales" are the actual distribution of the cost dollars. So $68.52 went directly to the customer's jobs and $49.88 went to the sales budget as an expense. There should be a sheet for each quote (your customer quote form) and a sheet for each job (the customer invoice).

Consolidated Log Sheets

Consolidate each separate sheet onto a daily log, which covers a month's activities. Then, at the end of each month, consolidate the daily log onto a monthly log. I'll use the job logs as an example, but you should consolidate all your logs the same way.

The daily job log— This is shown in Figure 4-6. I'm only showing three jobs. I hope you'll have 20, or 30, or more on your log.

The information in each row comes from the column totals on each job sheet. The last column is headed "Net $." You've already figured it on each

ACE PAINTING CO. — DAILY JOB LOG

JOB #	CUSTOMER	DATE	MATERIAL COST	MILEAGE COST	LABOR HRS	TRAVEL HRS	MATERIAL SELL	LABOR SELL	TRAVEL SELL	SALES TAX	TOTAL	NET $
10001	SMITH, B	4-1-87	52.00	24.00	8.5	2.0	73.84	169.58	63.90	4.43	311.75	142.07
10002	JONES, B	4-2-87	10.00	3.00	4.0	1.5	17.20	79.80	3.00	1.03	101.03	40.25
10003	HOWARD, S	4-2-87	15.50	3.00	4.0	1.5	22.01	79.80	3.00	1.32	106.13	39.56
	MONTH END TOTAL	—	77.50	30.00	16.5	5.0	113.05	329.18	69.90	6.78	518.91	221.88

TOTAL HRS = 21.5 AVG MAT COST = 25.83 AVG. M/U = 31% AVG. NET = 73.96 TOTAL DAYS = 2

AVG HR RATE = 18.56 AVG MAT SELL = 37.68 AVG GROSS = 172.97 GROSS vs NET = 42% PER DAY NET = 110.93

Daily job log
Figure 4-6

job sheet. Get it from "Net $" at the bottom of the job sheets. Across the bottom of the form is the summary of column totals. Here's what it all means:

- "Total Hrs" is the sum of the labor hours plus the travel hours.

- "Avg Hr Rate" is the average labor selling price. Find it by first adding "Labor Sell" and "Travel Sell," then dividing by "Total Hrs."

- "Avg Mat Cost" means the average material cost. It's the total material cost, divided by the total number of jobs.

- "Avg Mat Sell" is the average material selling price. It's the total dollars of "Mat Sell" (material sold at markup), divided by the number of jobs.

- "Avg M/U" is the average mark up percentage. It's "Avg Mat Cost" divided by "Avg Mat Sell," then subtracted from 1.00. Here, it's $25.83 divided by 37.68, or 0.69. Then, 1.00 minus 0.69 equals 0.31, or 31%. That's about standard.

- "Avg Gross" is the average gross. It's the dollar total of all the jobs divided by the total number of jobs. In our example, it's $518.91 divided by 3, or $172.97.

- "Avg Net" is the average net. It's the total from the "Net $" column, divided by the number of jobs.

- "Gross vs. Net" shows what percentage your net is of your gross. It's "Net $" divided by "Total." Here, that's $221.88 divided by $518.91, or 0.43 — 43%. And 43% is a reasonable figure.

- "Total Days" is the number of days worked on the jobs. Just add them up, from the "Date" column.

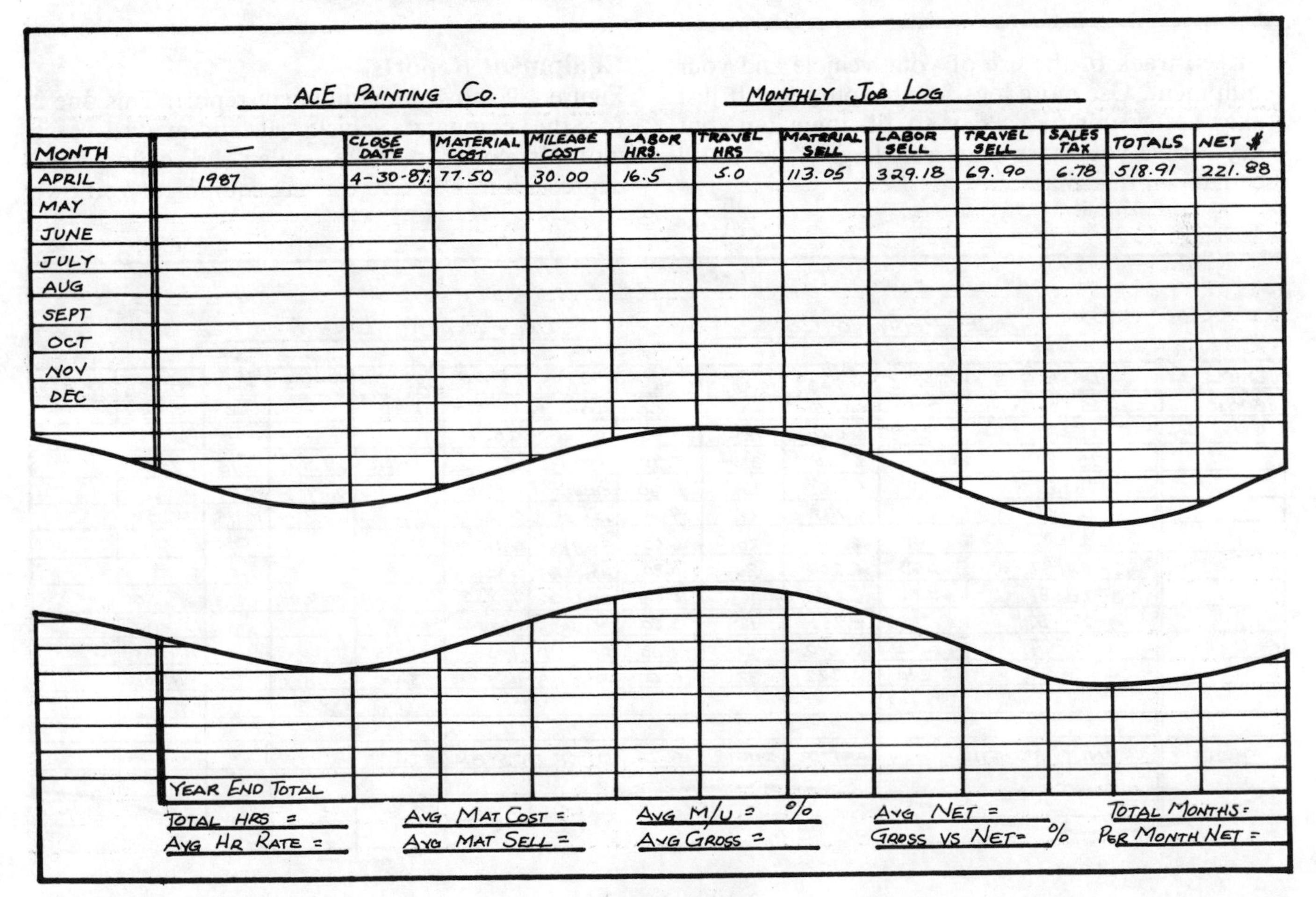

ACE PAINTING CO. — MONTHLY JOB LOG

MONTH	—	CLOSE DATE	MATERIAL COST	MILEAGE COST	LABOR HRS.	TRAVEL HRS	MATERIAL SELL	LABOR SELL	TRAVEL SELL	SALES TAX	TOTALS	NET $
APRIL	1987	4-30-87	77.50	30.00	16.5	5.0	113.05	329.18	69.90	6.78	518.91	221.88
MAY												
JUNE												
JULY												
AUG												
SEPT												
OCT												
NOV												
DEC												
	YEAR END TOTAL											

TOTAL HRS =
AVG HR RATE =
AVG MAT COST =
AVG MAT SELL =
AVG M/U = %
AVG GROSS =
AVG NET =
GROSS VS NET = %
TOTAL MONTHS =
PER MONTH NET =

Monthly job log
Figure 4-7

• "Per Day Net" tells you how much you made each day. Divide the "Net $" total by "Total Days."

If you study these figures you'll see a problem. The 43% in "Gross vs. Net" is good. It says that you are indeed making a fair return. But the average gross of $172.97, and the average net of $73.96 are too low. They're below your budgeted income level. This shows that you're getting the lower-priced jobs, not the higher-priced ones. You need to change your advertising, the market you aim at, or your sales pitch.

The monthly job log— Figure 4-7, the monthly job log, is a summary of the column totals from the daily job log of Figure 4-6. At the end of the year, calculate the bottom line the same way you did for the daily log.

Vehicle and Equipment Logs

Keep track of the use of your vehicle and your equipment. Use more logs for this purpose. It just takes a few minutes a day to fill them out, but they'll save you more than a little grief, as you'll see later in this chapter.

The Vehicle Log

Figure 4-8 shows your daily vehicle log. Keep it in the vehicle every working day. Record each day's mileage on the log. It breaks down the cost of running the vehicle and identifies what accounts are charged for what expenses.

In this example, I show the mileage sequentially, since the vehicle is used only for business. If it were a personal vehicle, there would be breaks in the beginning and ending mileage readings. At the end of the year you'd subtract the beginning mileage from the ending mileage to find total mileage. Then you'd subtract the total business mileage, to find total personal use mileage. (The IRS requires this information if you want to deduct your vehicle costs.)

Also, I show a materials pick-up. Charge this to FRT/IN. Show the rework of Job 10000 on its job sheet. And add the rework to the daily job log, Figure 4-6. Reworks change the cost of doing a job. They also change your markup and your profit. You must account for them.

Equipment Reports

Figure 4-9 shows an equipment report. This one is for the company vehicle. It summarizes each month's costs, mileage, miles per gallon, and depreciation. Use one for any equipment you own

ACE PAINTING CO. — DAILY VEHICLE LOG

VEHICLE #	DATE	BEGIN MILEAGE	END MILEAGE	TOTAL MILES	RATE X .30	CHARGE TO	JOB #	* CUST. PAID	* NON-PD.	* TRAVEL TIME	RATE	* TOTALS
180010	3-25-87	6,015	6,055	40	12.00	SMITH, B	10001	31.95	—	1.0	19.95	
	3-25-87	6,055	6,067	12	3.60	WEEKS, J	Q12501			.5		
	3-25-87	6,067	6,100	33	9.90	FRT	PICK-UP MATERIAL	—	9.90	.75	8.50	
	3-26-87	6,100	6,116	16	4.80	MENCO	Q12502			1.0		
	3-26-87	6,116	6,148	32	9.60	JAMES, B	Q12503	—	9.60	.75	14.50	
	3-27-87	6,148	6,168	20	6.00	MARKS	Q12504			.75		
	3-28-87	6,168	6,204	36	10.80	HOWARD	Q12505			1.0		
	3-28-87	6,204	6,240	36	10.80	BILL, K	REWORK 10000	—	10.80	.75	8.50	
	3-28-87	6,240	6,252	12	3.60	PETERS, M	Q12506			1.5		
	4-1-87	6,252	6,290	40	12.00	SMITH, B	10001	31.95		1.0	19.95	
	4-2-87	6,290	6,322	32	9.60	JONES, B	10002	3.00		.75	14.50	

*THESE CAN ONLY BE FILLED IN AFTER QUOTE IS DEAD OR JOB BILLED.

Daily vehicle log
Figure 4-8

that requires depreciation, or parts, or any other maintenance.

The numbers in the "MPG" column (miles per gallon) came from dividing the total miles driven each month by the total gallons of gas bought that month. See how the MPG went way down in July? Actually, it was on the down-swing back in May. You look for a problem and find a dirty air filter. But that alone shouldn't have caused such a big MPG drop. You check more and find only 10 pounds of air in a front tire. Inspection finds a nail. Repairs are made, and in August the mileage jumps back to normal.

The "Avg per day" column is the average miles driven per day. To find this number, divide the total miles by the total days of actual use.

Depreciation is shown in the "Deprec" column. We figured out the monthly depreciation in the previous chapter. The "Book value" (inventory value) column shows the last book value, minus the depreciation.

At the bottom of the report, "Tot Cost" means total cost; it's the sum of all the expenses. And "Tot Dep" is the total depreciation; the sum of all amounts in the "Deprec" column. "Cost Less Dep" is cost less depreciation. It's simply the total depreciation subtracted from the total cost. Here's an explanation of the other figures at the bottom of the report.

- "Cost Per Mile" is the total cost, divided by the total miles: $859.30 divided by 2,451.5, or 0.35. That's 35 cents per mile. And we haven't added in the interest charges on the loan, or the cost of insurance. So this cost is even higher. Guess what? You've been charging your customers 30 cents per mile, and losing about 10 cents per mile. At your projected yearly total of 39,000 miles, you'll lose about $3,900 over 36 months. Starting on September 1, raise the vehicle mileage markup rate to 40 cents per mile. Believe it or not, most painters don't even bother to charge for mileage! That's

ACE PAINTING CO. — EQUIPMENT REPORT
VAN S/N 358922 INV. # 180010

DATE	WASH & WAX	PARTS	OIL & LUBE	FUEL $	GAL.	TOTAL GAL.	TOTAL MILES	MPG	TOTAL DAYS USED	AVG PER DAY	DEPREC.	BOOK VALUE
4-15-87	7.50	—	—	18.00	12.85							9,500
4-30-87	7.50	—	—	18.00	12.85	25.70	539.70	21.0	21	25.7	111.11	9,388.89
5-15-87	7.50	—	—	21.00	15.00							
5-20-87	—	FAN BELT 8.50	—	—	—							
5-30-87	7.50	—	—	8.00	5.71	20.71	414.20	20.00	20	20.71	111.11	9,277.78
6-15-87	7.50	—	15.00	12.00	8.57							
6-30-87	7.50	—	—	21.00	15.00	23.57	447.83	19.00	21	21.33	111.11	9,166.67
7-15-87	7.50	—	—	21.00	15.00							
7-20-87	—	—	—	18.00	12.85							
7-30-87	7.50	—	—	21.00	15.00	42.85	492.77	11.5	22	22.40	111.11	9,055.56
8-4-87	—	AIR FILTER 3.25	—	—	—							
8-6-87	—	FIX TIRE 5.00	—	—	—							
8-15-87	7.50	—	—	18.00	12.85							
8-30-87	7.50	—	—	21.00	15.00	27.85	557.00	20.0	22	25.32	111.11	8,944.45
	75.00	16.75	15.00	197.00	140.68	140.68	2,451.50	17.43	106.00	115.46	555.55	—

COST PER MILE = $0.35 SALES TAX $11.82 COST LESS DEP = 303.75 AVG. MILES/DAY = 23.12 AVG. FUEL COST = $1.40 TOT. DEP. $555.55 TOT. COST $859.30

Equipment report, company vehicle
Figure 4-9

O.K., *if* you understand how much it truly costs and if you work that cost into your overhead markup. But ignoring this cost is downright foolish. Can you afford to give away your vehicles and equipment? That's what you'll be doing if you don't include their cost in every bid.

- Find the "Sales Tax" amount by adding up sales tax on the receipts. Or add another column just for sales tax.

- "Avg Miles/Day" is the average miles driven each day for the entire period covered by this log sheet. To find it, divide the total mileage by the total days.

- "Avg Fuel Cost" is average fuel cost, per gallon, for the entire period covered by this log sheet. Find it by dividing the "Fuel $" total by the "Gal." total.

There's a big advantage to keeping records like this. You can anticipate problems before they bankrupt your company. If that tire problem hadn't been fixed, it might have led to an accident. At the very least, you'd have lost a tire, needed a front-end alignment, and wasted money on gas.

Why You Need These Logs

You may say that you don't have time to keep logs. Well, let's see how much it's worth. It takes about five minutes a day to fill out this sheet. That's 19 hours a year. The tire you saved would have cost $78. The front-end alignment would cost $42. And fuel, at $20 a month, for four months equals $80. That's $78 plus $42, plus $80, or $200. And $200 divided by that 19 hours a year is $10.52 per hour. That's $10.52 per hour you just saved yourself, for doing this one log sheet. It pays to do paperwork.

In addition, we just found a $3,900 problem: when we realized that you weren't charging enough to cover your actual vehicle operating costs. That $3,900, divided by 19 hours a year spent filling out this log, is $205.26 per hour. Add it to your $10.52 an hour, and you just paid yourself $215.78 an hour. Just for doing the paperwork and studying the figures!

And when it comes time to settle up with Uncle Sam, good records are a must. Any deductible expenses you can't prove aren't deductible.

Expense Logs

Keep a log of your expenses for each month. It's an easy way to keep track of all costs and keep all cost information in one place.

Monthly Expenses

Now on to Figure 4-10, the monthly expense log. I show most of the expenses on one sheet. You'll probably need two or three sheets, for ease of accounting. Note that I logged both check numbers and cash out-of-pocket.

At the bottom of the form are the totals. The $32.00 amount in "Less N-D" is the nondeductible portion of the license plate fee. Subtract it from the total expenses to find your total deductible expenses.

Get the general idea from this example. Your account headings may be different. And, depending on how accurate or detailed you want to be, you may have more or fewer headings. I suggest more. The more categories you maintain, the more accurate and useful the monthly expense log becomes.

Yearly Expenses

Each month, transfer the column totals from the expense sheets to the yearly expense log, Figure 4-11. At the end of the year — I should say, at the end of your tax year — total the columns. Your tax year can end on December 31, but many painting companies choose another date. Talk to your accountant.

Now you have all your expense information for the IRS. Having it all on one sheet at tax time sure beats hunting through dozens of old shoe boxes for receipts.

Your total expenses— In Figure 4-11, the "Line Total" column is just that. Total across each line to get this figure. Add up each column; write the totals on the bottom line. Next, add up the "Line Total" column. Finally, add up all the totals on the bottom line. This answer should match the total of the "Line Total" column . . . Whoops! No match in our example. There's a mistake somewhere. Retotal in both directions. Are they all O.K.?, Sure, they are. We just forgot to add up the sales tax column. Do this, and everything balances. You see one reason I like a row-and-column form: it's easy to check for errors.

ACE PAINTING CO. — MONTHLY EXPENSES

ACC #	TO / ACC #	ELECT	GAS	VEH INSUR	SHOP INSUR	EQUIP REPAIRS	POSTAGE	MEMBER-SHIPS	UNI-FORMS	SALES TAX	PAID DATE	CK #
3504	PE 1013462	39.95									4-30	106
3505	CGAS		14.50								4-30	107
3506	ECO 57321			29.16							4-30	108
3507	ECO 57322				32.00						4-30	109
3508	— —					12.50				.68	4-30	110
3509	USPS						6.50				4-30	111
3510	—							—			—	
3511	—								DUST MASK 15.00	.90	4-21	101
ACC #		PHONE	Y.P. AD	ADVERT	OFFICE	OFFICE EQ REP	LIC FEE	FUEL + OIL	EQUIP EXP REP	SALES TAX	PAID DATE	CK #
3512	XXX · XX XX	25.50									1-30	112
3513			25.00								4-30	113
3514	NEWS DAILY			42.00							4-18	102
3515	OFFCO #1023				8.95						4-1	002
3516	OFFCO #1026					12.00					4-1	003
3517	DMY						VEHICLE 55.00				4-10	007
3518	SEE VEHICLE EX							25.70		1.54	4-30	CASH
3519	SEE EQUIP EX								24.00	1.44	4-30	120
TOP	SUB-TOTAL	39.95	14.50	29.16	32.00	12.50	6.50	—	15.00	1.58	—	—
BOTTOM	SUB-TOTAL	25.50	25.00	42.00	8.95	12.00	55.00	25.70	24.00	2.98	—	—
	NON-DEDUCTIBLE						<32.00>					
	TOTALS	65.45	39.50	71.16	40.95	24.50	61.50	25.70	39.00	4.56	—	—

TOT EXP = $372.32 LESS N-D = <32.00> TOT DED = $340.32
TOT CASH OUT = $27.24

Monthly expense log
Figure 4-10

Compare it to your budget— I show budget figures, and a delta (or difference). Remember our "target" budget for the year? Those amounts are used here, in the "Budget" row. But you should use your *actual* year-end budget figures.

How do you do that? First, figure your total yearly income from all sources. Then, mutiply it by each of the budget percentages we established. That's how to find out whether you're really within budget.

You could even find your actual monthly budget for each item. How? Just divide each of the individual target budgets by the number of months of the budget period. Then put those amounts in Figure 4-11. Now, you can easily compare dollars spent to dollars budgeted. You'll probably have to do some adjusting after a few months have passed.

Just remember this: Your target budget is a group of percentages of your target income. This target income changes from month to month. So, although your budget percentages may stay the same, the actual dollars you have available will change.

Look at Figure 4-11 again. The brackets [ ] in the example show you're under budget. That's a good sign. But you're not really under. The expenses are for 9 months; the budget is for 12 months. If you protect it out, I think you'll find yourself over budget on several items.

The gas (heat) account is already over budget. And look at the increase in electric cost, and the increase in gas cost. See the months when they happen? Most likely it's from increased use of air conditioning and heat. And why is vehicle insurance only $28.50 in December? Well, most installment contracts end up with "odd cents." That's the amount left over after the total premium was divided by the term.

Now, check the postage column. Business was slow in June and July. Also in October and November. So you did a few extra mailings to spur things on. In December? Well, it's always nice to say "Merry Christmas" and "Thank You" to customers. What's that in the memberships column? You joined the Chamber of Commerce and the Better Business Bureau.

ACE PAINTING CO. — YEARLY EXPENSES

		ELECT	GAS	VEHICLE INSUR.	SHOP INSUR	EQUIP REPAIR	POSTAGE	MEMBERSHIPS	UNIFORMS	SALES TAX		LINE TOTAL
	JAN	—	—	—	—	—	—	—	—	—		—
	FEB	—	—	—	—	—	—	—	—	—		—
	MAR	—	—	—	—	—	—	—	—	—		—
	APR	39.95	14.50	29.16	32.00	12.50	6.50	—	15.00	1.58		151.19
	MAY	46.50	11.50	29.16	32.00	—	6.00	50.00	—			175.16
	JUN	58.00	9.50	29.16	32.00	24.00	3.00	—	—			155.66
	JUL	65.00	8.00	29.16	32.00	32.00	21.00	50.00	—			237.16
	AUG	72.00	7.25	29.16	32.00	—	20.00	—	15.00	1.58		176.99
	SEP	60.00	14.75	29.16	32.00	14.00	6.00	—	—			155.91
	OCT	30.00	35.40	29.16	32.00	—	6.00	—	45.00	2.70		180.26
	NOV	24.50	42.60	29.16	32.00	18.75	20.00	—	—			167.01
	DEC	26.00	62.00	28.50	32.00	—	30.00	—	—			178.50
9-MONTH	SUB-TOTAL	421.95	205.50	261.78	288.00	101.25	118.50	100.00	75.00	=	1571.98	
12-MONTH	BUDGET *	510.00	200.00	360.00	360.00	250.00	340.00	150.00	100.00	=	2270.00	
	Δ	<88.05>	5.50	<98.22>	<72.00>	<148.75>	<221.50>	<50.00>	<25.00>	=	<698.02>	
	TOTALS	421.95	205.50	261.78	288.00	101.25	118.50	100.00	75.00	=	1571.98	1577.84

* THIS WAS PROPOSED 12 MONTH BUDGET — SEE TEXT

Yearly expense log
Figure 4-11

Know what you have here? History. And history of the last 12 months is the best predictor of what's going to happen in the next 12 months. Now you can *plan* for that needed advertising in June and October. You can plan for the increase in utilities during warm and cold periods. In short, you can use this sheet as a tool to plan next year's budget more accurately.

I could show you an endless variety of helpful forms. But you now have all the basics — with one exception. That's the balance sheet.

The Balance Sheet

A blank balance sheet is shown in Figure 4-12. This is similar to a form you might use when applying for a loan. But every balance sheet is similar in several important respects. It shows assets (what you own and what you're owed), liabilities (what you owe), and net worth. Think of a balance sheet as a financial snapshot. At some particular instant it summarizes your financial condition.

It's a good idea to fill out a balance sheet like this before you even go into business. You may find you're worth more than you thought.

Why You Need a Balance Sheet

Well, creditors want some sort of financial statement before they'll lend you money. This form tells it all. The IRS will require about the same information on your corporate tax return. If you decide to take in a partner, or to sell out, you'll need a balance sheet. If you decide to incorporate, you're going to need a current balance sheet. Are these enough reasons?

You'll want to draw one of these up annually at least — maybe even quarterly or monthly, to evaluate your progress and financial condition. I guess you're stuck with making these out. Fortunately, it's easy, if you've been keeping your records in order.

Financial Statement - Sole Proprietor

(Individual who is sole proprietor of a business)

______________________Office

Name______________________

Address______________________ Business______________________

FINANCIAL CONDITION AT CLOSE OF BUSINESS.., 19.......

ASSETS			
Cash on Hand and in Bank			
Accounts Rec. Customers—Ccrrent $			
Accts. Rec. Customers—Past Due $			
Total Accounts Receivable - $			
Less: Reserve Doubtful Accts. $			
Notes Receivable—Customers - $			
Less: Reserve Doubtful Notes $			
Trade Acceptances Receivable			
Merchandise—Finished			
Merchandise—In Process			
Merchandise—Raw Materials			
Readily Marketable Securities (Schedule 2)			
Net Cash Surrender Value Life Insurance (Sched. 1)			
TOTAL CURRENT ASSETS			
Real Estate and Bldgs. (Sched. 3) $			
Less: Reserve for Depreciation $			
Machy., Equip., Fixtures - $			
Less: Reserve for Depreciation $			
Automobiles and Trucks - $			
Less: Reserve for Depreciation $			
Interests in Controlled or Affiliated Cos. (describe):			
Other Securities Owned (Schedule 2)			
Mortgages Receivable (Schedule 5)			
Other Non-Current Receivables (Schedule 5):			
Deferred and Prepaid Items			
TOTAL			

LIABILITIES			
Accounts Payable—Not Due			
Accounts Payable—Past Due			
Notes Payable This Bank			
Notes Payable Other Banks			
Notes or Trade Acceptances Payable for Mdse.			
Other Notes Payable			
Portion of Equipment Contracts and Chattel Mortgages Due Within One Year			
Income Taxes Payable			
Other Taxes Payable			
Accrued Liabilities			
Other Current Debt (describe):			
Portion of Long Term Debt Due within One Year			
TOTAL CURRENT LIABILITIES			
Real Estate Encumbrances (Schedule 4)			
Non-Current Portion of Equipment Contracts and Chattel Mortgages			
Other Non-Current Debt (describe):			
TOTAL LIABILITIES			
Other Reserves (describe):			
NET WORTH			
TOTAL			

Profit & loss balance sheet
Figure 4-12

OPERATING RECORD FROM ______, 19___, TO ______, 19___:

If profit and loss statement does not fit your operations, please attach a statement on your own form.

Net Sales for Period - - - - $______
Cost of Goods Sold - - - - $______
Gross Profit - - - - - - - - - - - - - - - $______
Selling Expense - - - - - - - $______
Administrative Expense - - - $______
General Expense - - - - $______
Total Operating Expense - - - - - - - - - $______
Operating Profit - - - - - - - - - - - - - $______
Other Income - - - - - - - - - - - - $______
Total - - - - - - - - - - - - - - - - - $______
Federal & State Income Taxes $______
Other Deductions - - - - - $______
Total Deductions - - - - - - - - - - - $______
Net Income - - - - - - - - - - - - - - $______
Total Depreciation and Amortization included above $______
Deductions for Bad Accounts included above - - $______
Do above expenses include salary to yourself? ______
If so, how much? - - - - - - - - - - - - $______

Reconciliation of Net Worth:
Net Worth at beginning of Period - - - - $______
Net Income - - - - - - - - - - - - - $______
*Other Additions - - - - - - - - - - $______
Total - - - - - - - - - - - - - - - $______
Personal Withdrawals - - - $______
*Other Deductions - - - - $______ $______
Net Worth as of this statement date - - - - $______
*If Other Additions and Deductions involve important transactions please give details below:

______

MONTHLY SALES

Please enter here your approximate sales by months during the past fiscal period:

Jan.______ Apr.______ Jul.______ Oct.______
Feb.______ May______ Aug.______ Nov.______
Mar.______ Jun.______ Sept.______ Dec.______

Do the above figures include all of your income (both from your business and otherwise)? ______ or do they relate to your business only? ______
Please describe here any unusual factors influencing your volume or earnings during the past fiscal period ______

OPERATING RECORD FOR PAST FIVE YEARS IF NOT PREVIOUSLY FURNISHED

Year	19	19	19	19	19
Sales or Gross Income					
Net Income					
Personal Withdrawals					

STATEMENT: By whom was this statement prepared? ______
Fiscal year ends on ______ Regular time of taking inventory ______ Regular time of balancing books ______
Are your books audited by an independent accountant? ______ If so, by whom? ______
As of what date was his last audit made? ______ Was a separate signed report prepared by the accountant? ______
Does this statement include all of your assets and liabilities as an individual (i.e. personal non-business assets and liabilities as well as those relating to your business)? ______ If not, please also file a personal statement with us or describe here the extent and nature of your outside personal assets and liabilities ______
______ Have your personal income taxes for last year been paid? ______
Are you married? ______ If so, does this statement include any separate property of your wife (husband)? ______ Explain ______
Have you made a will? ______ Who is named executor of estate? ______
LEGAL STATUS: Do you do business under any style other than your own name? ______ If so, what? ______
______ Has a certificate of fictitious trade style been filed and published? ______
Have you ever gone through bankruptcy or compromised a debt? ______ If so, please give details ______

OTHER BANKS USED:

Name	City	Do you borrow there?	Maximum Debt Past Year
			$

Profit & loss balance sheet
Figure 4-12 (continued)

SALES AND RECEIVABLES: About what portion of your sales are for cash?________ Are a substantial part of your sales to any one customer?________ If so, approximately what portion?________ What are your usual selling terms on credit sales?________ Is any material part of sales customarily on longer terms?________ If so, please explain ________

About what amount of customers accounts receivable was past due on statement date?________ On what portion of credit sales do customers take discounts?________ Does any part of receivables represent merchandise out on consignment?________ Approximate amount $________

PURCHASES AND PAYABLES: What are your usual buying terms?________ Are a material portion of your purchases customarily on other terms?________ If so, explain________

Are you taking all available discounts?________ Please list principal suppliers on Schedule 6.

MERCHANDISE: Is this statement based on actual or estimated inventory?________ If actual, by whom taken?________

If estimated, on what basis?________

How is merchandise valued? (Specify clearly whether at original cost, replacement market, the lower of the two, or on what other basis)________

Is any portion of inventory consigned to customers?________ Approximate amount $________ Is any material part of the inventory excessive, slow moving or obsolete?________ If so, explain________

SEASONAL INFLUENCES: Are your sales seasonal?________ If so, when are they highest?________ Are your purchases seasonal?________ If so, when are they highest?________ Are your collections seasonal?________ If so, when are receivables at their low point?________ Does inventory fluctuate seasonally?________ If so, when is it highest?________ When lowest?________ Please do not omit the monthly sales figures requested above.

RENTAL: Does business rent?________ Present monthly rental paid $________ Date of expiration of lease________

SCHEDULE 1—INSURANCE

Has your present coverage been reviewed and approved by an insurance broker or adviser?________

Fire insurance:		Liability Insurance:	
On Merchandise	$____	Public Liability on Owned Autos	$____
On Mach'y, Equipt. and Fixtures	$____	Property Damage on Owned Autos	$____
On Buildings	$____	P.L. and P.D. on Non-owned Autos	$____
		Building & Elevator Pub. Liab.	$____

Please check those of following which are carried:

____ Explosion Ins. ____ Steam Boiler ____ Auto Fire, Theft ____ Business Interruption ____ Products Liability

____ Riot and Strike ____ Auto Collision ____ Workmen's Compensation ____ Robbery or Burglary ____ Machinery Breakdown

Is the extended coverage endorsement attached to fire policies?________ Do any policies contain a coinsurance clause?________ Basis____%

Is any insurance on a monthly reporting basis?________ Are employees having custody or control of property adequately bonded?________

Insurance on my life:

Beneficiary	Amt. of Policy	Cash Value	Amt. of Loans	Net Cash Value
	$	$	$	$
	$	$	$	$
	$	$	$	$

SCHEDULE 2—SECURITIES OWNED—Please attach separate schedule if needed.

Stock-Shares, Bond-Amounts	Description	Value at Which Carried on this Statement	Current Mkt. on Listed		Estimated Value on Unlisted		
			@	Amount	@	Amount	Yrly. Div.

In what name are the above securities carried?________

If in the name of yourself and co-owner are they joint tenancy?________

(Concluded on Page Four)

Profit & loss balance sheet
Figure 4-12 (continued)

SCHEDULE 3—REAL ESTATE AND BUILDINGS—Please give details of encumbrances on Schedule 4 opposite proper Parcel No.

	Location and Description Include Nature of Improvements	Title in Name of	Monthly Income	Valuation		Amount of Encumbrance (Details Below)	Assessed Valuation
				Land	Improvements		
Parcel # 1							
Parcel # 2							
Parcel # 3							
Parcel # 4							
Parcel # 5							

Are any properties in joint tenancy?__________ If so, which Parcel Nos.?__________ Has a homestead been filed?__________ If so, on which Parcel No.?__________ Are taxes or assessments delinquent on any of your properties?__________ If so, please give amount and details __________

SCHEDULE 4—REAL ESTATE ENCUMBRANCES

	Amount Owing per Sched. 4	Nature of Encumbrance and To Whom Payable	Int. Rate	Due Date	How Payable	Are Int.* and Prin. Current?
On Parcel #1 Above						
On Parcel #2 Above						
On Parcel #3 Above						
On Parcel #4 Above						
On Parcel #5 Above						

*If any payments of principal or interest are delinquent, please give details __________

Has foreclosure been instituted?__________ Details__________

SCHEDULE 5—MORTGAGES, AND OTHER NON-CUSTOMER RECEIVABLES

Name of Debtor	Amount Due	How Payable	Remarks (include description and value of any security)

SCHEDULE 6—PRINCIPAL SUPPLIERS—Please list concerns from which you buy large quantities and approximate amount due them on statement date.

Name and City	Amount Owed	Name and City	Amount Owed
	$		$
	$		$
	$		$

GENERAL REMARKS—Please explain here or in a supplementary letter any important differences between carrying values and actual values, any unusual receivables or payables of importance, or any other factors which have a bearing on interpretation of your financial statement.

____

____

____

For the purpose of procuring credit, from time to time, from the bank addressed, I furnish the foregoing as a true and accurate statement of my affairs on the date indicated. I have answered the questions on page two of this exhibit and certify that I have no liabilities except as reported therein and elsewhere in this exhibit. I agree to notify the said bank immediately, in writing, of any unfavorable change in my financial condition. I have carefully read the financial statement and supporting information on the four pages of this exhibit, both printed and written, and I solemnly declare and certify that this is a true and correct account of my financial condition on the date stated.

FOR BANK USE
BRANCH CERTIFICATION: This is a copy of the original statement, properly signed, in the credit files of this Branch.

Date__________ __________ Manager

__________ (Date Signed) __________ Signature

Profit & loss balance sheet
Figure 4-12 (continued)

Inventory Control

Do you think inventory control is only for Sears, K-Mart, and McDonald's? Wrong — you'd better do it too. It's not hard. "Inventory control" simply means keeping track of all the things you have, or use, in your business. It's part of your overall management plan. And it's probably one of the most neglected parts of any painting company. That neglect can be costly.

How costly? Well, inventory control lets you know if you have too much or too little of any item. I once worked for a company with inventory and equipment valued at $350,000. After controls were established, we found that nearly $65,000 of this was obsolete stock that would never be used! Even worse, some obsolete items were still being ordered — why, I'll never know. That $65,000 could have been in the bank earning interest. Instead, it was invested in inventory that was useless. Worse, for several years the company paid property tax on the obsolete stock.

I use the term *inventory* to mean everything you own for the operation of your business. All the tools, equipment, vehicles, forms, furniture, disposable materials, and materials to be re-sold. Each item belongs to one of these groups:

- Capital equipment
- Disposable materials
- Resale materials

We covered most of these items in the chapter on accounting to the IRS. What you need is a system for controlling them. What are they worth? Where did they come from? Where did you use them? How much did you pay for them? What did you sell them for? Can they be sold? For a profit? Do you have too many of one item, not enough of another? What are the carrying charges? Should you re-order? When should you re-order? It just makes good sense to know where your money is and what it's doing. That's the purpose of inventory control.

Vendor File

The first control, shown in Figure 4-13, is the vendor file. The form is pretty much self-explanatory. It simply puts in one place all the information you need on each vendor. Add to it each vendor you use regularly.

Note that I used a five-digit vendor number. This gives you up to 99,999 vendors, more than enough for years of doing business. If you want, you can use the first two digits for vendor type: paint, tools, paper goods, etc.

Equipment Inventory Log

Now, on to the equipment inventory log shown in Figure 4-14. It shows where you're spending the big dollars. It gives you tax information. It shows the cash equity of ownership in the equipment. And it shows the total cost of credit that's been extended to you.

Also, it's an important record of what you own. A record you can use for insurance claims, for loan security, or for determining what you and your company are worth, should you want to sell out. Keep a second copy of this form in a safe deposit

ACE PAINTING CO — VENDOR FILE

VENDOR	NAME	STREET ADDRESS				PHONE	#	CONTACT	TERMS	DUE DATE	LIMIT	
00010	WV MOTORS	29605	LOST HILLS, CITY			XXX	XXXX	JAY	CONTRACT	30TH	10,000	
00015	EQCAL	10135	LAKE,	CITY		XXX	XXXX	BILL	CONTRACT	22ND	2,500	
00025	SMC	123	OAK,	CITY		XXX	XXXX	MARY	CASH	A/R	150	
00030	PAINTCO	5348	LIBERTY,	CITY		XXX	XXXX	ED	1% 10 NET 30	15TH	1,500	

Vendor file
Figure 4-13

ACE PAINTING CO. EQUIPMENT INVENTORY LOG

INV #	DESCRIPTION	S/N	PUR.	COST	CASH DOWN	BALANCE OWED	MONTHLY PAY	TERM	TOTAL	SALES TAX	DEPREC MONTH	1ST YEAR DEPREC
180010	VAN, ½ TON	358972	4-1-87	9,500	2,500	7,000	236.80	36 MO	11,024.80	570.00	111.11	1335.32
180020	AIRLESS	2538	4-1-87	1,500	—	1,500	37.50	48 MO	1,800.00	90.00	23.43	210.87
180030	DESK, OFFICE	—	4-1-87	350	350	—	—	36 MO	350.00	21.00	8.33	74.97
180040	CHAIR, OFFICE	—	4-1-87	100	100	—	—	36 MO	100.00	6.00	2.22	19.98
180050	COMPUTER, IBM	167829	5-6-87	3,000	500	2,500	80.55	36 MO	3,399.80	180.00	55.55	440.40
180060	CALCULATOR	26543	5-8-87	60	60	—	—	—	60.00	3.60	—	—
				14,510.00	3510.00	11,000.00	354.85		16,734.60	870.60	200.64	2,083.54

TOTAL INTEREST = $2,224.60

Equipment inventory log
Figure 4-14

box or other safe place. Don't keep it in the same location as the original.

Look at the log again. Note that inventory item #180060 wasn't depreciated. That's because it's worth less than $100. We'll take the tax write-off when it's sold or scrapped.

Did I leave anything out of the equipment inventory? I probably left out several things you'll want to include. But, most important, I left out the model numbers of the equipment. Maybe we should add a second line for each item. Then we could include warranty terms, service contract information, and who sold it. It's up to you, but here's my theory: the more detailed, the better.

Inventory Log

Figure 4-15 is the inventory log for items other than equipment. The vendor code you see here directs you to the vendor file in Figure 4-13.

I've assigned a "house" part number for each item. Of course, manufacturers have model and part numbers. But you'll be buying from many different sources, so you'll have a conglomeration, and sometimes an overlap, of source part numbers. By assigning your own numbers, you can avoid the possible duplication of part numbers for two or more items that are *not* the same. Standard part numbers for paint type, texture and color are contained in Federal Standard #595a. Chapter 10 tells you where to send for a copy. If you're considering buying a computer, find out what kind of number system will work best, and start using it now. It'll prevent problems when you switch to a computerized system.

Inventory Use Report

Now let's look at Figure 4-16, the inventory use report. It's a single sheet. Make one up for every major item of equipment or category of supplies you stock. (If you buy paint by the job, your inventory will be much smaller.) Because everything shown is tax-deductible, it's a valuable report.

Note that on June 15 your delivery cost went up. Why? Did you change vendors? Did the vendor raise his prices? Better have a talk with him — he just cost you $3.00. And, since the cost went up, so did the sales tax. He also charged you $5.00 freight. The total increase is $8.20. Almost the cost of a gallon of good latex.

ACE PAINTING CO — INVENTORY LOG

P/N	DESCRIPTION	UNIT	SIZE	VENDOR	MFG	MFG. P. N.
10001	PAPER TOWELS	ROLL	8" X 11"	00025	SMC	—
10002	PAPER DROPS	CA	9'X12'	00025	SMC	—
10003	RAGS	BDL	STD	00025	SMC	—
↓						
10101	ENAMEL, TINT BASE	EA	GAL	00030	PAINTCO	03547
10102	ENAMEL, TINT BASE	EA	QT	00030	PAINTCO	03548
↓						
20100	COLOR, YEL	TUBE	15 OZ	00030	PAINTCO	106531
20101	COLOR, RED	TUBE	15 OZ	00030	PAINTCO	106532

Inventory log
Figure 4-15

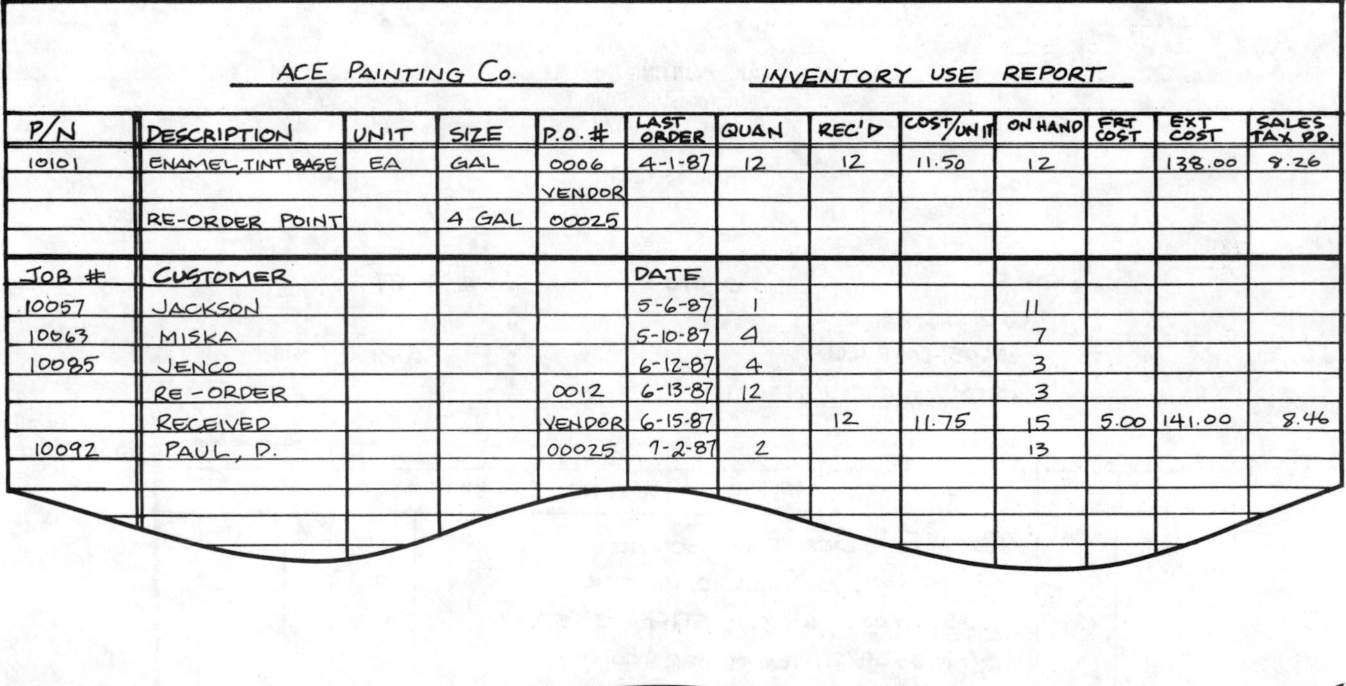

ACE PAINTING CO. — INVENTORY USE REPORT

P/N	DESCRIPTION	UNIT	SIZE	P.O. #	LAST ORDER	QUAN	REC'D	COST/UNIT	ON HAND	FRT COST	EXT COST	SALES TAX PD.
10101	ENAMEL, TINT BASE	EA	GAL	0006	4-1-87	12	12	11.50	12		138.00	8.26
				VENDOR								
	RE-ORDER POINT		4 GAL	00025								
JOB #	CUSTOMER				DATE							
10057	JACKSON				5-6-87	1			11			
10063	MISKA				5-10-87	4			7			
10085	JENCO				6-12-87	4			3			
	RE-ORDER			0012	6-13-87	12			3			
	RECEIVED			VENDOR	6-15-87		12	11.75	15	5.00	141.00	8.46
10092	PAUL, D.			00025	7-2-87	2			13			

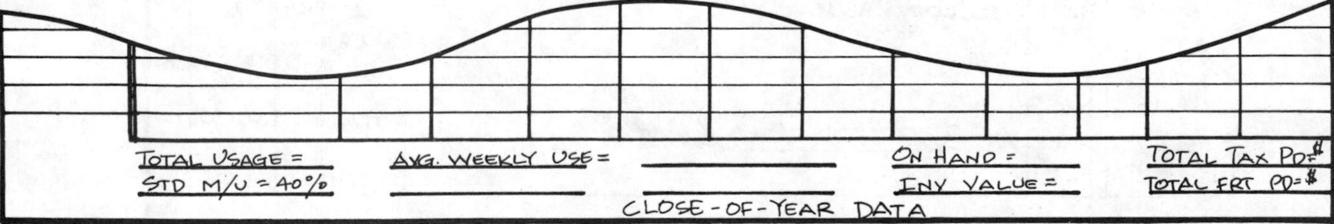

TOTAL USAGE =
STD M/U = 40%
AVG. WEEKLY USE =
ON HAND =
INV VALUE =
TOTAL TAX PD= $
TOTAL FRT PD= $
CLOSE-OF-YEAR DATA

Inventory use report
Figure 4-16

If the increase is legitimate, raise your selling price. If it isn't, use this sheet to show the vendor his mistake. Maybe he forgot to give you your discount. Or maybe he hired new help who didn't know that you don't pay a delivery charge.

This sheet is also your use history. Show it to your dealer to prove you deserve a bigger discount.

Re-order points— Note that we've established a re-order point for this item. A re-order point is the date when you must order more materials. Maybe you have to buy a minimum amount to get your discount. Or maybe you must get an item in before the current stock is used up.

It's best to stock high-use items like this one. Don't count on the vendor having it in stock when you need it for your customers. There's nothing worse than dropping in at the paint store at 7 AM to find they don't have an item you need for the job at 8 AM. Running out of stock can cause you to miss promised delivery dates. The result is lost time, employees paid to do nothing, and the loss of customers and your good reputation.

Average use— At the bottom of the form is some year-end data, and one piece of information you should figure out more than once a year — the average weekly use. Do it for each part number. It tells you whether some item, like primer paint, is getting stale. It tells you whether you're buying too much or too little in a given time period. Here's how to figure average use:

1) Subtract the current date from the last purchase date to find how many weeks have passed.

2) Subtract the amount ordered from amount on hand to find how much you've used.

3) Divide how much you've used by the number of weeks that have passed, to find how much you used each week.

Why keep inventory use files?— Here are the advantages. It lets you project future needs so you can budget your money better. It lets you see where your dollars are invested. And it gives you your markup (M/U), and the potential market value.

ACE PAINTING CO
2468 Industrial Road
Glidden, WA 20030

(206)555-1234

No. 0001 Customer's Order 00010 Date 4-1-87

SOLD TO: WV MOTORS
29605 LOST HILLS
CITY, ZIP

SHIPPED TO: SAME

DATE SHIPPED	SHIPPED VIA	TERMS	F.O.B.	SALESMAN
	PICK-UP	ABC BANK LOAN	DEALER	BILL

PURCHASE ORDER FOR ZT-80
CARGO VAN, 1/2 T, RADIO, HEATER
PASS. SEAT. 6 CYL, STICK 3 SPEED
S/N 358972. TERMS PER CONTRACT,
COLOR BLUE

SALE PRICE	$9,500.00
SALES TAX	570.00
FRT	100.00
TOTAL	$10,170.00

x Bill Jones Acct. Rep.

Sample purchase order
Figure 4-17

Poor inventory control ties up cash that you should be using to make money. I don't care if it's $250 or $350,000. If something's not useable, or not selling, get rid of it. Think of inventory as fertilizer. Sitting idle in a pile it's doing nothing for anyone. Spread it around and put it to work and it does wonders.

Purchase Orders

So how do you order all these items you use? With a purchase order (PO), of course. Look at Figure 4-17. Note that it's the same general-purpose form we used for the customer quote and invoice. The PO number, at the top left, directs you to your purchase order log. Or, if you wish, to your checkbook and the number of the check you paid with.

Purchase Order Log

Keep a log of all purchase orders you issue. Look at Figure 4-18. You can see at a glance that PO #0002 is still open. No stock has come in yet. And PO #0003 is on account, with monthly payments due. Also, there were some back orders (B.O.'s). Better check to see if you need that stuff for a job. If so, can you get it from another vendor?

Charts

It helps if you can see the numbers. Do this with charts. I love charts. They show me at a glance where my business has been, where it is, and where it's going. They point out the changes faster than a list of numbers ever could. I keep 30-day charts on the following:

- Quotes given versus quotes turned into jobs
- Income versus total expenses
- Each expense item, such as advertising, phone, postage, shop supplies, and inventory dollars

It's easy to do. Just lay out a separate sheet of 8½" by 11" graph paper for each item. Chart dollars upwards along the left-hand edge. Chart time — in days, weeks, or months — along the bottom horizontal line. Get the figures from the files you created. Plot these figures on the graph. Finally, it's connect-the-dots time.

What will you see? You'll see whether your business is growing. You'll see when expenses are about to get out of hand. You'll see peaks and dips as you go through holidays, the seasons, the vacation months, and other time periods. (This lets you adjust your schedules, and your material purchases, in the future.)

You'll see business drop, and know it's time to spend more time and money on promotion. You'll see business boom, and know it's time to raise prices or add more help. You'll see the best time for a vacation for you or your men. You'll be able to spot trends in your business. You'll know when something's wrong, like a sharp drop in quotes converted. (This usually means a fly-by-night in the area.) You can spot overall trends in your business area, like a change in the makeup of the population.

Charts can show you a lot. But you have to make them and use them. They're a tool for your success. Try it. Look at Figure 4-19 for some samples.

ACE PAINTING CO — P.O. LOG

P.O. #	VENDOR #	DATE	DATE REC	BACK ORDER	B.O. REC	INVOICE DATE	DATE PAID	CK #	MATERIAL COST	SALES TAX	FRT IN	TOTAL
0001	00030	4-1	4-3	—	—	4-15	4-20	128	50.00	3.00	2.00	55.00
0002	00030	4-1		4-3		—						
0003	00015	4-2	4-2	—	—	4-15	4-30	131/55.00	650.00	39.00	12.00	701.00
0004	00025	4-3	4-3	—	—	4-3	4-3	CASH	10.00	.60	—	10.60
0005	00030	4-4	4-7	4-7	4-9	4-30	5-5	146	28.50	1.71	—	30.21

Purchase order log
Figure 4-18

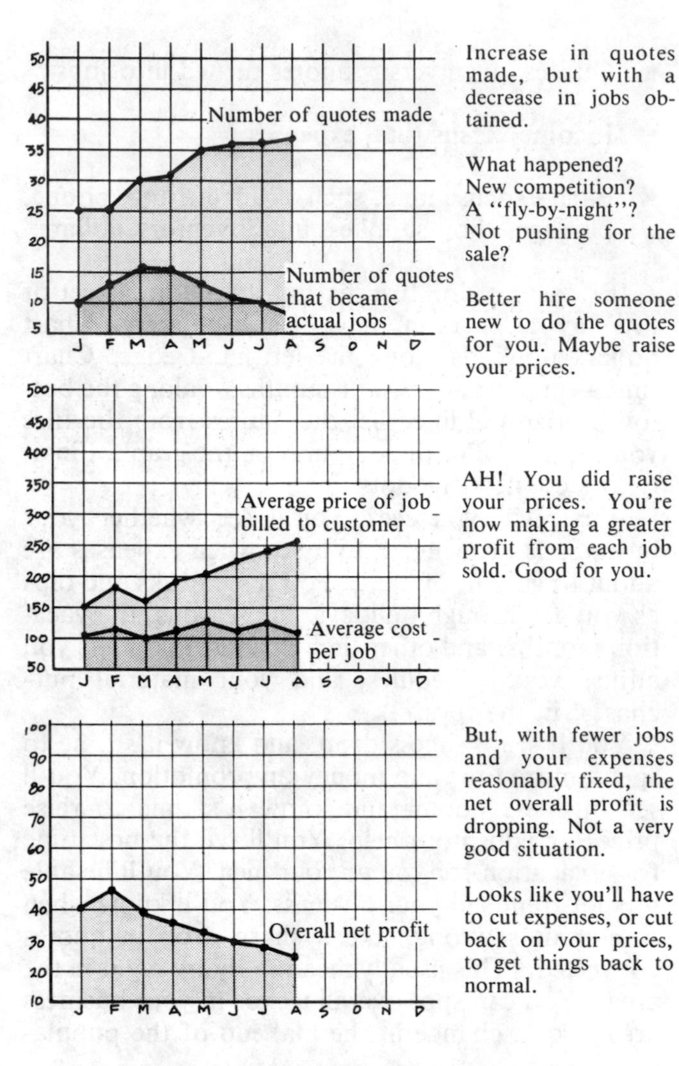

Using charts for management control
Figure 4-19

Computer Accounting

Finally, I'm going to jump into the future for a moment or two — to the time when your business is so prosperous that it just takes too long to do your accounting by hand. Is this the time to get a computer? Most businesses these days either have a computer or have access to one. After reading the last two chapters you can see how much paperwork is required. A computer can help you sort out all these figures and facts.

Computers are great for things like letter writing, spreadsheets and even routine accounting. Here's why. Nearly every company needs about the same program. My paint company can use the same letter writing program your insurance agent uses — and that program costs only a few hundred dollars.

But get much more complex than that and you're into the area of custom programming. No two paint companies are exactly alike. Each has slightly different needs for inventory control, billing and management reports. No standard program is going to fit your requirements exactly. That leaves only two choices: either spend several (perhaps many) thousand dollars to have custom programs written and maintained, or try to adapt your procedures to a program already written by someone who knows nothing about the way you do business. That's seldom an easy choice.

I've seen million-dollar companies try to computerize their "by-hand" systems. It was an enormous job. Number systems (part numbers, invoice numbers, vendor numbers, and so on) had been set up at random. Many had been changed several times over the years. It took months of program conversions and de-bugging to bring it all under computer — and management — control.

No painting business is going to fail because it didn't have a computer. But some will go belly up because the owner spent too much time and money trying to get a computer to do what could have been done just as easily by hand. If you have a good office system but can't handle the volume, a computer may help. If your office records are a mess, getting a computer will just make them worse. Here's my advice: Stick to painting and stay out of computer stores as long as possible.

If You Can't Resist

If you're determined to try using a computer, be advised at the outset that selecting a computer is an adventure — and seldom is as easy and painless as the computer ads imply. Start by doing your research. Talk to contractors who already have computers. Buy some computer books and magazines. One book that should be relevant to your operation is *Computers: The Builder's New Tool.* You can order a copy from the order form in the back of this book. Have a clear idea of what you need before you talk to any salespeople. You'll need to buy both hardware and software, but software is the key.

Hardware— This is the machine itself. All computers have a CPU (Central Processing Unit). That's what does the number-crunching. The rest of the box supports the "peripherals" you'll need:

- A screen. It's also called a CRT (Cathode Ray Tube), or a monitor. Get one with as large a screen as you can afford. Be sure it has a glare-free screen, or else get a screen cover.

• A keyboard. Get one that feels comfortable to your hands. A number pad to the right side of the keyboard is a must.

• The storage media. There are two types: disk storage and memory storage. Buy as much storage capacity as you can afford. A practical minimum for memory storage is 512 kilobytes. If you can afford it, get a 10 megabyte hard disk. It holds a lot: All your programs and information can be on one disk. It'll take a long time to fill up. And hard disks are faster than floppy disks.

• You'll also need a printer. A "daisy-wheel" letter quality printer is essential for letter writing. Its printing looks like it was done on a typewriter, not by a computer.

Software— This means programs. Without software, the computer is just a box of parts. Don't even consider buying a computer until you've found the key programs you want to use. Talk to people who are actually using the ones you're interested in. Test those programs. Look for reasons why they won't work in your office. There's usually no practical way to modify most of the programs sold in computer stores. Either they work "as is" or they'll never work at all.

If you don't love trying to figure things out by reading directions from a manual, you're going to hate computers. Learning to use almost any computer program can be a chore. My rule of thumb is to allow one hour of learning time for every $5 in program costs. So for a $250 program, expect to spend 50 hours with the computer before you're really using the program as it's intended to be used. If you've got better things to do with your time, leave computers to those who really love them.

We've covered some pretty important points in this chapter. I hope you'll refer back to this information many times over the next few months. As your company grows, you're going to need all, or nearly all, of the forms and records I've covered. When your company is bigger than mine, you'll need lots more. And that's the subject of the next chapter: bringing in the new customers you need to keep growing and to stay prosperous.

chapter five

Finding Your Gravy Train

We've covered setting up your business, some of the legal and tax considerations, and accounting and record keeping. The next question is: "What are you going to paint?" Your answer: "Houses of course — what else?"

Okay, but what if . . .? What if there just isn't enough business? What if your competition is strongly established? What if your Yellow Pages already have 12 pages of house painters? What then?

Fortunately, there are *many* other choices — and most will be more profitable than house painting. But most *will* require an investment in special equipment. I call these *specialty* paint contracting businesses.

Whatever you want to paint, there's a market for you out there. You just have to find it. This chapter will help you do just that: It's a catalog of paint contracting specialties. One of these specialties may be just what you're looking for. If you don't find it here, don't give up. I've only scratched the surface. There are more paint specialties than you can imagine — and many are far more profitable than any house painting company you're likely to run into.

Paint Contracting with Specialty Equipment

To become a specialist you need equipment and skills that the average general painting contractor doesn't have, and that the newcomer certainly doesn't have. In fact, in some areas they can't even rent the equipment. This equipment can cost thousands of dollars — but has a reasonably fast payback if it's used regularly. (Payback is the time it takes for earnings to pay off the purchase price.) This equipment is also tax deductible, as capital equipment.

What kinds of equipment are we talking about? Sandblasters, waterblasters, texture coating hopper spray systems, fiberglassing spray systems, portable airless spray systems, electrostatic spray systems, heated hopper spray systems, bucket lifts, scaffolding, and road striper systems. Most of these units cost from $1,200 to $4,000 new. Rebuilt, they'll cost about 50% of the new price. Used and worn, they'll cost about 25% of the new price.

Which ones do you need? A full-line general painting contractor needs them all — or, at least, access to all. This isn't practical for you since it would be too big an investment. Or is it, in your case? Do you have the money or the credit to swing the purchase? Much of this equipment isn't readily available in rental yards. This is where you can fit in.

Rentals

How about you renting the equipment? You wouldn't be doing any painting yourself. Instead, you'd rent the equipment to others. Of course, your investment will be large, since you'll need an assortment of good-quality equipment. To start with, you should have:

- At least two airless units
- One waterblaster
- One sandblaster
- One fiberglass spray outfit
- One texture hopper spray system
- At least two different types of sanders
- Several scaffolds
- Assorted ladders, including a few 30-foot ones
- An assortment of brushes, rollers, covers, sandpapers, edge guides, painting tarps, spray equipment and replacement tips. These you'll sell.

Your total investment in equipment will be about $16,000, plus tax. With your other start-up expenses, you can easily invest $20,000 or more.

How much can you earn? Daily rental income, if everything is rented, can be about $350. Payback could happen in 50 to 60 rental days. Rental days? They're not like week days: rental days are the days each unit is actually rented. For some equipment, payback could take as long as a year. Before you invest all that money, check with local contractors. Can you really keep all the equipment rented out? An item isn't doing anything but depreciating if it sits in your shop.

Be sure to have a lawyer draw up a good rental contract. And get a deposit: One day's rental plus 20% is common. Also, get the contractor's license number, a credit card number, or some other satisfactory ID and security.

Be sure to maintain the equipment in good condition. You must have insurance against theft and liability. Make sure the user — who may not be the person who rents from you — knows how to use the equipment, including safety precautions.

You don't need a store: You can rent equipment from your home or truck. If you rent from a truck, add the extra service of delivery and pickup at the job site. Of course, a store is a definite plus. Another possibility is to work with a local paint store.

There are several other ways to make a good living if you own the right painting equipment. You buy the equipment and rent it, and your services, to homeowners or contractors. Later sections in this book offer more details.

Sandblasting Service

Many surfaces have to be sandblasted — there's no other economical way to clean them. Stucco homes, heavy equipment, farm equipment, cement walls and floors, block walls, tanks and pools are examples. These all need sandblasting to completely remove the old finish before re-coating.

Besides a sandblaster, you'll need a belt or disk sander, a block and tackle, ladders, and maybe scaffolding and a heat stripping unit.

Waterblasting Service

This is similar to sandblasting but uses a high-pressure jet of water instead of sand. Waterblasting strips old coatings from wood sidings, fences, and barns. It also works on most concrete and cement.

You'll need a waterblaster, ladders, and scaffolding.

Textured Wall Coatings Application

You need a hopper gun spray system and a sander for this. It can be a good business in areas with lots of new construction. Textured wall coatings make life easier for the building contractors, since less care is needed during the drywall hanging process. And most painting contractors don't have the equipment needed for this type of work.

Acoustical Ceiling Application

You need a hopper gun spray system and a ladder for this, too. You apply the textured coating to ceilings instead of walls. (These ceilings are common in tracts in California.) Again, your big market is new construction. There is, however, also a big market for the re-spray of water-damaged ceilings, or of ceilings in homes built before 1974. Few, if any, homes were done with textured ceilings before this. Now homeowners want to upgrade. You can get a more fancy effect by adding sparkles with a small hand sprayer.

Fiberglassing

You'll need a fiberglassing outfit, sanders, ladders, and maybe a sandblaster. Fiberglassing is mainly used for lawn furniture, boat hulls, auto and recreational vehicles, and pool areas. Because it's used mainly for boats and other marine uses, fiberglassing businesses are well-established on the coasts and inland lake areas. You can also do fiberglassing work for OEM's (original equipment manufacturers). It's new to the home industry, so the market is basically untapped. And fiberglassing is one of the only practical materials for large monolithic (one-piece) structures. The potentials are unlimited.

Electrostatic Spray Painting

This system uses an electrical charge to direct the paint coating onto and around a metal surface. It's the only hand-type spray system that will work on many objects. Your electrostatic work will be mostly for OEM's, but it's not limited to the OEM market.

You'll need electrostatic spray outfits for both liquid and powder, sanders, ladders, and a sandblaster.

Road Striping

This is a good little specialty business. You use your equipment to apply special reflective paint to roads, highways, and parking lots.

Since you're dealing mostly with government agencies, this business brings in top dollars. It might be hard to get those first few contracts, but once you do, you're in. And road maintenance work has a spin-off: painting rails and bridges. The rail in Figure 5-1 needs paint, and someone will get the contract.

Consider road striping, rail and bridge painting
Figure 5-1

In the meanwhile, there's always the private-sector market to get you started. Parking lots are everywhere, and they need repainting every few years. Figure 5-2 shows a typical parking lot job.

That covers several small specialty businesses, using most of the major equipment. One more item comes to mind, but it's expensive.

Tank Truck Bulk Sprayer

This is a small truck, on which is mounted a 500-gallon tank, an air compressor, and a spray system. This equipment costs at least $20,000. You'll also need ladders.

Parking lot striping is always in demand
Figure 5-2

What do you spray? Paint, of course. Also, coatings for shake roofs, coatings for mobile home roofs, and insecticides for homes and lawns. If you add a heater to thin out the material, you can spray asphalt coatings for flat roofs, fences, and driveways.

Lower-Cost Specialty Painting Business

But what if you who don't happen to have several thousand in your wallet to invest in capital equipment? You can still find your spot in the marketplace — your niche. Here's a group of lower-cost small specialty painting businesses to look into. (Later sections of this book tell you just how to do most of these jobs.)

Bathtub and Shower Enclosure Refinishing

You need a small portable air compressor and a conventional spray gun. They'll cost under $200. You'll use either polyurethane, or epoxy, or both. It depends on whether you're covering fiberglass or enamel. Check with your local paint store.

Customers are willing to pay up to 70% of the cost of a new tub or shower. Why? Since most tubs and shower enclosures are installed before the drywall is applied, replacing them is a major project. Your prime markets are older neighborhoods and rental neighborhoods. Besides refinishing, you can also offer a complete color change to match a customer's decor.

Wood Floor Refinishing

You'll need drum and disc sanders for this. You can rent both from most equipment rental shops.

First, you completely strip all the old wax and finish down to the bare wood. After sanding, you may restain the floor. Finally, clear-coat with a polyurethane formulated for use on wood floors. The operation is nearly the same for all wood floors. This is one of the few times you can probably charge by the square foot. If you're good, and if you've practiced (or apprenticed) on floor refinishing, then accept parquet floor refinishing jobs. If not, forget even trying to do parquet floors. It takes a lot of skill and time to do them right. Stay with straight grained plank floors.

Cabinet Refinishing

This is another good specialty niche. But it does take a lot of practice and experience to do it right. Once again, your customers are usually willing to pay a good price — up to 60% of replacement cost. You can sell new hardware, too, if you have a good supplier. You need sanders, a compressor, a small step ladder, and a heat stripping unit.

Some companies have grown very quickly by offering a two-day cabinet refinishing system. It works like this: First, remove all the doors and hardware. Next, sand the doors. Finally, recover them with wood or plastic laminate. Doing the job this way, you only have to stain and clear-coat the framing.

Sign and Mural Painting

Are you artistic? Then here's your niche. Signs are needed everywhere. You'll need to learn silkscreening and need specialized equipment — a compressor, ladders, and a silkscreen setup. Everywhere you look, you sees signs like the one in Figure 5-3 that need repainting.

Sign painting can be profitable
Figure 5-3

Mural painters aren't much in demand, but if you can get into the real high-priced home market, and if you're good, and I mean *good,* you can demand prices that will make your head spin. Stores and offices may also be clients, paying you for decorating the insides and outsides of otherwise dull buildings. I've seen murals done in miniature tiles, and murals done in pre-painted tile. Then, again, there's always mural wallpaper.

Speaking of wallpaper, there's another niche. Not in installing it, but in removing it.

Wallpaper Stripping

This job is time-consuming. And it isn't easy. Most paint contractors and homeowners would rather pay someone else to do it. Your tools are a wallpaper steamer, a ladder, scrapers, a razor knife, some sponges, some solvents, and a lot of plastic bags. Again, this is a job you can generally price by the square foot.

Small Specialty Businesses

Do you like the idea of turning something old and worn into something of beauty? Then read on for a few interesting, and maybe profitable, ideas.

Vinyl Repair and Coloring

Everyone has some vinyl object — a chair, a couch, a car seat, a car top — that needs repair. To get into this business, write to Mohawk Products for their catalog. It has all the information, products, tools and kits you'll need to get started in the field. What about those ads in the trade magazines? The ones that want hundreds of dollars for training and kits? Better check Mohawk out first. Write to:

Mohawk Finishing Products, Inc.
Rt 30 N
Amsterdam, NY 12010

Antique and Furniture Refinishing

This is an art that can be very satisfying, if done right. Though not a big-volume business, it's a good sideline. It's interesting work — you'll venture into ''antiquing,'' bronzing, staining, and applying decals. You'll also have to do some repair work.

One caution: What may look like junk when it comes in may be priceless to your customers. Practice on used items you pick up at garage sales.

You need sanders, a compressor, and a heat stripping unit. You can get other materials and supplies from Mohawk Products. Also, check out Albert Constantine and Sons, Inc. These two companies can supply just about everything you'll need.

Another question: What's an antique these days? It's anything over 40 years old. A few years ago it was anything over 100 years old, but the nostalgia fad over the last decade has changed that. Now it's 40? I guess that includes me, and many homes. Which brings us to another group of specialty businesses.

Handyman and Painting Businesses

Are you a good handyman as well as a good painter? If so, here are some more business possibilities for you.

Old Home and Building Restoration

The restoration business is booming. Your main tools are an airless unit, sanders, ladders, a heat stripping unit, a hopper gun, scaffolding, a wallpaper steamer, and a waterblaster. Buy or rent other tools as they're needed.

You'll probably work three to six months on each house, bringing a run-down mess up to code and modernizing it. You can do the work on a contract basis or buy the home, live in it while doing the remodeling, and then sell it when finished.

Try buying buildings in foreclosure sales or by making some arrangement with a delinquent owner. You can usually get excellent terms (no money down) if the building is really a derelict. Restore the building while you live in it. You'll pay at least $20,000 for most of the materials you use. But some material yards will sell on credit or accept payment on sale of the building.

The very first day you take ownership is the day you put the place up for sale. Then work hard, and hope that by completion you'll have it sold and have your profit. A good profit is 10% to 15% of the sale price, so you'd better be good at estimating your costs before you buy. One thing I can guarantee is that you'll get your share of painting — and of painting problems. So study the chapter on common problems and their solutions: You'll run into almost everything I've included.

Water-Damage Repair

This is another handyman/painting specialty business. Almost all buildings suffer water damage at some time. Your job is to learn to spot the problem, to know the solution, and to do the prep work for the homeowner or general painting contractor. You'll need sanders, a heat stripping unit, ladders, scaffolding, and a waterblaster. A waterblaster to repair water damage? Certainly. A waterblaster not only cuts through old paint, it can also cut away failed stucco, block, brick, and concrete.

This business will lead to a lot of work, and to the chance to bid the actual painting. The following section describes a possible sub-specialty of your water-damage repair business.

Basement and retaining wall waterproofing— This applies to swimming pools, too. This service is especially needed in the east and north sections of the country. The techniques you need to know, and the products for you to use, are available from several companies. The best source of information I know of is:

Thoro Systems Products
7800 NW 38th Street
Miami, Florida 33166

Concrete Preparation

You're not afraid to play with acids? Then maybe "concrete prep" work is your niche. Pools, walks, slabs, walls, driveways, and a host of other items are made of concrete. They must all be prepared and cleaned before being painted. Most do-it-yourselfers, and many contractors, just don't like to handle the acids that must be used. They're probably smart — acids aren't exactly toys to be played with by people who don't know what they're doing. And many don't understand the proper acid solutions and neutralizer requirements.

If you're prepared to devote the time and effort to study what acids do what in what solutions, what safety precautions must be taken, how they must be stored, what neutralizers to use — you can be the local acid expert. Besides acids, you'll need a heat stripping unit, a sandblaster, and a waterblaster.

Cleaning and Prep Business

Another specialty market is found in retirement or lower-income areas. These are the areas with many pre-fabs, trailers, and mobile homes. Most have aluminum roofs and siding that oxidizes or chalks. You can start your business with as little as a bucket of soapy water, a hose, and a mop. Later, when business picks up, get yourself a waterblaster. Not only is simple cleaning good business; if your customers want to repaint, your service becomes a must. Just remember that you're dealing with people on low or fixed incomes, so price your work accordingly. Many times a $200 wash job can save (or put off) a $800 paint job.

Do the high work — like the top of this barn
Figure 5-4

The Over-30 Market

Not afraid of heights? A lot of painters are, and so are most do-it-yourselfers. Most homeowners don't own ladders long enough to reach the roof of a two-story house. More important, they know that moving a big ladder several dozen times takes too long and is dangerous. And many painting contractors don't much like the idea either — especially when there are plenty of one-story homes to work on. So here's your niche. You can specialize in the over-30 market. Over 30 feet high, that is. Like finishing off the top of the barn in Figure 5-4.

Work with the contractor or the do-it-yourselfer. You do what they won't. For equipment, you'll need scaffolds, ladders, and possibly even roof jacks or block and tackle to do this safely. You should have a spray unit, sanders, and a sandblaster or waterblaster, too. Just by advertising that you'll do work over 30 feet high, you'll get lots of complete painting jobs.

Businesses in the Great Outdoors

Do you like to be outdoors? O.K. Here are some outdoor specialties that don't include painting houses. They involve all sorts of structures and items that require special coatings or techniques.

Farm Building and Fence Painting

This is a good one if you live in a rural area. You can get into it for low to moderate cost. A waterblaster is a plus, but not always necessary. A portable power generator and an airless rig can make life easier. Again, they aren't essential. What *is* essential is a special selection of coatings that are safe for the animals that will be in contact with them — and will try to eat them. Most latex paints will do the trick, though not always. On horse ranches, for example, you'll usually be required to paint the fencing with hot tar.

Hot tar coatings— For this, you'll need a portable melting pot unit and spray gun. On pig farms you have a special problem: pigs eat walls. Most rural building supply houses can fill you in on what types of coatings are acceptable for various farm buildings and animals.

Tree Painting

You apply insect-repellent paint to tree trunks, from ground level to about five feet above ground level. And you paint all the stubs left after limbs

Tanks and pipe need regular repaints
Figure 5-5

have been trimmed off. Otherwise, insects get into the tree and kill it.

A portable airless rig and power generator are a plus for this type of business. And you'll need several different kinds of ladders to do it safely. Potential customers are homeowners, orchards, tree trimming companies, local communities, and power companies.

Painting Outdoor Furniture and Planters

This is a good market for a specialist. Most wood decks, and many homes, are built of redwood. Homeowner's association parks — and many local government parks — have redwood benches, tables, buildings, and playground equipment. All these items require a specially-formulated coating. And they all need recoating once per season, either in late fall or early spring. Besides ladders and sanders, a portable airless rig and a power generator are a plus.

Pipe, Tank, and Bridge Painting

These items always need painting. This market can be hard to break into, since you're dealing with government agencies, or with government-size private industry. But the pay is excellent. And one good contract can have you set for life.

Your primary markets are the oil and utility companies. They can be hard to please. You'd better know the safety rules. You also need to know government and OSHA (Occupational Safety and Health Association) specifications. Many of the coatings used are marine-type enamels and polyurethanes, containing insecticides and fungicides. Tanks like the one in Figure 5-5 need to be repainted regularly.

You'll have to estimate material requirements for odd shapes. (See Chapter 7, on estimating.) And you'll need equipment like airless sprayers, sanders, a tank truck, ladders, scaffolding, and a sandblaster. If this market's a little over your head, look into the next one.

I know a painter who has only one customer. Don't feel sorry for him. His customer is one of the biggest oil companies in the country. He paints nothing but gas stations — day after day, week after week. Monotonous? Maybe. But he seems to tolerate the monotony of buying himself a new Cadillac every October also. But in all honesty, he's pretty good at it. Year after year his contract is renewed.

Construction and Farm Equipment Painting

Equipment like this always needs to be re-painted. Check Figure 5-6. Most farm equipment isn't quite this bad, but you can see there's a market here. You'll need a portable airless sprayer and a sandblaster to do the work profitably.

Specialty Businesses Indoors

So much for the great outdoors. Say you're an inside person and don't like to travel. Then let the work come to you. Here are a few businesses where exactly that happens.

Painting for Original Equipment Manufacturers (OEM's)

Many OEM's have the facilities for building their product, but not the facilities for painting what they build. You can offer that service. You'll need a building (owned or rented), a conventional spray and air compressor outfit, drying racks or heat lamps, and a spray booth. Depending on the size, shape, and material of the items, you may also need a pickling tank, a dip tank, and an electrostatic spray outfit.

You can see it's not a low-cost business to get into. But that works in your favor. Most manufacturers and general painting contractors won't in-

Farm equipment needs paint, too
Figure 5-6

vest in their own system. They'd rather use your services. Look at Figure 5-7. The painter who does the industrial painting for this one company grosses over $130,000 per year.

You'll have two big problems in setting up this business. One is money. You need about $40,000. The other is your local building department. You must have a building that meets ventilation and fire safety code requirements.

A Painter's Bookkeeping Service

No painting involved, but related to it. Most small paint contractors need a bookkeeping service that understands, and specializes in, service to the small contractor. Most painting contractors are far happier handling a paintbrush than an accounting ledger. You can do it for them. While they sweat it out with a scraper and paint fumes, you sit comfortably in your office. You'll need a small computer to do this properly.

Your Own Store

Then there's always this alternative: own your own paint store. Sell the products to the people who go out there and do the painting. You could also rent them the equipment they need. But again, this is a business that takes money. If this sounds interesting, contact major paint companies for their dealer price list and possible franchise opportunities.

Consulting and Estimating Service

Do you enjoy meeting people? There's a market for a consulting and estimating service. Your potential customers are do-it-yourselfers, small general contractors, and governments. You do the legwork, the troubleshooting, the color selection recommendations, and the estimating. You could even set up a weekend school for do-it-yourselfers and trainees. The only thing I have to say here is: You'd better be right at least 95% of the time.

Another possibility is helping newcomers with their business set-ups, financing, paperwork, equipment selection, and advertising.

Touch-up Work

Are you good at detailed work? Do you have good eyesight and color recognition? If so, there's profit in touch-up work.

Everything, at one time or another, gets a nick or two that detract from its looks. Most of the time, a

Industrial painting is lucrative
Figure 5-7

complete refinishing isn't necessary. Automobiles, furniture, appliances, walls, and equipment fall into this group. Once more, check out Mohawk Products. They supply touch-up materials for most anything you'll encounter.

"Specific General" Contracting

Specific general contracting? By this I mean that you stick to certain aspects of the market, like painting and decorating storefronts and lobbies. Or office and factory painting and decorating.

If you're near the water, marine, boat, and equipment painting is good. Automotive painting is needed everywhere. Silkscreen painting is needed in most industrial park areas. Apartment painting is needed in urban areas. A small-job painter is needed everywhere to take the jobs that the big boys won't handle — one door, one window, part of a house. Of course, the customer knows he's going to pay a higher rate for one room than he would for the whole house.

Painting contractors often need extra help. Their workers don't always show up. They've promised to start a job on Wednesday and here it is Tuesday and there's two days of work left in their current job. Here you come to the rescue to save their reputation for reliability. It's going to cost them, of course. Larger contractors often find themselves behind schedule. Or maybe they just underestimated the manpower required. Now they need a pro. You contract to be that pro. Or take advantage of the fact that larger contractors often like to get in, do the work, and get out fast. Prep work slows them down — so you contract to do their prep work.

There's a chapter later in this book on some of the alternatives to painting. There may be a good niche for you there. Even if you're a large paint contractor, some of these options could be a profitable sideline for your business.

Well, if nothing else, I hope you now have some idea of the potential market for your services. The paint and coatings field is big. Finding that niche, that place you fit into, can be rewarding. Not only with profits, but also with personal accomplishment. If you don't find that niche, you're going to be battling it out with every guy in town who owns a paintbrush.

Remember my friend who paints gas stations. He started by painting just one station, taking a chance. It turned into a gravy train for him. I hope this chapter has offered some ideas on how to find your gravy train.

Selling

If you've followed the steps in the first five chapters, you're ready to go. Your business is set up. Your paperwork and equipment are in shape. You've decided what share of the market to make your own. You're ready to start bringing in the profits — except for one minor point. You don't have any jobs yet. You haven't got any work to do.

Unfortunately, getting plenty of the right kind of work isn't always easy. For most paint contractors it takes some selling. Your ads in the Yellow Pages and classified section of the local paper will bring in inquiries. You'll get calls and requests for bids. But those aren't jobs. Every prospect that calls you probably calls two other painters. And some will get a dozen bids and end up doing the job themselves. You've got to convert those calls into jobs if you're going to survive and thrive in the painting business.

There's more to making money in the painting business than just being ready and able to do the job. There's more to getting work than just quoting the lowest price. Every paint contracting company needs someone who is good at creative selling, getting work at a fair price by meeting the owner's needs and requirements — even when some cut-throat operator has quoted a much lower price.

If you're a better painter than salesman, this chapter is for you. It may offer more valuable information than all the other chapters in this manual combined. If you're an experienced salesman, you should still be able to pick up some tips that will help you close more sales. The most important point in the chapter is this: *Don't underestimate the importance of salesmanship.* It can make a big difference to your company, to you, and to your family. Some very mediocre painters have developed outstanding incomes simply because they were good at selling themselves and their services. Even the best painter in the world is going to be a starving painter until someone hires him.

Learning to Sell

You'll meet many painting salespeople who aren't really salespeople at all. Sure, they have a sales job. But they don't know even the basic principles and techniques of selling. Fortunately, the basic principles are easy to learn and use. The best-paid salespeople working for the most professional companies in the industry use these three basic principles:

1) Know your service.

2) Overcome the objections.

3) Shut up!

Sounds simple, doesn't it? It *is* simple. But poor salespeople don't follow these three easy rules.

Step 1: Know Your Service

Try this. Sit down in front of a mirror. Now, talk as long as you can about painting. Keep talking until you can't think of anything else to say. Time yourself.

Did you talk for more than twenty minutes? If not, you don't know enough. How can you expect to talk to potential customers? How can you answer their questions? How can you be the expert they trust? Remember, the first rule of selling is *know your service.*

Know what the customer wants and needs— These aren't always the same thing. They might want the wrong kind of paint for a specific application. Or they want you to paint a wall in a color that doesn't complement their decor. Or they want their home's exterior painted in a color that doesn't fit in with the other houses on the street. Guide them into making the right decisions. You'll both benefit, and your customers will thank you for advising them.

Remember, they called you because you're the expert. Good selling means using the facts to your advantage. Be precise and accurate, not pushy. Your job is to guide, not to demand.

Step 2: Overcome the Objections

Next, learn how to overcome objections. Here's an exercise that will help. Find someone who can spare some time. (At least 30 minutes.) Both of you sit down. Now, try to sell a paint job to your victim. Keep trying — but have them offer objections every time you ask for the order. No matter what you say, they should have a reason for not buying from you. See if you can answer every objection. If you can't, you don't know what you're selling. Go back to Step 1.

If you answered every objection, you're halfway to success. Try this with a couple of different people, so you have to answer a wide range of objections.

Here's something else to remember: The customer came to you. He wants to buy; there's a job he wants to have done. If he's balking, and objecting to buying from you, then there must be a reason. *You have to find that reason.*

Some typical objections— The list below is just a sample of the kinds of objections you're likely to hear.

- I'm only getting estimates right now, I'll call you later.
- How do I know you'll do a good job?
- Gee, I didn't realize it would cost that much.
- Boy, I really hoped to get the hallway painted too. (This person is after a "freebie.")
- I don't like that brand of paint.

Start with yourself. Ask them, flat out, if it's *you* they object to. They'll always say no. Keep asking questions. Each time they say no, you're that much closer to making the sale. You're getting to the *real* reason they don't want to buy from you. But save the money question (their objection to your price) for last.

Find the real objection— Sooner or later, they'll say "Yes, that's why I'm not ready to buy yet." When they do, start thinking. Is this the real objection? What can you do to overcome this objection? What are the alternatives? If you don't know your product and your business, you may be stuck. Then you can only say, "Thank you," and beat a retreat. But if you *do* know your product and service, try this.

Repeat the objection back to them— Sometimes you can overcome an objection by selling it back to them. It goes something like this:

"Mr. Smith, are you saying that the *only* reason you don't want to have me paint your house is because I use El Cheapo brand paint?" (or whatever their objection was).

Say it several times, in different ways. Once they admit that this is the only reason, you're set up to deliver the clincher. "Would I get the job if I used Rolls-Royce Superior brand with a 50-year unlimited warranty?" . . . (whatever your answer is to the objection).

Remember that list of objections a few lines back? Here's an example of a good way to deal with the first one.

"Mr. Smith, I know you're just getting estimates now. That's fine with me. I'd do the same thing. I'll leave this *written* estimate with you so you can compare it to the others you get. You do, though, understand that I work on a first-come, first-served basis. From our conversation, I'm sure you realize that I'm a professional and will give you a professional job." (It's tempting to cut down your competition at this point. Don't do it, it reflects badly on you). "If you accept my offer today, we'll start Monday. When you've compared the other bids for service, preparation work, paint quality, and workmanship, you'll find my estimate quite competitive. That's the only reason you're delaying your decision — to get other estimates? Correct?"

You've done several things here.

- You complimented their intelligence.
- You submitted a professional *written* estimate.
- You told them that delaying a decision delays the starting date on their job. You may not be available later.
- You asked them to accept you as a professional.
- You offered prompt service.
- You stressed quality. And, by doing so, inferred that someone else may not be as good.
- Most important, you sold their objection back to them and requested an answer.

All this plants a seed of doubt in their mind: Do they really want more estimates? If they answer "yes" to your "Correct?" question, counter with "Why?" This puts them in the difficult position of trying to justify their delay to themselves. If they answer "No," and you counter with "Why?" they have to give you the *real* objection. Now you have one you can answer.

If they object to the price— What if their real objection is the price? This is probably the most common — everyone wants a bargain. Find out whether it's the price itself, or just that they don't have enough cash on hand. If it's price alone, counter with something like this:

"Mr. Smith, my prices are competitive for the quality of work I do. I could give you the job at, say, half price. But do you *really* want a half-price paint job? Your home deserves the preparation work that's in my bid and the quality materials I plan to use."

Do you think they'll say "Yes"? Not likely. But, once again, you planted a seed. That low estimate in their desk drawer just became questionable. Is *that* painter going to splash on another coat without any prep work? Will the paint peel the day after the check clears?

If the reason is that your prospect doesn't have enough cash available, counter with this:

"Mr. Smith, you know you can trust me to do a good job for you. A quality job. A long-lasting job. Would you hire me if I could show you a way to finance the work?"

I'm assuming, of course, that you have financing alternatives available. You may accept credit cards, or you may have your own monthly payment plan. Whichever answer they gave, you've shown your professionalism. And you've maneuvered them into having to make a decision.

If the customer is going to finance the work, he has to receive a *right of rescission* notice from you. A sample contract you may copy and use, and a copy of this federally-required notice, are in Figures 6-1 and 6-2.

Step 3: and Then, Shut Up

When your prospect gives the O.K., you have the job. Hand him the contract and a pen, and then shut up. *Don't say another word.* They'll either sign or they won't. If they sign, thank them, set a date to start the job, and leave. If they don't sign, you still haven't found the real objection. Go back to Step 2 until you find the true reason they're not buying from you.

By the way, never, ever ask a customer to "sign" the contract. People have been taught not to sign anything, especially not contracts. Ask them to "O.K." it. Or to "authorize" it. Or just put it, and a pen, in their hands.

Why do I insist that you shut up after the clinching question? Here's why. You're about to make the sale. You've worked hard on it because they've been objecting. It's likely they just plain ran out of ideas on what to object to. If you open your mouth and say anything, you give them the chance to start the objections again. I don't care if you have to sit in dead silence for an hour. Make them be the first to speak.

Before I go on, let me show you a classic mistake made by inexperienced salespeople:

Customer asks: "Do you have Royal Blue?"

You answer: "Yes."

Customer asks: "Do you have Tutti-Frutti Green?"

"Yes."

"Do you have Passion Pink?"

"Yes."

Proposal and Contract

For Residential Building Construction and Alteration

Date________________19____

To ____

Dear Sir:

We propose to furnish all material and perform all labor necessary to complete the following:

____

____

____

____

____

Job Location:

All of the above work to be completed in a substantial and workmanlike manner according to the drawings, job specifications, and terms and conditions for the sum of

Dollars ($________________)

Payments to be made as the work progresses as follows:____________

____

____

the entire amount of the contract to be paid within________days after substantial completion and acceptance by the owner. The price quoted is for immediate acceptance only. Delay in acceptance will require a verification of prevailing labor and material costs. This offer becomes a contract upon acceptance by contractor but shall be null and void if not executed within 5 days from the date above.

By____________________

"YOU, THE BUYER, MAY CANCEL THIS TRANSACTION AT ANY TIME PRIOR TO MIDNIGHT OF THE THIRD BUSINESS DAY AFTER THE DATE OF THIS TRANSACTION. SEE THE ATTACHED NOTICE OF CANCELLATION FORM FOR AN EXPLANATION OF THIS RIGHT."

You are hereby authorized to furnish all materials and labor required to complete the work according to the drawings, job specifications, and terms and conditions on the back of this proposal, for which we agree to pay the amounts itemized above

Owner ______________________________

Owner ______________________________ **Date**________________

Accepted by Contractor________________________ **Date**________________

Sample contract
Figure 6-1

Notice To Customer Required By Federal Law

You have entered into a transaction on________________which may result in a lien, mortgage, or other security interest on your home. You have a legal right under federal law to cancel this transaction, if you desire to do so, without any penalty or obligation within three business days from the above date or any later date on which all material disclosures required under the Truth in Lending Act have been given to you. If you so cancel the transaction, any lien, mortgage, or other security interest on your home arising from this transaction is automatically void. You are also entitled to receive a refund of any down payment or other consideration if you cancel. If you decide to cancel this transaction, you may do so by notifying

____
(Name of Creditor)

at______________________________________
(Address of Creditor's Place of Business)

by mail or telegram sent not later than midnight of____________. You may also use any other form of written notice identifying the transaction if it is delivered to the above address not later than that time. This notice may be used for the purpose by dating and signing below.

I hereby cancel this transaction.

____________________ ____________________
(Date) (Customer's Signature)

Effect of rescission. When a customer exercises his right to rescind under paragraph (a) of this section, he is not liable for any finance or other charge, and any security interest becomes void upon such a rescission. Within 10 days after receipt of a notice of rescission, the creditor shall return to the customer any money or property given as earnest money, downpayment, or otherwise, and shall take any action necessary or appropriate to reflect the termination of any security interest created under the transaction. If the creditor has delivered any property to the customer, the customer may retain possession of it. Upon the performance of the creditor's obligations under this section, the customer shall tender the property to the creditor, except that if return of the property in kind would be impracticable or inequitable, the customer shall tender its reasonable value. Tender shall be made at the location of the property or at the residence of the customer, at the option of the customer. If the creditor does not take possession of the property within 10 days after tender by the customer, ownership of the property vests in the customer without obligation on his part to pay for it.

Right of rescission notice
Figure 6-2

Well, you get the idea. A thousand colors later, you still haven't made the sale. Your first answer should have been, "Do you want Royal Blue? Are you going to have me do this job in Royal Blue?"

After the *customer* says yes, you say, "O.K., I can do this in Royal Blue." Hand over the pen and the contract and *shut up.*

When Customers Say No

Suppose you've used every trick in your bag, but didn't get the job. Don't give up. You can still turn a profit from that job that seems lost. And there *will* be some lost jobs. You'll run into a do-it-yourselfer who just wants to bounce ideas and figures off you. You'll run into someone whose budget truly can't handle the cost. Many sales calls are wasted on people who, for one reason or another, have no intention of hiring anyone to do the work.

How can you turn these lost hours into a profit? First, never assume a no-sale. Sell, sell, sell. Sell yourself and your knowledge of the product. But if that doesn't work, you can still salvage something for your time. Here's how:

Offer to sell the materials— You are (or should be) buying at a discount. Of course, you're not in the business of selling paint at retail; you're in the business of selling paint jobs. But, by offering to sell materials, you keep the sales pitch going. Now you can talk up the quality of the materials you use and compare them with the bargain brands. You can explain why you use quality brushes and rollers, not the bargain brands and why you use proper patching materials, not off-the-shelf materials.

What are you doing? Planting that seed of doubt again. Planting a suggestion that the job may fail if the materials are wrong. And you just put yourself right back into the picture. You know the right materials, and how to apply them. You're back to selling your services.

If they actually accept your offer to sell them materials only, sell to them at retail and pocket the difference. Just remember to collect the sales tax, and make sure that there's enough markup to cover your time and mileage.

Offer to rent equipment and tools— Do this only if you have a separate set of tools for this purpose. Don't rent your own tools.

Offer to split the work— They do part of the job, you do the rest. How about this: you do the high work on the scaffold; they do the low part? Splitting the work is accepted more often than you might think. Many people feel that they can swing a brush or roll a roller as well as you can. It may be true. Anyone can paint. Of course, you do it better, faster and more efficiently. But a homeowner doing his own house isn't usually in any hurry.

The key difference between a professional paint job and the homeowner's is the prep work. (There's more about this in the following chapters). Prep work is where most customers fold. They don't understand why it's necessary, or how to do it properly. Therefore, you have a valuable service to offer. You also have the tools and equipment for doing the job right. They'd have to buy or rent the right tools or spend ten times as long doing it with elbow grease alone.

Also, many paints and coatings (epoxies, polyurethanes, stains) require special care or special application methods. Offer to do those parts of the total job.

Offer to be their consultant— For, say, $20, give them a detailed, written report. Tell them exactly what they must do to complete the job successfully. You may think, for $20, why waste my time? There's a reason other than the $20. Read on.

Selling Insecurity

Those last two ideas work, but maybe not the way you would expect. Here's what usually happens. The customer realizes how much work is involved, and wonders if he can do it. Insecurity and a lack of technical knowledge keep the service industries in business. Insecure people wonder if their decision to do it themselves is the right one. They'll think it over. And, most of the time, they'll call you back to accept your bid. After all, they have your written report, which shows that you're the expert.

If they do the job themselves— Sometimes, of course, your prospect will actually do the work himself. If so, one of two things happens. Either they do the work correctly, or they get into trouble. If they have problems, they'll try to get help. Who's the expert? You. Where do they go for help? To you, maybe. I say, "maybe" because they might be embarrassed to tell anyone that they messed up the job. But if you were accepted as a friend even though you didn't get the job, if you advised and encouraged them, the chances are that they'll turn to you when in trouble. Now's your chance to resell your labor.

If they did the job right, they see you as the expert. You gave them good advice. You boosted their confidence. You showed them you're a friend and a professional. They'll give your name to anyone who asks them to recommend a painter. And, next time they have a project, they'll remember how much work the last one took. This time, they may decide it's easier to hire you.

Most people have some insecurity— By "insecurity" I mean "fear of the unknown." People don't like to open new doors. It's tough to get a person to do business with you the first time. But once they hire you — once they've gone through that door for the first time — they're relaxed and at ease. If they trust you, they'll come back to you before going to the next unopened door — your competitor. So, even if you lose the job, smile and be friendly. Maybe they won't like the painter they do hire. Next time, they'll remember you, the one with the smile.

If I haven't convinced you, just think for a moment of your own shopping habits. Chances are, you use the same paint store, the same gas station, the same drug store — because it's easy and comfortable.

Contractor Warranties

There's another sales clincher you can use: the warranty. Sell it right along with the service.

Most building contractors warranty their work for one year from completion. But few painting contractors warranty the job for more than a few weeks. There are just too many variables: customer damage and abuse, customer-supplied or poor-quality paints, customer prep work, roof or pipe leaks, externally-caused damage, misunderstandings of what the contract called for, color changes due to variations in light and surface, and normal aging of the paint.

Your Responsibilities

Within the warranty period, you are responsible for poor prep work you did, or for spots you missed. Offer a written warranty, with *exclusions* and *coverages* clearly listed. Cover yourself against unjustified reworks.

Double-check your work— Look at Figure 6-3. I'm not sure whether this job was done by the homeowner or by a hired pro. Anyway,

Figure 6-3
Painter's holiday

someone has a rework job. Note the holidays under the eaves. If it isn't repainted pretty soon, the color match will be impossible and the whole under-eave will have to be repainted. This can be a costly second trip. Add up the time spent driving to and from the job, getting set up, and redoing the work. Add to that the fact that you'll probably never be asked to do any future work for this owner or anyone he knows. *Double-check all your work before you leave the job site.*

A knowledgeable painter rarely does a bad paint job. It's usually the do-it-yourselfer, or the painter hired for $5 an hour, who causes problems. But, when rework is justified, do it fast, do it right, and do it with a smile. Most people realize that mistakes sometimes happen. Just try to minimize reworks as much as possible.

How to Handle Reworks

When call-backs happen, the way you handle the situation makes the difference, the difference between a satisfied customer and an unhappy one. Do the job grudgingly and you've made an enemy. Admit your error, and chances are you'll be forgiven. By the way, since you're back in the customer's home, you have the chance to sell more work. At the very least, ask for the names of others who need a painter. Sell, sell, sell. You won't make any money if you don't.

Before you request final payment, be sure the workmanship satisfies *you.* After all, it's your reputation that's on the line.

Presenting Yourself

Now you know the basics of selling paint jobs. But before leaving this chapter, I want to offer some more tips: the best times for appointments, how to make a good impression, the sales tools you should bring, and more.

When to Make Sales Calls

Would you say that the time to make a sales call is whenever the potential customer phones? You'd be wrong. That may be the worst time. It cuts into your job time.

What time should you set for the appointment?— Well, if you have a full crew of painters, and a full or part-time salesperson, normal work hours are best for sales calls. But most painters work alone, or in small partnerships. Their time is valuable. Set appointments for times you're not likely to be painting. Does the job call for indoor painting? Set the appointment for after hours (between 5 and 9 PM) on the day the customer calls. Why? Because indoor work can be estimated in the evening hours. Does the job call for outside painting? Set the appointment for the next available Saturday, or for a work day you set aside just for sales calls. Why? Because outdoor estimating should be done during daylight hours when you can see what you're estimating.

The key to sales calls— It isn't the time of day; it's *being there.* If you say you'll be at their home at 7 PM, then be there at 6:55 PM and wait at the curb for five minutes before ringing the doorbell. Don't show up at 7:30. Being late for an appointment is inconsiderate and rude, and reflects badly on you. Set a time, and be there on time. If you *have* to be late, call to say you're going to be late.

Dress to Sell

Your appearance affects how people react to you. People want to know they're hiring a quality person. If you dress like you're going to a 1960's antiwar rally, you likely won't get in the door. But dressing like a banker doesn't help either. In the first case, your prospect will assume that anyone who dresses like a bum couldn't possible do quality work. In the second instance, if you're too well-dressed, they'll assume you overcharge for your services or are just a high-power salesperson who knows very little about painting.

What should you wear? I find that a pair of clean painter's whites or slacks and a semi-dress shirt work best. Look like a painter, but a clean and decent one. Be clean-shaven, with a neat haircut. Look the way your customers look. They'll be more comfortable, and so will you.

Sales Tools

What do you bring with you to close the sale? First of all, you need your estimate sheets, a contract, and a pen.

Bring some paint chip samples, and a few pictures. But not, as many people recommend, pictures of past jobs. Why? Because everyone has a different idea of what looks good. By showing a picture of a home you recently completed, even if that customer was satisfied, you may kill this sale. This customer might not like yellow with red trim. Instead, bring pictures of how you protect furniture, bushes, walks, cabinets, and so on. This is what I found works best.

Estimating Tools

Remember that you'll need a tape measure, a scratch pad, and a pencil. Add a hammer, a screwdriver, and a flashlight. Since most estimating requires some calculations, bring a small battery-operated calculator. And have extra batteries for both the flashlight and the calculator. Mess up on the initial sales visit and most people will assume you'll mess up on the job too.

Other Display Aids

There are some other "tools" you might find useful. They help you show the customer what they're getting. People about to spend hundreds (or thousands) of dollars want to see what their money is buying.

Sample boards— If you do stucco, ceiling texture, wall texture, or shake/shingle restoration work, make up a few "sample boards." These are nicely-framed panels, at least one foot square. Cover each panel with one of the actual materials and textures that you apply. Colors should be off-whites or nice pastels. Along with your samples, bring a book with manufacturers' specifications and warranties.

The painter who shows actual samples will be way ahead of the painter who shows nothing — or one who shows only photos.

You can also use sample boards to help sell mobile home roof coatings, industrial coatings, and similar coatings. If you plan to do industrial or government painting, however, get the "Federal Standard" paint chips. (For more on this, see Chapter 10 on paint color.)

Color books— Some painters arm themselves with color books when selling to homeowners. This is optional. You can kill a sale by giving a customer too many color choices — they just can't decide. I carry a full set of sample colors, but use them only if necessary.

Most homeowners already know the colors they want before they call you. Many even have color chips. This can be a problem if you use a commercial paint. You'll have to cross-match to their color, and check with them that your paint is the right color. It's often easier to simply buy the color and brand of paint the customer wants. That puts the responsibility on them, not on you.

I won't claim that this chapter has covered everything a paint salesman can know about selling paint jobs. But it has covered the basics. Anyone who understands and uses the principles I've discussed here should have no trouble converting plenty of prospects into buyers. Then it's up to you to do work that's as good as your sales pitch promised.

If you want to pick up more selling techniques than I've provided here, order a copy of *How To Sell Remodeling*. An order form is bound into the back of this book. It covers selling all types of work, not just painting, but selling is selling.

Estimating

The hardest thing most painters must do isn't the painting, or even the selling. It's the estimating. In this chapter, we'll discuss estimating methods, and rules for good estimating. Finally, I'll offer some charts you can use when compiling estimates.

Estimating Labor Costs

I can tell you, to the ounce, how much paint it takes to cover a given surface. Later in this chapter I do just that. But the hard part is estimating how long it will take to cover that surface. An accurate labor time must be included in your price to the customer. How do you do it? Read on.

Manhour Estimating

This means simply adding up how long it takes to do each part of a job. Sounds easy, doesn't it? In some businesses it is easy. The auto repair industry, for example, has books that list all common auto repair work and the time that each repair should take. These references are generally accepted by most insurance companies and are used to price work done at most repair shops. This system works because all cars of the same make, model and year are pretty much the same. Replacement parts are made to fit perfectly. Work is done at the repair shop under controlled conditions. The time needed won't vary much from job to job.

There's no such guide for painters, and probably never will be. Why not? There are just too many variables. Every 1965 four-bedroom house isn't the same. The work is always done at the owner's site, not in the painter's shop. That makes every painting job a custom job. No one estimate will cover all cases. But that doesn't mean that an accurate estimate is impossible. It just means that you have to study the variables more closely.

Variables in Estimating

What do I mean, too many variables? To see what I'm talking about, let's paint a standard interior door. Here's where we start asking questions.

General— Who's doing the job — a pro or a novice? Are you using a brand of paint you know, or an unknown brand supplied by the customer? How lax, or how critical, is your customer?

Condition and preparation— Is the door flat, or does it have raised panels, molding, or louvers? Will panels require repair or replacement? Are panels real wood, or metal, or plastic? Are there scuff marks, lipstick marks, or other hard-to-remove imperfections? Is the wood grain smooth, or will it need filler and sanding? Do the jamb and molding need repair? Do they fit properly? Are screws, nails, and glue joints secure, or will they require repair? Are the stops in place, or do you have to install them? Is the door bottom clear of the floor, or does it rub on the carpet? Are there one, two, three, or twelve coats of prior finish on it? Was the last coat applied properly, or will you have to strip it? Does it have to be washed, scrubbed, or cleaned with chemicals?

Hardware— Can the hardware be easily removed? Is the hardware coating worn off? Will it need to be replaced or refinished? Are the hinges clean and free-swinging?

Painting— Will it be painted or stained? Will it be painted in latex, enamel, lacquer, or polyurethane? Will both sides be the same color? Is either side to be done in more than one color? Should you brush, roll, or spray this door? Will the new finish need one, two, or more coats to cover properly? If multiple coats, what's the drying time between coats? Do you have to sand between each coat? Will you have to make a second, or third, trip to finish the job? Can you remove the door for painting? If you can't remove it, can you work undisturbed, or will people be in and out all day?

Did I miss anything? Every variable affects how long it takes to paint this door. And there are as many questions to ask about every part of the rest of the house. Colors? Water damage problems? Type of paint? Trim? Furniture to be moved? Pictures to be moved? Carpeting to be protected? Bookshelves to be painted? On and on.

And then you have the biggest variable of all: your customers. Every customer is different. Some will be very demanding. That alone may increase your labor time by 25%.

Standard-Situation Estimating

You want standard manhours for painting? I wish I could give it to you. Believe me, I tried to find estimates that apply to most jobs. I'm a good estimator. I've been doing it for years, both on paint jobs and in factories. In factory work I can estimate the manhours required to assemble hundreds of parts and be accurate within a few percent. In the factory, though, conditions are controlled, and each step is repeated over and over. I can time each step being done by many people, and take an average.

Under factory conditions and with proper equipment, I can paint a door in 15 seconds. In a home, under good conditions, it may take me 35 minutes. But it could take you five hours, and require three trips to the job. So whatever labor estimate I offered would be right for some of my jobs and wrong for most of your jobs.

I'm going to exercise an author's prerogative and omit any manhour estimates from this manual. If you have to have them, several books of manhour estimates for painting have been published. Some are listed on the last pages of this manual. Use these manuals if you must. But I'm going to offer a better service than tables of manhours. I'm going to explain how to compile your own labor standards — figures that will probably be wrong for every painter you'll ever meet, *but just right for you.*

Square-Footage Charges

Some painters don't even worry about manhours. They just charge by the square foot of floor or wall, say 55 cents. This is great, if it works. That is, if it works for you, in your location, with your customer base, with your competition. And if you can pay the bills and make a profit.

Out of curiosity, I've averaged some jobs, *after* they were completed. It worked out to anywhere from 35 cents to $1.50 a square foot of floor, depending on the variables. I can make money at both rates, and at anything in between. It all depends on the job.

Why such a big difference in square-footage rates? Some jobs used little paint, but required lots of labor. Other jobs needed little labor, but used gallons and gallons of paint. Most painters can't bid by the square foot and expect to make money consistently. You'd get too many jobs with little or no profit and lose a lot of work that might show a good profit.

Notice that we've talked a lot about labor times but we *still* haven't estimated that interior door. The next section, though, shows you a reasonable way to figure labor costs, one that will help you cover the variables.

Half-Day Estimating

Finally, here's an answer. Half-day estimating is similar to manhour estimating, but leaves you a lot more room for those variables. It works like this: Figure all jobs as a half-day or full-day work for one man rather than by the hour. Here are some standards you can use:

- A kitchen — with degreasing, moving refrigerator and stove, and trimming-in around cabinets — takes a full day.

- A full bath and a half bath together take a full day.

- A living room and dining room combination takes a full day.

- Two bedrooms take a full day.

- Closets are deceiving. They look like small, quick jobs. You'll find that most take half to a full gallon of paint, require semi-gloss, scuff-mark cleaning, and take half a day.

- Your estimating, selling, and collecting time will add one-half day on most jobs. Including this overhead time, an average three-bedroom home takes five working days to complete. This is for a good-quality job in a furnished home.

- The exterior of the same home takes one to two days prep; one day of large surface rolling or spraying; and at least two days of trim and window work. Figure a minimum of five days. That makes painting the interior and exterior of a typical three-bedroom house a two-week job.

So with the previous few pages in mind, what's the answer? How much do you charge your customers? In my forty-odd years, I've sold services and service-related products, wholesale goods, and tons of assorted retail goods. It all comes down to one-easy-to-learn fact:

You don't have full control of the price you charge. Your customers control it.

Why? If they aren't willing to pay the price you ask for the quality you offer, you don't work. Once you determine what they're willing to pay, you're back in control. You either supply at that rate, or not at all. But if you're priced right for your marketplace, and still aren't making a profit, you have two choices: You can find a new marketplace, or you can start cutting expenses.

One job sold out of each three bids you make is a good average to shoot for. Are you averaging less than that? Chances are, you're priced too high. Averaging more than that? You're probably giving the jobs away.

Estimating Tips

Here are some tricks you can use to find out the rates for your area. Ask a few of your competitors to bid repainting your (or a friend's) home. Do your own best estimate first, so you have something to compare with theirs. Take their estimate, deduct the material cost, and divide by a good wage for your area. This will give you an idea of the manhours they figured into the job.

Use your competitors' figures or quotes— Use this trick whenever you're out bidding a job. It's based on the fact that most people get more than one bid. If your potential customer quickly accepts your bid, chances are you're the low bidder. If your potential customer says "Well, I'll let you know," then you're probably priced higher than the competition.

Whatever they answer, do this: Start for the door, turn, and shrug a little. Then say "Mr. and Mrs. Smith, I'm happy I had this chance to meet nice people like you, and that I could be of service. Oh, by the way: just out of curiosity, what did you figure this job was worth?"

Most times, they'll figure there's no harm in telling you. But the amount they mention isn't what *they* thought: It's what the competition offered. Add 10% to it (they'll usually understate), and you'll be pretty close to the best quote they got.

If you were the low bid, and got the job, you learned something. If you were high, and didn't get the job, here's your second chance. Say something like this: "Gee, you really think it can be done for that amount? If I take another look around, and sharpen my pencil a bit, do I have the job?" Now they think it might be bargain time. Come back with their figure, plus about 5%. You'll likely get the job. Just be sure you want it at that price. In either case, you learned a little more about your competition, and how they figure prices and manhours.

Protect Your Estimate

Now about that low bid you made. There's a way to both protect yourself and allow for changes if they're needed. Write up a formal contract that specifies what you will do for the price you bid. No more, no less. Then, at the bottom, insert this clause.

> *This contract is for work and materials as indicated. It does not include additional materials or labor for additional, customer-requested repairs or painting; nor does it cover repairs or painting not listed. All additional materials and labor will be rebid or billed at an additional, agreed-upon rate.*

This clause does several things for you. First, it gets rid of the "But it will only take you a minute to fix it." That little "extra minute" job usually runs into another and another and another, until you've lost any profit you could possibly have made. It also gets rid of the "Gee, I didn't know," and "Is it really that bad?" These are usually problems the customer knew about, lived with, and figured that it's your tough luck if you didn't include it in your bid. You fix it, or you can't finish the job and get paid.

This clause will usually turn an almost-honest person honest. Now they know you're too smart to be taken in. Now they'll start pointing out all the hidden defects you didn't see. Now you have a chance to add to your low bid, getting your price up to where you make a profit.

Rules for Good Estimating

Many new estimators make two mistakes when estimating labor costs. First, they don't get the manhours right, usually estimating far fewer hours than are really needed. I can understand that. It's a mistake anyone can make. With experience, errors like that should happen less and less.

But the second mistake is inexcusable. That's not knowing your actual hourly labor cost. Don't make that error. It's easy to figure your cost per manhour and every paint estimator should know it. Start with the base wage. Let's say that's $10 per hour. Taxes and insurance usually add about 35% to that. Here's the breakdown on taxes and insurance.

Your state unemployment insurance will be about 4% of wages. F.I.C.A. tax will add about 8%. Federal unemployment insurance adds about 1% more. All states require worker's compensation insurance. Your cost will be between 10% and 15% of wages paid to your painters. And even the most basic bodily injury and liability insurance policy will add about 6% more. Please note that these figures will vary from state to state and from year to year. Your accountant will have the exact figures you should use.

There's no legal way to avoid paying these taxes. Both your state and the federal government have very effective ways of imposing penalties on paint contractors who try to avoid paying their employer taxes or fail to buy the insurance that's required.

Adding 35% to our $10 base wage, our hourly cost is now $13.50. But don't stop yet. We're just getting started. Next, add the fringe benefits and apprentice charges you pay. Fringes include vacation and holiday pay, medical and dental plans, etc. For many employers these fringes will add 10% to the base wage. That brings our hourly cost to $14.50.

Many paint contractors add the cost of supervision and overhead to their hourly labor cost. Others include supervision and overhead as a separate item in their estimates. Whichever method you use, be sure to include these important costs in your estimate. Let's say we decide to include supervision and overhead in the hourly labor cost, and estimate these expenses at $50,000 for the year. A lot of that figure may be the wage you draw as owner of the business. How do you load that cost into each productive painter hour? Easy! Say you expect to keep five painters busy nearly all year. Figuring 2,000 manhours in a year, that will be 10,000 painter hours this year. Dividing the $50,000 overhead cost by the 10,000 hours, we get $5 per painter hour for supervision and overhead. Add that $5 to our $14.50 and we've got $19.50 for an hourly labor cost.

Notice that at $19.50 we haven't made a dime yet. We've just covered our costs. Let's add a little profit so we can buy new equipment some day and carry a few slow pay accounts when necessary. How about a 20% profit? Does that seem like too much? O.K., but don't go less than 10%. That brings the hourly selling price for labor to nearly $21.50. I doubt that your hourly labor price will ever be less.

Finding your true selling price for labor is an important key to better estimates. It's just as important as making accurate manhour estimates. After your first few jobs, you'll start to get the feel of things. After a few months, you'll be able to spot problem jobs and price yourself accordingly. But remember these rules of good estimating:

Rule 1) Know your cost of doing business.

Rule 2) Keep accurate records of your estimates.

Rule 3) Price your work so it sells, and sells at a profit.

Estimating Materials

The first step is knowing what type of paint or coating to use. So in this section you'll find some charts on application methods, types of paint, and drying times. This is essential information for all estimates.

I hate math and formulas. But I hate even more losing money because I underestimated the material required to finish a job. So I've included some charts to help you figure out the square footage of many different surfaces: rafters, doors, window trim, finished lumber, and more. Best of all, I've done the calculations for you. In the third part of this section you'll find paint coverage charts.

Area to be painted	Latex	Oil/alkyd	Epoxy	Urethane	Enamel	Stain	Lacquer
Molding	x	x		x	x		x
Doors	x	x		x	x		x
Walls	x	x					
Bath	x	x			x		
Kitchen	x	x			x		
Toys		x		x	x		
Furniture				x	x	x	x
Window frames	x	x		x	x	x	x
Cement or block	x	x	x				
Wood siding	x	x				x	
Automotive					x		x
Appliances			x	x	x		x
Ornamental iron		x	x		x		
Laundry	x	x			x		
Kid's room	x	x			x		
Marine				x			
Stucco	x						
Floors			x	x			

Types of paint for various applications
Figure 7-1

Paints for Different Surfaces

First, look at Figure 7-1. It shows you what types of paint to use for different surfaces.

You not only have to decide the type of paint, you also have to decide whether it should be glossy, flat, or in between. Figure 7-2 shows the appropriate finishes.

Paint Application Methods

Now that you know what kind of paint to use, how do you apply it? Figure 7-3 shows paint application methods. Of course, many times you'll have to use a combination of methods.

In Figure 7-3, the spray column assumes that you have the right guns and compressor. Also note that *new* roof shingles should be dip-coated before installation. But if they're already installed, use an airless at the low-pressure setting. Better still, use a pump-type garden sprayer. It will cause a minimum of overspray, and the paint will penetrate better.

Paint Drying Times

Once you've applied the paint, how long do you have to wait for it to dry? Well, latex will dry to the touch in 5 to 30 minutes, when applied with an airless sprayer. Let urethanes dry at least five days between sanding coats. The chart in Figure 7-4 gives you more information. But remember that drying times vary, depending on brand, coat thickness, humidity, and how porous the surface is.

Figuring Surface Areas

You've decided what paint to use for the job and how to apply it. The next step is to figure out how much surface area you must cover with that paint. It's easy for flat areas like walls. But what about odd-shaped areas like rafters, or trim and moldings? This section tells you how to do it.

Flat Surface Area

It's easy to figure the area of any two-dimensional surface. Simply divide it into rectangles and multiply each height by the appropriate width. The measurements can be in inches or in feet, whichever applies. To change square inches to square feet, divide square inches by 144.

Area to be painted	Flat finish	Semi-gloss	Full-gloss	Satin
Moldings		x	x	x
Doors		x	x	
Walls	x	x		
Bath		x	x	
Kitchen		x	x	
Toys		x	x	x
Furniture		x	x	x
Window frames		x	x	
Cement or block	x	x		
Wood siding	x			
Automotive			x	
Appliances			x	x
Ornamental iron		x	x	
Laundry		x		
Kid's room	x	x		
Marine			x	
Stucco	x			
Floors		x	x	

Paint for various applications
Figure 7-2

Area to be painted	Brush	Roller	Spray	Airless	Glove	Aerosol
Plaster ceiling	x	x	x	x		
Acoustical ceiling		x	x	x		
Walls		x		x		
Woodwork, trim	x	x	x			
Toys	x		x			x
Furniture	x		x			x
Window frames	x	x				
Pipes	x				x	
Wire fence		x			x	
Ornamental iron			x		x	
Under eaves	x	x		x		
Stucco siding		x		x		
Rough wood siding		x		x		
Cement or block		x		x		
Automobile			x			x
Appliances			x			x
Roof shingles				x		

Best application methods
Figure 7-3

Type paint	Tack free	Hard	Full cure
Latex	1 hour	6 hours	30 days
Oil/alkyd	6-8 hours	24 hours	30 days
Lacquers	1-2 minutes	1 hour	--
Epoxys	1 hour	2 hours	--
Urethanes	1-2 hours	24 hours	30 days

Paint drying times
Figure 7-4

Measuring walls— Here's a trick for measuring standard walls: Since standard wall height is between 7 and 8 feet, just measure the width and multiply by 8. This way, you'll be sure to have enough paint. Actually, it's a good idea to round up measurements (round 4½ feet up to 5 feet, for example) just to be sure you buy enough paint.

Speaking of walls, you'll need to deduct the areas of windows, patio doors, and regular doors from the surface areas of the walls they're in. Here's how to do it:

Deducting window areas— Windows are easy: Measure the height and the width, multiply them, and subtract that answer from the total square footage of the wall. Again, if the measurement is in inches, remember to divide total square inches by 144, to find square footage.

Deducting patio door areas— If the patio door is the standard 6'8'' high, all you have to do is measure the width, then look up the surface area to deduct on this chart:

Width (feet)	Deduct (sq. ft.)
4	26.75
5	33.25
6	40.00
7	46.50
8	53.25
10	66.50
12	80.00

Deducting standard door areas— Most doors are 6'8'' high. Measure the width, then use the following chart to find the deduction:

Width (inches)	Deduct (sq. ft.)
24	13.50
26	14.50
28	15.50
30	16.75
32	17.75
34	19.00
36	20.00
40	22.25
48	26.75

Deducting garage door areas— A standard garage door is 7 feet high. Measure the width, then deduct according to the following chart:

Width (feet)	Deduct (sq. ft.)
8	56
10	70
14	98
16	112
20	140

Finished Lumber Surface Area

Sometimes you'll need to paint one, two, three, or all four sides of pieces of finished lumber. How much paint do you need? Well, look at Figure 7-5. It has a drawing and several charts. Use this figure to find out how much square footage you'll have to cover. Here are some examples of how to use Figure 7-5:

Example 1: How much surface area is in 50 linear feet of 2 x 6, on side A only? Look at the chart. Find the multiplier for 2 x 6, side A. It's 0.468. Multiply 50 (the number of linear feet) by 0.468: the surface area is 23.4 square feet.

Example 2: How much surface area is in 500 linear feet of 2 x 4, on sides ABC? Look at the chart to find the multiplier for 2 x 4, sides ABC. It's 0.723. Multiply 500 by 0.723 to get 361.5 square feet.

Example 3: How much surface area is in 185 linear feet of 4 x 10, on sides ABCD? Find the multiplier for 4 x 10, sides ABCD, on the chart. It's 2.176. And 185 times 2.176 equals 402.56 square feet.

Side A or C only			
2 x 4	.296	4 x 4	.296
2 x 6	.468	4 x 6	.468
2 x 8	.625	4 x 8	.625
2 x 10	.791	4 x 10	.791
2 x 12	.958	4 x 12	.958
Sides AB, AD, BC, or CD			
2 x 4	.426	4 x 4	.593
2 x 6	.598	4 x 6	.765
2 x 8	.755	4 x 8	.921
2 x 10	.921	4 x 10	1.088
2 x 12	1.088	4 x 12	1.255
Sides B or D only			
2 x --	.130	4 x --	.296
Sides ABC or ADC only			
2 x 4	.723	4 x 4	.890
2 x 6	1.066	4 x 6	1.233
2 x 8	1.380	4 x 8	1.546
2 x 10	1.713	4 x 10	1.880
2 x 12	2.046	4 x 12	2.213
Sides ABCD			
2 x 4	.853	4 x 4	1.186
2 x 6	1.196	4 x 6	1.530
2 x 8	1.510	4 x 8	1.843
2 x 10	1.843	4 x 10	2.176
2 x 12	2.176	4 x 12	2.510

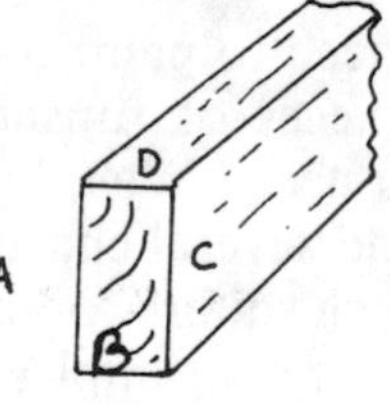

Finished lumber surface area
Figure 7-5

The most useful calculations are probably those for sides ABC, since this is your basic exposed beam or exposed roof rafter. The side A calculations are good for exterior wood trim.

Remember that unpainted new wood will need at least two coats. So double the calculated value to find actual square footage of coverage needed.

Roof Gable Surface Area

A gable is the triangular end of a building, measured from the eaves to the ridge. Use Figure 7-6 to find the square footage of the gable area, if you know the roof pitch. *Pitch is the roof rise in inches for each foot of width at the eaves.* The 4/12 pitch is the most common, followed by the 8/12 pitch.

This table is handy, since you can measure width at the eaves at ground level. Or you can take both the width and pitch off the plans, doing the estimate without actually measuring the building. If you're estimating a half-gable, just divide the figures in the columns in half.

Remember: *These figures do not include overhangs.* You must calculate them separately, and add them in. I'll show you how just a bit further on.

Roof gables where pitch is not known— If you don't know the pitch, you'll have to measure. Measure width from the ends of the eaves, across the width of the building. Then, measure height from a line across the building, as shown in Figure 7-6, to the ridge peak. Now you have two ways to find square footage:

1) Multiply the height by the width, then divide by two.

2) Find the pitch, then use the table in Figure 7-6. To find the pitch, divide the height, in feet, by the width, in feet.

Roof Overhang Surface Area

Most homes have roof overhangs. They direct water away from the sides of the house and shade the house from the sun. You'd think that you could find the square footage by simply measuring the length and width of this overhang, then multiplying the two figures. But consider the rafters. Maybe they look like they won't take much paint. But if you spread out the three sides of a 2 x 6 rafter, you have over a square foot of surface for each foot of length.

Figure 7-7 is a table of coverages for several common rafters, rafter spacing, and overhang lengths. Here's how to use it: Let's say a house has overhangs on two sides. The rafters are 2 x 6, spaced 24 inches apart. So *Y* equals 24.

1) Measure *straight out* from the wall to the end of the overhang. Let's say it's 30 inches. *X* equals 30.

2) Count the number of rafter spaces. Let's say there are 18.

3) Look in the 2 x 6 rafter table to find that the square footage for each section is 9.25 square feet.

4) Multiply this 9.25 by the 18 spaces. The total is 166.5 square feet for the overhang on one side of the building.

Width at eaves (in feet)	Pitch							
	3/12	4/12	5/12	6/12	7/12	8/12	10/12	12/12
2	.25	.50	.50	.50	.75	.75	1.00	1.00
3	.50	.75	1.00	1.25	1.50	1.50	1.75	2.25
4	1.00	1.50	1.75	2.00	2.50	2.75	3.50	4.00
6	2.25	3.00	3.75	4.50	5.25	6.00	7.50	9.00
8	4.00	5.50	6.75	8.00	9.50	10.75	13.50	16.00
10	6.25	8.50	10.50	12.50	14.50	16.75	20.75	25.00
12	9.00	12.00	15.00	18.00	21.00	24.00	30.00	36.00
14	12.25	16.50	20.50	24.50	28.50	32.75	41.00	49.00
16	16.00	21.50	26.75	32.00	37.50	42.75	53.50	64.00
18	20.25	27.00	33.75	40.50	47.25	54.00	67.50	81.00
20	25.00	33.50	41.75	50.00	58.50	66.75	83.50	100.00
22	30.25	40.50	50.50	60.50	70.50	80.75	100.75	121.00
24	36.00	48.00	60.00	72.00	84.00	96.00	120.00	144.00
26	42.25	56.50	70.50	84.50	98.50	112.75	140.75	169.00
28	49.00	65.50	81.75	98.00	114.25	130.75	163.25	196.00
30	56.25	75.25	94.00	112.50	131.25	150.00	187.50	225.00
32	64.00	84.50	106.75	128.00	149.25	170.75	213.25	256.00
34	72.25	96.50	120.50	144.00	168.50	192.75	240.75	289.00
36	81.00	108.25	135.25	162.00	189.00	216.25	270.00	324.00

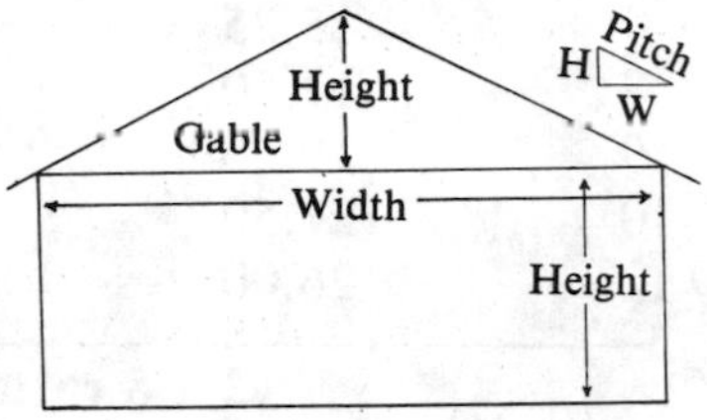

Pitch
H
W
Height
Gable
Width
Height

Roof gable surface area
Figure 7-6

5) Finally, since there are overhangs on two sides, double this figure. There will be 333 square feet to cover.

This table works for pitches of up to 8/12, for lengths (dimension *X*) up to 36 inches; and for pitches of 4/12, with lengths (*X*) from 4 feet to 12 feet. The 4-foot and up lengths in the table are for carports, patio covers, beamed ceilings, and similar areas. Also, I've added a little extra to cover the end side and the rafter ends. But remember, this is area for one coat only. Double the figures if two coats are needed.

Baseboard and Molding Surface Area

One of the hardest estimating jobs is figuring how much paint you need for covering trim. Figure 7-8 has multipliers to make estimating trim area much easier. For simplicity, I used decimals (0.25 = 1/4", 0.50 = 1/2", 0.75 = 3/4"). Round up measurements of 1/8 inch or less to the nearest 1/4 inch, to assure proper coverage.

For instance, let's say you have to paint the front surface and one edge of 750 linear feet of baseboard molding. You'll use one coat of paint. The molding is 2¼ inches high and 1/8 inch thick. Here's how to figure the surface area:

1) Look at the measurements. You know that 2¼ is the same as 2.25. Round up the 1/8 to 1/4, which is the same as 0.25.

2) Add the two together, for a measured surface of 2.50 inches.

3) Look in the chart for 2.50. The multiplier is 0.208.

4) Multiply the 750 linear feet you have to paint by 0.208, for an answer of 150. You must buy enough paint to cover 150 square feet of molding.

2 x 4 rafter table

X	Y=16"	Y=24"	Y=48"
14 inches	3.00 sq. ft.	4.00	6.75
18	4.00	5.25	9.00
22	4.75	6.50	11.00
24	5.25	6.75	11.75
30	6.25	8.25	14.25
36	7.25	9.50	16.50
4 feet	9.00	11.50	20.00
5	10.75	14.25	24.75
6	12.75	17.00	29.50
7	15.25	20.00	35.00
8	17.50	23.00	40.00
9	19.50	25.75	44.75
10	21.50	28.50	49.50
12	26.00	34.50	59.75

2 x 6 rafter table

X	Y=16"	Y=24"	Y=48"
14 inches	3.50 sq. ft.	4.50	7.25
18	4.50	6.00	9.25
22	5.50	7.00	11.75
24	6.00	7.75	12.75
30	7.25	9.25	15.25
36	8.50	10.75	17.75
4 feet	10.00	13.00	21.50
5	12.50	16.00	26.50
6	15.00	19.00	31.50
7	17.75	22.50	37.50
8	20.00	26.00	43.00
9	22.50	29.00	48.00
10	24.00	32.00	53.00
12	30.00	38.50	64.00

2 x 8 rafter table

X	Y=16"	Y=24"	Y=48"
14 inches	3.75 sq. ft.	4.75	7.75
18	5.25	6.50	10.25
22	6.25	7.75	12.50
24	6.75	8.50	13.50
30	8.00	10.00	16.00
36	9.50	11.75	18.75
4 feet	11.50	14.25	22.75
5	14.00	17.50	28.00
6	16.75	20.00	33.50
7	19.75	24.75	39.50
8	22.75	28.50	45.50
9	25.50	31.75	50.75
10	28.00	35.00	56.00
12	33.75	42.25	67.50

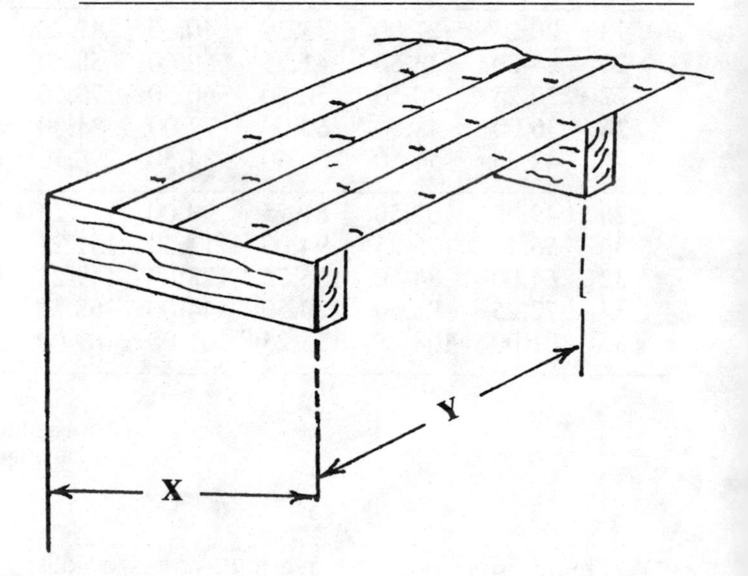

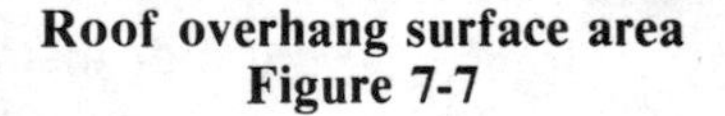

Roof overhang surface area
Figure 7-7

Window sill trim— Since most window sills are made of 1 x 4 lumber, of fairly standard widths, I've developed a table of multiplying factors for finding the surface area.

2'	3'	4'	5'	6'	7'	8'	9'
.82	1.23	1.64	2.05	2.46	2.87	3.28	3.69

Here's an example of how to use this table: A home has several windows. Two are 4 feet wide; their multiplier (from the table) is 1.64. Three windows are 6 feet wide, so their multiplier is 2.46. One window is 8 feet wide and has a multiplier of 3.28. Therefore, to find total square footage:

(2 x 1.64) + (3 x 2.46) + (1 x 3.28) = 13.94 sq. ft.

Round it up and you have 14 square feet of surface to cover. This is for a single coat, of course.

You'll need at least two coats for durability. So double this figure and buy paint to cover 28 square feet. Use full gloss or semi-gloss for painted sills. Why? Because gloss protects the surface by reflecting some of the sun's rays. It also makes cleanup easier.

Varnish clear coats generally require three coats. If you're varnishing these windows, buy enough varnish to cover 42 square feet. For clear coats, use a spar varnish, or a polyurethane with ultraviolet light (U/V) inhibitors added. Use exterior paints and varnishes for both interior and exterior window sills. Why? Because both are exposed to ultraviolet light from the sun, to dust and to other contamination.

Measured	Multiplier	Measured	Multiplier
1.25"	.104	8.75"	.729
1.50	.125	9.00	.750
1.75	.146	9.25	.771
2.00	.167	9.50	.792
2.25	.188	9.75	.813
2.50	.208	10.00	.833
2.75	.229	10.25	.854
3.00	.250	10.50	.875
3.25	.271	10.75	.896
3.50	.292	11.00	.917
3.75	.313	11.25	.938
4.00	.333	11.50	.958
4.25	.354	11.75	.979
4.50	.375	12.00	1.000
4.75	.396	12.25	1.020
5.00	.417	12.50	1.041
5.25	.438	12.75	1.062
5.50	.458	13.00	1.083
5.75	.479	13.25	1.104
6.00	.500	13.50	1.125
6.25	.521	13.75	1.145
6.50	.542	14.00	1.166
6.75	.563	14.25	1.187
7.00	.583	14.50	1.208
7.25	.604	14.75	1.229
7.50	.625	15.00	1.250
7.75	.646	15.25	1.270
8.00	.667	15.50	1.291
8.25	.688	15.75	1.312
8.50	.708	16.00	1.333

Trim square-footage multipliers
Figure 7-8

Door Surface Area

Standard doors are 6'8" high but come in many widths. To find the surface area of one side of a door, use the table already given in the section on deducting doors from wall surface area. To find surface area of both sides and edges of a standard door, see Figure 7-9. Again in this figure, I've used decimals for ease, and rounded square footage up to the nearest 1/4 (0.25).

Width (inches)	Both sides and edges (sq. ft.)
24	28.50
26	30.50
28	32.75
30	35.00
32	37.25
34	39.50
36	41.75
40	46.25
48	55.00

Door surface area, both sides and ends
Figure 7-9

Doors get a lot of abuse. They need either a clear coat or a semi-gloss or full-gloss enamel. On exterior doors, use exterior paint or varnish. Clear coats should have a U/V inhibitor.

New doors need at least two coats of enamel. So double the square footage shown. Many new doors are pre-primed. Be sure to degrease them after installation. Apply the paint or varnish within 60 days of installation. After 60 days, you'll need a new coat of primer.

For clear coats, use three coats. Triple the square footage shown. And remember to add a sealer coat to your estimate for new, clear-coat doors.

Garage Door Surface Area

The standard garage door is 7 feet high. Most single garage doors are 8 feet wide and most double doors are 16 feet wide. To figure the surface area of one side of a garage door, see the table in the section on deducting garage doors from wall surface area. To find the surface area of both sides of a garage door, see Figure 7-10.

Width (feet)	Both sides (sq. ft.)
8	112
10	140
14	196
16	224
20	280

Garage door surface area, both sides
Figure 7-10

Finding Surface Areas of Odd Shapes

Architect-designed homes usually have non-rectangular surfaces. You can, of course, guess at the surface area. If you want to estimate more accurately, use the following formulas, tables, and suggestions. Many irregular shapes are just a combination of rectangles, triangles or circles. Estimate the surface area of each and then accumulate the totals.

Surface Area of a Cylinder

You'll often have to paint cylindrical objects: round columns or posts, tanks, pipes, tank cars or trucks, round barns, silos, or houses. There's an easy way to calculate the surface area of a cylinder:

1) With a cloth tape, measure around the girth.

2) Multiply the girth by the height (or length) to find the surface area.

If your measurements are in inches, remember to divide the total square inches by 144.

To find the surface area of a cylinder end, first determine whether it's circular (a fairly flat surface) or spherical (a ball-shaped surface). Then use either Figure 7-11 or 7-12.

Diameter	Area	Diameter	Area
1	1.00	16	201.00
2	3.25	17	227.00
3	7.00	18	254.50
4	12.50	19	283.50
5	19.75	20	314.25
6	28.25	21	346.50
7	38.50	22	380.25
8	50.25	23	415.50
9	63.75	24	452.50
10	78.50	25	491.00
11	95.00	26	531.00
12	113.00	27	572.50
13	132.75	28	615.75
14	154.00	29	660.50
15	176.75	30	707.00

Surface area of a circle
Figure 7-11

Surface Area of a Circle

All you have to do is measure the diameter of the circle. Then use the chart in Figure 7-11. Diameter is the distance from edge to edge, through the center. If you measured in inches, divide the figures in the columns by 144 to get square feet.

By the way, if you have a circle with a diameter that's not on the chart, the formula is:

$$\text{Area of circle} = 0.7854 \times d^2$$
(d = diameter)

Surface Area of a Sphere

You might need to know the surface area of a sphere to paint a spherical tank, or the spherical end of a cylindrical propane tank. If you know — or can measure — the radius of a sphere, Figure 7-12 will tell you the surface area. *Radius is the distance from the center to the outer edge of a sphere.*

But you can't just put your tape against the tank and measure from one side to the other. That would measure the curve, too, so it would be a false measurement. How do you measure it? One way is to find the diameter, then divide it in half to get the radius. To find the diameter, extend two parallel lines out from the tank, from top and bottom (or from each side), and measure the straight distance between these lines. Now you have the diameter of the tank. Or, if the tank's supporting structure touches it at opposite sides, you can measure from one support to the opposite one to find the diameter.

As with the previous figure, divide the column figures by 144 to get square feet. And don't forget this: If you're using Figure 7-12 to find the surface area of one end of a cylindrical tank, you must divide by two. Figure 7-12 gives the total area of a ball. The two tank ends are two half balls. One tank end is one half ball.

What if the radius doesn't measure in even feet? You can still use the chart. For example, a spherical tank has a radius of 4½ feet. The chart shows 4 feet as 201, and 5 feet as 314.25. The difference between them is 113.25. Divide 113.25 in half, and add the answer to 201.00. The surface area is 257.62 square feet. That's the surface you have to paint.

If you need it, the formula is:

$$4 \pi r^2$$
($\pi = 3.1416$, r = radius)

Many industries today use spherical tanks to store liquids and gases under pressure. Be careful when painting these tanks. Pressures are high and the contents may be highly flammable. The next chapter covers painting safety.

Radius	Area	Radius	Area
1	12.50	11	1520.50
2	50.25	12	1809.50
3	113.00	13	2123.75
4	201.00	14	2463.00
5	314.25	15	2827.50
6	452.50	16	3217.00
7	615.75	17	3631.75
8	804.25	18	4071.50
9	1018.00	19	4536.50
10	1256.75	20	5026.50

Surface area of a sphere
Figure 7-12

Paint Coverage Chart

Now that you've figured out what paint to use, how to put it on, and the area you have to cover, how much do you buy? Figure 7-13 suggests paint coverage on different surfaces.

Paint-Remaining Charts

You'll often find that there's some paint left in a can. Or you'll wonder whether you've still got enough paint left to do some small area. Either way, you need to know how much area the paint left in the can will cover. Here's where the paint-remaining charts come in. Figures 7-14 through 7-17 are charts for pint, quart, and gallon cans, and for five-gallon plastic buckets. I've done all the volume calculations for you. And I've rounded down the answers to compensate for possible losses.

You just measure the height of the paint left in the can, then look it up on the chart. The charts give coverage in square feet. As you can see, the area a given amount of paint will cover depends on the coverage stated on the can's label. Note that *coverages are for a single coat, over a reasonably smooth surface.* Since many factors influence how much coverage you'll actually get, use these charts as a guide only.

How to Use the Paint-Remaining Charts

Here's an example. There's 3½ inches of paint left in a one-gallon can. The label says a gallon covers 400 square feet. Look up 3 in the "Inches remaining" column on the one-gallon chart in Figure 7-16. Now, look across that row to see what's under the "400" column heading. You see 172 square feet. Next, look across the "1/2" row. Under 400, you see 28 square feet. Add these together. The paint remaining in this can will cover about 200 square feet.

Use the "350" column when spray painting with a paint that normally covers 400 square feet per gallon. Why? Because you'll lose paint in spray lines and in overspray. How much? Between 10% and 15%. I didn't include aerosol cans in my charts because it's hard to determine the paint remaining in an aerosol can. Most aerosol cans, when full, have enough paint to cover 40 square feet.

One-Pint Coverage Chart

This means a one-pint (16 fluid ounce) container. The standard size is 3½ inches high by 3¼ inches wide. If the label lists coverage by square feet per pint, you can still use the chart. Just multiply by 8 to get the square feet per gallon. For instance, the label says a paint covers 25 square feet per pint. Multiply 25 by 8 to get a coverage of 200 square feet per gallon.

One-Quart Coverage Chart

This means a one-quart (32 fluid ounce) container. The standard size is 4⅞ inches high by 3⅞ inches wide. If the label lists coverage for square feet per quart, just multiply by 4 to get the coverage per gallon. For example, if the label says a quart of paint covers 100 square feet, multiply that by 4 to get a per-gallon coverage of 400 square feet.

One-Gallon Coverage Chart

This is a one-gallon (four-quart) can. Standard size is 7 inches high by 6½ inches wide.

Five-Gallon Coverage Chart

This refers to a five-gallon plastic paint container. Standard size is 12 inches high by 11¼ inches wide. The full level, though, is 11¾ inches.

	Quart (sq. ft.)	Gallon (sq. ft.)	Exterior stucco (gallon) (sq. ft.)
Latex			
flat	100	400	300
semi-gloss	100	400	300
ceiling			
smooth surface	100	400	
ceiling tiles	40	160	
acoustical ceiling	25	100	
exterior stain			
smooth surface		250	
rough surface		150	
full gloss	50	200	(Two coats usually needed.)
Oil/alkyd			
flat	100	400	
semi-gloss	100	400	
full gloss	100	400	
exterior stain			
smooth surface		350	
rough surface		150	
Liquid plastics	125	500	
Concrete sealer		75	
Semi-transparent oil stain			
new smooth surface		250	
new rough surface		150	
recoat		250	
Aerosol, 13 oz. spray	40 sq. ft. (usually less - several coats are often needed)		

Paint coverage chart
Figure 7-13

Inches remaining	Normal coverage per pint, at per-gallon rate 100	200	350	400	500	600
¼	.5	1.5	3.0	3.5	4.0	5.0
½	1.5	3.5	6.0	7.0	8.5	10.5
¾	2.5	5.0	9.0	10.5	13.0	16.0
1	3.5	7.0	12.0	14.0	17.5	21.0
2	7.0	14.0	25.0	28.5	35.5	42.5
3	10.5	21.0	37.0	42.5	53.5	64.0
3½ (full)	12.5	25.0	43.5	50.0	62.5	75.0

One pint coverage chart
Figure 7-14

Extra-Coverage Surfaces

Many surfaces you'll be painting aren't smooth. They'll take more paint than the coverage listed on the can. From experience, let me suggest some percentage increases for your surface area calculations.

- Shake shingles require an extra 7% to 10% if edges are to be coated.
- Cement block requires 5% to 8% more.
- Lap-type siding requires 5% to 7% more.
- Corrugated metal requires 15% more.
- Rough plywood will take 2% to 3% more.

Why should you use these figures? Here's an example. A cement-block wall is 8 feet high and 100 feet long. Length times width gives 800 square feet of surface area. If you bought two gallons of a paint with 400-square-foot coverage, you'd run out 8 feet short of the wall's end. To increase the coverage figure by 8%, use a multiplier of 1.08. And 1.08 times 800 is 864 square feet.

Buy enough to do the job. The few dollars extra are nothing compared to the cost of an extra trip to the paint store, even ignoring the time spent blending new paint into the paint already dried on the wall. Also important, your customer's confidence will slip a notch or two if your paint can runs dry 8 feet from the end of the job and you have to go scampering off to the paint store.

Inches remaining	**Normal coverage per quart, at per-gallon rate**					
	100	**200**	**350**	**400**	**500**	**600**
¼	1.5	3.0	6.0	6.5	8.0	10.0
½	3.0	6.5	11.5	13.0	16.5	20.0
¾	5.0	10.0	17.0	20.0	25.0	30.0
1	7.0	13.0	23.0	26.0	33.0	40.0
2	13.0	27.0	46.0	53.0	66.0	80.0
3	20.0	40.0	70.0	80.0	100.0	120.0
3¾ (full)	25.0	50.0	87.0	100.0	125.0	150.0

One quart coverage chart
Figure 7-15

Inches remaining	**Normal coverage per gallon, as listed on label**					
	100	**200**	**350**	**400**	**500**	**600**
¼	3.5	7	12.5	14	18	21.5
½	7	14	25	28.5	36	43
¾	10.5	21.5	37.5	43	53.5	64.5
1	14	28.5	50	57	71.5	86
2	28.5	57	100	114.5	143.5	172
3	43	86	150.5	172	215	258.5
4	57	115	201	230	287	344.5
5	71.5	143	251	287	359	431
6	86	172	301.5	345	430.5	517
7 (full)	100	200	350	400	500	600

One gallon coverage chart
Figure 7-16

Inches remaining	Normal coverage per gallon, as listed on label				
	100	200	350	400	500
¼	10	20	37	40	50
½	21	42	75	82	103
¾	32	64	113	128	160
1	43	86	150	172	215
2	86	172	301	344	430
3	129	258	452	516	645
4	172	344	602	688	860
5	215	430	753	860	1075
6	258	516	904	1032	1290
7	301	602	1054	1264	1565
8	344	688	1205	1376	1720
9	387	774	1356	1548	1935
10	430	860	1505	1720	2150
11	473	946	1657	1897	2370
11¾ (full)	500	1000	1750	2000	2500

Five gallon coverage chart
Figure 7-17

Estimating Worksheets

All right, you've learned how to estimate manhours and estimate materials. You know what your labor costs are. You're all set, right? Where could you possibly go wrong? I'll tell you where the danger lies, then offer a solution.

No matter how accurately you estimate the labor and materials for the work you include, you're bound to lose your shirt if you *forget to include* something that needs to be done. To help prevent that, use an estimating worksheet for every job you estimate. Make up some of your own, or borrow mine. I use Figure 7-18 for estimating interior paint jobs and Figure 7-19 for exteriors. For interior jobs, make up a different worksheet for each room. Use the form as a checklist, to make sure you include every part of every job.

I've also included a copy of each worksheet filled out, showing how I use it for a typical residential estimate (Figures 7-20 and 7-21). Of course you'll have to substitute your own labor and material costs, and estimate the manhours for your crew.

If you've never used an estimating worksheet, give these a try. They can't hurt. And you may find that you're turning out more accurate — and more profitable — estimates.

In Summary

The end of this chapter brings to a close the first part of this book. We've covered the basics of every painting business. I hope it's convinced you to do your homework before spending money on equipment, inventory and advertising. I hope it's convinced you to set your business up right the first time, to know your product, your goals, your limitations and your costs. I hope it's convinced you to keep accurate records and use them to help your new business grow.

If you've found these first few chapters dull and irrelevant, maybe you shouldn't be running your own company. Not everyone wants to and not everyone should. If you enjoy painting but not the painting business, stick to the brushwork.

Whether you're in business for yourself or not, the rest of this book is for you: the nuts and bolts of painting. It will help you understand and solve all the common painting problems. It's bound to be valuable reading for anyone who wants to do quality painting with as few problems as possible.

We'll begin where every good painter should begin. Safety first.

Job #: ____________ **Customer:** ____________ **Date:** ____________ **Room:** ____________ **Sheet** ____ **of** ____

Furnished______ **Unfurnished**______

X = DAMAGE
WINDOW
DOOR
FP FIREPLACE
SH OR CB SHELF OR CABINETS

A
B
C
D

Item	Dimensions		S.F.	Prep	Prime	# Coats	Color	Type
1) Ceiling	(A x C)	=	______ S.F.	Prep ______	Prime ______	# Coats ____	Color ______	Type ______
2) Wall A	(A x H)	=	______ S.F.	Prep ______	Prime ______	# Coats ____	Color ______	Type ______
3) Wall B	(B x H)	=	______ S.F.	Prep ______	Prime ______	# Coats ____	Color ______	Type ______
4) Wall C	(C x H)	=	______ S.F.	Prep ______	Prime ______	# Coats ____	Color ______	Type ______
5) Wall D	(D x H)	=	______ S.F.	Prep ______	Prime ______	# Coats ____	Color ______	Type ______
6) Window 1	(L x W)	=	______ S.F.	Prep ______	Prime ______	# Coats ____	Color ______	Type ______
7) Window 2	(L x W)	=	______ S.F.	Prep ______	Prime ______	# Coats ____	Color ______	Type ______
8) Window 3	(L x W)	=	______ S.F.	Prep ______	Prime ______	# Coats ____	Color ______	Type ______
9) Window 4	(L x W)	=	______ S.F.	Prep ______	Prime ______	# Coats ____	Color ______	Type ______
10) Door 1	(W x H)	=	______ S.F.	Prep ______	Prime ______	# Coats ____	Color ______	Type ______
11) Door 2	(W x H)	=	______ S.F.	Prep ______	Prime ______	# Coats ____	Color ______	Type ______
12) Door 3	(W x H)	=	______ S.F.	Prep ______	Prime ______	# Coats ____	Color ______	Type ______
13) Door 4	(W x H)	=	______ S.F.	Prep ______	Prime ______	# Coats ____	Color ______	Type ______
14) Cabinets		=	______ S.F.	Prep ______	Prime ______	# Coats ____	Color ______	Type ______
15) Shelves		=	______ S.F.	Prep ______	Prime ______	# Coats ____	Color ______	Type ______

16) Trim: Top of walls ______ bottom of walls ____ doors ____ windows ______ Total LF _______

Notes:

Subtotals:

Type _______ Color _______ Gal _______ Manhours _______

Type _______ Color _______ Gal _______ Manhours _______

Type _______ Color _______ Gal _______ Manhours _______

Type _______ Color _______ Gal _______ Manhours _______

Type _______ Color _______ Gal _______ Manhours _______

Interior paint estimating worksheet
Figure 7-18

Job #: ________ **Customer:** ________ **Date:** ________ **Room:** ________ **Sheet** ____ **of** ____

WINDOW
DOORWAY
FP FIREPLACE
X = DAMAGE

A B C D H W

One story: ________ **Two story:** ________ **Dormers:** ________ **Overhang:** ________

1) Wall A	=	______ S.F.	Prep ______	Prime ______	# Coats ____	Color ______	Type ______
2) Wall B	=	______ S.F.	Prep ______	Prime ______	# Coats ____	Color ______	Type ______
3) Wall C	=	______ S.F.	Prep ______	Prime ______	# Coats ____	Color ______	Type ______
4) Wall D	=	______ S.F.	Prep ______	Prime ______	# Coats ____	Color ______	Type ______
5) Wall E	=	______ S.F.	Prep ______	Prime ______	# Coats ____	Color ______	Type ______
6) Wall F	=	______ S.F.	Prep ______	Prime ______	# Coats ____	Color ______	Type ______
7) Dormers	=	______ S.F.	Prep ______	Prime ______	# Coats ____	Color ______	Type ______
8) Overhangs	=	______ S.F.	Prep ______	Prime ______	# Coats ____	Color ______	Type ______
9) Window	=	______ S.F.	Prep ______	Prime ______	# Coats ____	Color ______	Type ______
10) Doors	=	______ S.F.	Prep ______	Prime ______	# Coats ____	Color ______	Type ______
11) Garage doors	=	______ S.F.	Prep ______	Prime ______	# Coats ____	Color ______	Type ______

Notes:

Other damage repair:

Totals:

Type ________ Color ________ Gal ________ Manhours ________

Type ________ Color ________ Gal ________ Manhours ________

Type ________ Color ________ Gal ________ Manhours ________

Type ________ Color ________ Gal ________ Manhours ________

Type ________ Color ________ Gal ________ Manhours ________

Exterior estimating worksheet
Figure 7-19

Job #: I-235 **Customer:** M. J. SMITH **Date:** 2-25-87 **Room:** LIVING ROOM **Sheet** 1 **of** 5

Furnished_____ **Unfurnished**_____

A = 15
C = 12

			S.F.	Prep	Prime	# Coats	Color	Type
1) Ceiling	(A x C)	=	180	BLEACH	NO	1	WHT	TEXTURE
2) Wall A	(A x H)	=	120	PATCH	NO	1	RUST	FLAT
3) Wall B	(B x H)	=	120	PATCH	NO	1	WHT	FLAT
4) Wall C	(C x H)	=	96	NO	NO	1	WHT	FLAT
5) Wall D	(D x H)	=	96	PATCH	NO	1	WHT	FLAT
6) Window 1	(L x W)	=	48	REMOVE	YES	3	CLEAR	POLY
7) Window 2	(L x W)	=						
8) Window 3	(L x W)	=						
9) Window 4	(L x W)	=						
10) Door 1	(W x H)	=	21	YES	YES	2	WHT	SEMI
11) Door 2	(W x H)	=						
12) Door 3	(W x H)	=						
13) Door 4	(W x H)	=						
14) Cabinets		=	12	NO	NO	1	WHT	FLAT
15) Shelves		=						

16) Trim: Top of walls NO bottom of walls YES doors YES windows YES Total LF 100

Notes: WINDOW MUST BE STRIPPED – PLUS STAIN & POLY – ALLOW 3 WORKING DAYS FOR POLY TO DRY BETWEEN COATS. ADD 6 HRS TRAVEL.

Subtotals:

Type	Color	Gal	Manhours
TEX	WHT	2	4
FLAT	RUST	1/4	1.5
FLAT	WHT	1	2.5
SEMI	WHT	1/4	1.0
POLY	CLEAR	1/2	8.0 + 6.0

23 HRS @ 25.00 $ 575.00
MATERIALS
4 GAL @ 15.00 60.00
$ 635.00

Completed interior worksheet
Figure 7-20

Job #: A-25 **Customer:** R.J. JONES **Date:** 1-20-87 **Room:** ______ **Sheet** 1 **of** 1

A = 8'H x 30'L
B = 8'H x 30'L
C = 8'H x 12'W
D = 8'H x 12'W
OH = 1' x 60'

WINDOW
DOORWAY
FP FIREPLACE
X = DAMAGE

A
B
C
D
FP

H
W
H = 2' = 2Y
W = 12'

One story: ______ **Two story:** ______ **Dormers:** ______ **Overhang:** ______

1) Wall A	=	240 S.F.	Prep Y	Prime Y	# Coats 2	Color BLU	Type SEMI
2) Wall B	=	240 S.F.	Prep Y	Prime Y	# Coats 2	Color "	Type "
3) Wall C	=	96 S.F.	Prep Y	Prime Y	# Coats 2	Color "	Type "
4) Wall D	=	96 S.F.	Prep Y	Prime Y	# Coats 2	Color "	Type "
5) Wall E	=	— S.F.	Prep	Prime	# Coats	Color	Type
6) Wall F	=	— S.F.	Prep	Prime	# Coats	Color	Type
7) Dormers	=	— S.F.	Prep	Prime	# Coats	Color	Type
8) Overhangs	=	60 S.F.	Prep Y	Prime Y	# Coats 2	Color BLU	Type SEMI
9) Window	=	33 S.F.	Prep Y	Prime NO	# Coats 2	Color WHT	Type FULL
10) Doors	=	82 S.F.	Prep Y	Prime Y	# Coats 2	Color WHT	Type FULL
11) Garage doors	=	— S.F.	Prep	Prime	# Coats	Color	Type

Notes: X WATER DAMAGE — REPAIR WALLS 3 HRS @ 25/HR = $75.00

FP EXTERIOR IS WOOD — TO BE PAINTED

Other damage repair:

Totals:

Type SEMI	Color BLU	Gal 1	Manhours 8
Type PRIMER	Color WHT	Gal 2	Manhours 8
Type PREP	Color —	Gal —	Manhours 4
Type FULL	Color WHT	Gal 1 QT	Manhours 20
Type	Color	Gal	Manhours

20 + 3 HRS @ 25/HR = $575.00
1 GAL SEMI BLU = 18.00
2 GAL WHT PRIMER @ 15/GAL = 30.00
1 QT WHT GLOSS = 10.00
$633.00

Completed exterior worksheet
Figure 7-21

Painting Safety

When you see that this chapter is about safety, you'll probably skip right on to Chapter 9. But before you do, let me ask you to read the next three paragraphs. You won't have an opportunity to read about painting safety very often. In fact, this may be the only chapter you'll ever read on the subject. Let me suggest that you make the most of it. I'm going to explain everything my years as a paint contractor have taught me about safety on painting jobs. If you want to ignore that, O.K., it's your choice. But if you read this chapter and believe what I say and follow my suggestions, you'll qualify as the type of safety-conscious professional painter that every owner, lender, insurance carrier and employee would prefer to work with.

But, you say, you're already a safe painter. You run safe jobs. You haven't had anyone hurt on a job for at least the last six weeks. Why waste time on anything that doesn't end up as a bank deposit?

You're right. Safety is the last thing most paint contractors worry about, especially smaller, less experienced paint contractors. The big boys tend to take safety very seriously, and for a good reason. Safety pays. Any serious accident on a job is almost sure to wipe out the profit on that job. Accident prone contractors pay higher insurance rates, if they can get insurance at all. And the loss of one good man — even for a short time — may be the most expensive loss of all.

The most profitable, most professional paint contracting companies that I know have good safety records. Most of the serious painting accidents I can recall didn't happen in the larger, better-managed painting companies. They happened in smaller, marginal companies. Maybe it's no coincidence that many of those companies aren't around any more.

Truth in Advertising

When it comes to paints, manufacturers have thrown the book away. It's a competitive business: Paint manufacturers go to great lengths to protect their formulas. But to the user, this protection can be downright confusing, if not hazardous. To see what I mean, try reading the labels of a few paint cans. It's hard to tell whether you have a harmless coating or a lethal weapon on your hands.

As an example, let's take the label of a can of stucco paint. (I'm not trying to pick on this particular product — it's just typical of a majority of labels.) The label lists no ingredients. The label contains no safety warnings. It does say that the product is water clean-up, and water-repellent. To my way of thinking, water clean-up means latex. And water-repellent usually means oil-based. The color-chip card says that the paint gives "acrylic protection." The written description says the paint is an "oligomer resin," a "modified polysaccharide."

Here's the fun part. What is this paint made of? Let's try to break it down. If it's a latex, it contains natural or synthetic rubber. If it's an oil, it probably contains a drying oil. Many drying oils are poisonous and flammable. Now, "oligo" means few. And "mer" means belonging to the class of. So we can translate "oligomer resin" to "a resin that is in a class of few resins." Next, "modified" means changed. And "poly" translates to many. Finally, a "saccharide" is something containing sugar. So we have "many changed sugars"? About "acrylic protection": an acrylic is a form of acrylic acid. But maybe they're adding acrylic acid as the modifier?

I could go on with this analysis of the words, but you get my point. You figure it out: Is this product harmful or dangerous? Is it toxic, poisonous, or flammable? Or is it harmless? As I said, this is only one example of the way paint manufacturers try to bury their recipes.

Decoding the Labels

Even though it's a smokescreen, always read paint can labels. They won't always tell you what's in the can, and they may list ingredients that you can't identify. Here's a clue, though. Look at the last few letters in the words describing the ingredients:

- If the word ends in ...*ic* you can figure there's an acid present. Acids aren't flammable, but are highly corrosive. They'll burn eyes and skin. Acids "eat" into old paint so the new paint will adhere better.

- Words ending in ...*ol* indicate alcohol. Two examples are phenol and methanol. Alcohols are poisonous and highly flammable. They're generally used to reduce the drying time.

- Words ending in ...*ene* mean aromatic hydrocarbons. Most of these are also poisonous and highly flammable. Xylene is an example. They're usually added as the "carrier" ingredient.

- Words ending in ...*ide* are chemical compounds, mainly salts and acids. Chloride is one. These are usually caustic or bleaching by nature and will burn eyes and skin.

The next chapter goes into more detail on paint components and why they're used. If you want even more information, if you want to be a *real* expert on paint, Appendix A has some detailed information on the chemistry and physics of paints.

The Ideal Paint Label

I can understand the reasons for keeping paint formulas secret. But it seems only sporting to let painters know what they're up against. I'd like to see a label similar to the one in Figure 8-1 on all paint cans. And, speaking of labels, the manufacturer who adds a paint-remaining scale to the side of his cans will be a hit in my book. Especially if it's embossed on the can so I can read it from the inside.

With a label like this, you don't need to know the ingredients. What you will know is how to use this paint correctly and safely. There have been too many injuries, accidents, and fires involving paints. If you're ever in a position to influence a paint manufacturer, please explain that we painters need labels like Figure 8-1.

I hope paint manufacturers will establish good labeling standards. If they don't recognize their responsibilities, government will probably get into the act sooner or later, forcing them to adopt a safety label.

Before you jump up and say "Hey, the label in Figure 8-1 doesn't say anything about . . . " read on. *This proposed label covers only safety.* The usual information on spreading rates and application methods would still be listed elsewhere.

Building Codes and Paint Standards

The International Conference of Building Officials (ICBO) publishes the Uniform Building Code (UBC). It's a building code recommendation that your local building department can follow. If you live west of the Mississippi river, they probably do. Among other things, it establishes paint standards for fire safety in different types of building construction and building occupancy. Many communities change some sections of the code, however. You need a copy of the code in force in your area. Your local building inspector's office should be able to sell you a copy, or tell you where to get one.

Reason for the Codes

The building code is like a law; it's intended to be enforced, not read and understood. That makes it harder to follow what the code requires. But the code still provides protection, because your inspector knows what the code demands, even if *you* can't figure it out. It protects you, your employees, your customers, and innocent bystanders from harm.

Caution: Read Before Use

This product meets or exceeds the following minimum classifications:

Occupancy group ________________ Flash-point ________________

Flame spread ________________ Construction group ________________

Dry coat flame spread is ____ feet per minute

Flash-point of ____ °F ____ is/ ____ is not considered dangerous.

____ Keep away from open flame, extinguish all pilots lights, and use only in well-ventilated area.

Vapors are ____ toxic ____ nontoxic. Vapors will ____ rise ____ fall.

Liquid is ____ flammable ____ nonflammable.

Liquid is ____ poisonous ____ is not poisonous by ____ handling, and/or ingestion. If swallowed: ____

____

____

Product ____ does ____ does not contain asbestos and/or lead.

Fumes from burning of dried film are ____ toxic ____ nontoxic.

Product ____ can ____ cannot be used on plastics without harmful effects.

Use ____________ for thinner. Use ____________ for solvent.

Is approved by the following government agencies. ________ NFC ________ ASTM ________ ICBO.

Product contains ____ acids ____ bases ____ neither.

Intended use of this product is for ____________ of ____________ type surfaces. It is especially formulated to have the following attributes: ________________

____

____

All paint products are potentially hazardous if used or disposed of carelessly. Dispose of cans and/or liquid in hazardous-material waste.

Figure 8-1
The ideal paint safety label

We live in a litigation-happy society. If your failure to follow the codes results in property damage, completion delays, or personal injury, you can be liable for damages. You could lose not only money and time, but also your home and your business. Judges won't accept "Well, I didn't know" as an answer. As a professional painter, it's your business to know. I'll outline the basics for you.

Residential use— In residences there's usually no problem with using any paint. But there are rules for wall coverings and construction materials. Consult your local building inspector if you plan to use any plastics, or other nonstandard products.

Industrial, institutional, military or government use— There are rules governing paints for these uses. Paints must meet, or exceed, the requirements for construction, occupancy, and flame-spread. The National Fire Code (NFC), and the Uniform Building Code give full instructions.

I'll cover the basic UBC and NFC requirements. If you're covering any wall with more than a 28th of an inch of material, buy and read the codes. The UBC is available from:

The International Conference of
Building Officials
5360 S. Workman Mill Rd.
Whittier, CA 90601

Order the NFC from:

The National Fire Protection Association
470 Atlantic Ave.
Boston, MA 02210

These books aren't cheap, but the information is invaluable. If you want to open a paint store, a paint storage area, or a paint spray area, you better read UBC chapters 5 and 9 (section 901) very carefully.

Using the UBC

The Uniform Building Code divides occupancy into several classifications and subclassifications. They're listed alphabetically, from A to R. Group H is the most strict. The R and M occupancies are residences and garages. They're the least strict. Look at Figure 8-2 (UBC Table 5-A) for descriptions of the occupancy groups. This figure contains the Uniform Fire Code reference tables from the Uniform Building Code.

First, refer to the UBC to determine the building type. "Type" refers to the kind of construction

TABLE NO. 5-A—WALL AND OPENING PROTECTION OF OCCUPANCIES BASED ON LOCATION ON PROPERTY
TYPES II ONE-HOUR, II-N AND V CONSTRUCTION: For exterior wall and opening protection of Types II One-hour, II-N and V buildings, see table below and Sections 504, 709, 1903 and 2203.
This table does not apply to Types I, II-F.R., III and IV construction, see Sections 1803, 1903, 2003 and 2103.

GROUP	DESCRIPTION OF OCCUPANCY	FIRE RESISTANCE OF EXTERIOR WALLS	OPENINGS IN EXTERIOR WALLS
A See also Section 602	1—Any assembly building with a stage and an occupant load of 1000 or more in the building	Not applicable (See Sections 602 and 603)	
	2—Any building or portion of a building having an assembly room with an occupant load of less than 1000 and a stage 2.1—Any building or portion of a building having an assembly room with an occupant load of 300 or more without a stage, including such buildings used for educational purposes and not classed as a Group E or Group B, Division 2 Occupancy	2 hours less than 10 feet, 1 hour less than 40 feet	Not permitted less than 5 feet Protected less than 10 feet
	3—Any building or portion of a building having an assembly room with an occupant load of less than 300 without a stage, including such buildings used for educational purposes and not classed as a Group E or Group B, Division 2 Occupancy	2 hours less than 5 feet, 1 hour less than 40 feet	Not permitted less than 5 feet Protected less than 10 feet
	4—Stadiums, reviewing stands and amusement park structures not included within other Group A Occupancies	1 hour less than 10 feet	Protected less than 10 feet
B See also Section 702	1—Gasoline service stations, garages where no repair work is done except exchange of parts and maintenance requiring no open flame, welding, or use of Class I, II or III-A liquids 2—Drinking and dining establishments having an occupant load of less than 50, wholesale and retail stores, office buildings, printing plants, municipal police and fire stations, factories and workshops using material not highly flammable or combustible, storage and sales rooms for combustible goods, paint stores without bulk handling Buildings or portions of buildings having rooms used for educational purposes, beyond the 12th grade, with less than 50 occupants in any room	1 hour less than 20 feet	Not permitted less than 5 feet Protected less than 10 feet

From the Uniform Building Code, ©1985, International Conference of Building Officials

Uniform Building Code occupancy categories
Figure 8-2

TABLE NO. 5-A—Continued
TYPES II ONE-HOUR, II-N AND V ONLY

GROUP	DESCRIPTION OF OCCUPANCY	FIRE RESISTANCE OF EXTERIOR WALLS	OPENINGS IN EXTERIOR WALLS
B (Cont.)	3—Aircraft hangars where no repair work is done except exchange of parts and maintenance requiring no open flame, welding, or the use of Class I or II liquids Open parking garages (For requirements, See Section 709.) Heliports	1 hour less than 20 feet	Not permitted less than 5 feet Protected less than 20 feet
	4—Ice plants, power plants, pumping plants, cold storage and creameries Factories and workshops using noncombustible and nonexplosive materials Storage and sales rooms of noncombustible and nonexplosive materials	1 hour less than 5 feet	Not permitted less than 5 feet
E See also Section 802	1—Any building used for educational purposes through the 12th grade by 50 or more persons for more than 12 hours per week or four hours in any one day 2—Any building used for educational purposes through the 12th grade by less than 50 persons for more than 12 hours per week or four hours in any one day 3—Any building used for day-care purposes for more than six children	2 hours less than 5 feet, 1 hour less than 10 feet[1]	Not permitted less than 5 feet Protected less than 10 feet[1]
H See also Sections 902 and 903	1—Storage, handling, use or sale of hazardous and highly flammable or explosive materials other than Class I, II, or III-A liquids [See also Section 901 (a), Division 1.]	See Chapter 9 and the Fire Code	
	2—Storage, handling, use or sale of Classes I, II and III-A liquids; dry cleaning plants using Class I, II or III-A liquids; paint stores with bulk handling; paint shops and spray-painting rooms and shops [See also Section 901 (a), Division 2.] 3—Woodworking establishments, planing mills, box factories, buffing rooms for tire-rebuilding plants and picking rooms; shops, factories or warehouses where loose combustible fibers or dust are manufactured, processed, generated or stored; and pin-refinishing rooms 4—Repair garages not classified as a Group B, Division 1 Occupancy	4 hours less than 5 feet, 2 hours less than 10 feet, 1 hour less than 20 feet	Not permitted less than 5 feet Protected less than 20 feet
	5—Aircraft repair hangars	1 hour less than 60 feet	Protected less than 60 feet
H (Cont.)	6—Semiconductor fabrication facilities and comparable research and development areas when the facilities in which hazardous production materials are used are designed and constructed in accordance with Section 911 and storage, handling and use of hazardous materials is in accordance with the Fire Code. [See also Section 901 (a), Division 6.]	4 hours less than 5 feet, 2 hours less than 10 feet, 1 hour less than 20 feet	Not permitted less than 5 feet, protected less than 20 feet
I See also Section 1002	1—Nurseries for the full-time care of children under the age of six (each accommodating more than five persons) Hospitals, sanitariums, nursing homes with nonambulatory patients and similar buildings (each accommodating more than five persons)	2 hours less than 5 feet, 1 hour elsewhere	Not permitted less than 5 feet Protected less than 10 feet
	2—Nursing homes for ambulatory patients, homes for children six years of age or over (each accommodating more than five persons)	1 hour	
	3—Mental hospitals, mental sanitariums, jails, prisons, reformatories and buildings where personal liberties of inmates are similarly restrained	2 hours less than 5 feet, 1 hour elsewhere	Not permitted less than 5 feet, protected less than 10 feet
M[2]	1—Private garages, carports, sheds and agricultural buildings (See also Section 1101, Division 1.)	1 hour less than 3 feet (or may be protected on the exterior with materials approved for 1-hour fire-resistive construction)	Not permitted less than 3 feet
	2—Fences over 6 feet high, tanks and towers	Not regulated for fire resistance	
R See also Section 1202	1—Hotels and apartment houses Convents and monasteries (each accommodating more than 10 persons)	1 hour less than 5 feet	Not permitted less than 5 feet
	3—Dwellings and lodging houses	1 hour less than 3 feet	Not permitted less than 3 feet

[1]Group E, Divisions 2 and 3 Occupancies having an occupant load of not more than 20 may have exterior wall and opening protection as required for Group R, Division 3 Occupancies.

From the Uniform Building Code, ©1985, International Conference of Building Officials

Uniform Building Code occupancy categories
Figure 8-2 (continued)

used in building. Type is used in the occupancy classification section (Table 5-A) of the code.

The top part of Figure 8-3 (Table 42-A) gives flame-spread classes I, II, and III. These are the results of actual tunnel tests of how fast the flame will spread. The label on the paint can should show either the flame-spread or the classification. If not, contact the manufacturer. Or use a brand that's labeled.

The bottom part of Figure 8-3 (Table 42-B) shows the maximum flame-spread class you can use for each occupancy group in Figure 8-2. You must use all three of these tables to find the final minimum classification of paint to use.

Using unrated or underrated material can cause the entire building to be downgraded in classification. Why? Because buildings take on the lowest classification of the materials used in them. A Class I building, if painted with Class III paint, becomes a Class III building. You can use a higher grade paint than necessary, but beware of using a lower class than the code requires. It could result in an expensive rework, and possibly a lawsuit.

TABLE NO. 42-A—FLAME-SPREAD CLASSIFICATION

MATERIAL QUALIFIED BY:	
Class	Flame-spread Index
I	0-25
II	26-75
III	76-200

TABLE NO. 42-B—MAXIMUM FLAME-SPREAD CLASS[1]

OCCUPANCY GROUP	ENCLOSED VERTICAL EXITWAYS	OTHER EXITWAYS[2]	ROOMS OR AREAS
A	I	II	II[3]
E	I	II	III
I	I	II	II[4]
H	I	II	III[5]
B	I	II	III
R-1	I	II	III
R-3	III	III	III[6]
M	NO RESTRICTIONS		

[1]Foam plastics shall comply with the requirements specified in Section 1712.
[2]Finish classification is not applicable to interior walls and ceilings of exterior exit balconies.
[3]In Group A, Divisions 3 and 4 Occupancies, Class III may be used.
[4]In rooms in which personal liberties of inmates are forcibly restrained, Class I material only shall be used.
[5]Over two stories shall be of Class II.
[6]Flame-spread provisions are not applicable to kitchens and bathrooms of Group R, Division 3 Occupancies.

From the Uniform Building Code, ©1985, International Conference of Building Officials

Uniform Building Code flame-spread classifications
Figure 8-3

An example of using the UBC— You're contracted to paint and wallcover a school building. The job entails doing a lab in ceramic tile, putting up cork bulletin boards in two classrooms, painting five classrooms, paneling a library, putting up plastic sheet in a hall, and carpeting the walls of a first grade playroom. The school is used for grades K through 12, five days a week. The average class size is 45 students.

Looking at UBC Table 5-A, we find that the occupancy is Group E-2. Table 42-B shows that occupancy group E must have classrooms rated Class III, exitways of Class II, and vertical exitways of Class I. You might think that Class III materials will work in all the classrooms. Here's where "hunt and read" comes into play. UBC chapter 9, section 4204 (b), states that all carpet on walls must be Class I.

Chapter 17, section 1712 (a), covers plastics. Plastic sheet has to be marked and most have a flame spread of greater than 75 and a smoke density rating of 450, or Class III. According to Table 42-B, we need Class II materials for the hallways. But read section 1712 (a): *The interior of the building shall be separated from the foam plastic by an approved thermal barrier having an index of 15* . . . and chapter 42, section 4201: *Requirements . . . shall not apply to . . . materials which are less than 1/28 inch in thickness cemented to the surface of walls or ceilings* . . . . By using very thin plastic, we can cover the hall walls.

Now what about the ceramic tile in the lab? Turn to UBC chapter 30, sections 3001 and 3005. Section 3001 says that with building inspector approval, we can tile up to 48 inches high from the finish floor. Over that, we go to section 3005 (d), which specifies the type and thickness of the lath and mortar. Flame spread should be Class I. We can install the tile with no problem.

In the library, you're pretty safe to put up any Class III paneling. But if the wall is an exterior wall, it's required to be fire resistive. Then chapter 42, section 4203 applies. You must fire-stop every 8 feet, and if you use furring it can be no more than 1¾ inches thick. The intervening spaces must be filled with a Class I or inorganic material. Chapter 25, section 2518 (f), describes a fire-stop.

Finally, let's consider the bulletin board installation. Chapter 42, section 4204 (a) states that bulletin boards covering not more than 5 percent of the gross wall area of the room may be installed.

By now, some of you are saying "Oh, boy!" I agree it's a pain — but there are reasons for these codes and classifications. They're trying to buy time: time to beat the fire to the exits. Class III materials in classrooms allow students to get into the hallway. Class II materials in the halls slow the spread of the fire there. Class I in the stairwells slows the spread even more, preventing students from being trapped in burning stairwells. Stairs make a good natural path for fire spread.

I've just touched on a very few of the UBC codes. You must also look in the NFC. Sometimes it might just be simpler to take your plans to the local building inspector and let him figure out if you're in compliance. All codes are open to interpretation, and it's the interpretation of the building inspector that counts.

Think Safety

When painting, think safety. Take precautions against spillage, splatters, kids and pets, fumes, and odors. Most — though not all — paints are only dangerous when wet. Whether a dried paint is dangerous depends on what's in it and where it's used.

Poisonous Paints

Paint that's around children or pets often gets chewed on. On farms, animals chew on paints. There are many paints specially made for nursery and farm use. Read the label.

White paint looks good on fences — but it doesn't last. The sun, wind, and rain soon destroy it. In addition, horses love to chew on fences. The black horse-ranch fences shown in Figure 8-4 solved the problem. The coating is hot liquid tar. The tar sinks into the wood. When warmed by the sun, it softens and reseals any breaks. And it tastes real bad, so the horses have stopped nibbling.

Additives to watch out for— Some paints have poisonous additives. Watch out for tributyltin (a mildewcide), chlorpyrifos (an insecticide), pentachlorophenol (a preservative), and copper napthenate (a preservative). These poisons remain in the dried paint. Other paints contain lead and arsenic, both of which accumulate in the body and are highly poisonous.

Many dried paints give off toxic fumes when burned. Don't use these types of paints for interior applications. Most of the paints that contain the chemicals I just mentioned fall into this group.

Black fences at horse ranch
Figure 8-4

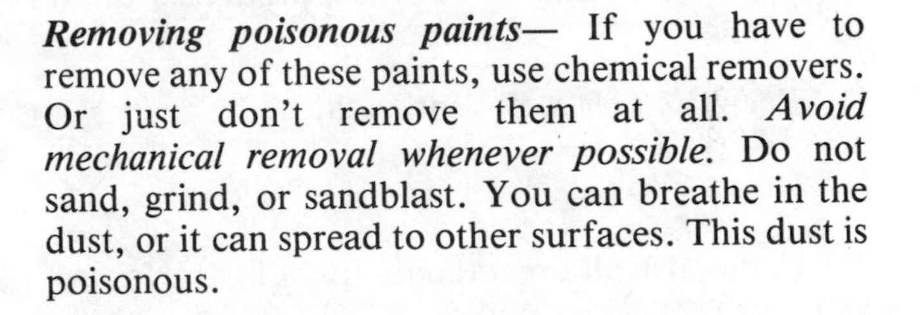

Removing poisonous paints— If you have to remove any of these paints, use chemical removers. Or just don't remove them at all. *Avoid mechanical removal whenever possible.* Do not sand, grind, or sandblast. You can breathe in the dust, or it can spread to other surfaces. This dust is poisonous.

Paint Fumes

Not long ago in Los Angeles, police stopped a man for drunk driving. He was all over the road and nearly caused several accidents. He wasn't intoxicated from drinking, but from painting. It happens to all professional painters at one time or another.

The fumes from most paints are toxic. They may not be poisonous, but they starve the brain of oxygen. When your brain doesn't get the oxygen it needs, you become lightheaded, dizzy, faint, and make careless mistakes. Climbing a ladder, driving a car, or using motorized equipment while in this state is dangerous to you — and to others around you. Any time you're painting indoors, think about ventilation.

Flammable Paints

Some paints, and their fumes, are flammable. On my ideal paint-can label, I used the term *flash-point.* What is it? It's the minimum temperature (in degrees Fahrenheit) at which a flammable liquid will give off flammable vapors. That means vapors that can catch fire, if a spark or flame is present. A *flammable liquid* is one with a flash-point below 140 degrees F. As you can see, we're talking about pretty low temperatures.

Most manufacturers don't give a flash-point on their labels. You can be sure, though, that if it says "combustible" — it is. Many painters don't know it, but even some water-based paints are combustible.

Another Los Angeles painter found himself up a tree, both literally and figuratively. He didn't put out the pilot light on a gas range when he was painting a kitchen. Blew the house into kindling, and he landed in the branches of a tree some 20 feet away. His next stop was the hospital, and when he got out, in the defendant's dock in court.

Rules for flammable paints— It does happen, so take care. Fumes of any paints containing alcohol, aromatic hydrocarbons, petroleum distillates, acetates, or turpentines are flammable and potentially explosive. This includes most oil-base paints, varnish and oil stains, as well as many commercial cleaners. When using one of these products, follow these rules:

1) Keep the area well ventilated.

2) Don't smoke.

3) Extinguish all open flames (pilot lights, for example).

4) Don't flick light switches on and off, or turn on any electrical motors while you're painting. Unplug coffee pots, refrigerators, automatic fans, and other electrical appliances before you start.

I know, you never did all this before, and never had any problems — but it only takes once.

Lead-Based Paints

Lead-based paint was used in older homes, usually pre-1960. It creates a special problem. It's difficult, if not impossible, to remove. *So don't remove it.* Trying to completely remove old lead-based paint by sanding, sandblasting, chemically stripping, or burning is hazardous to your health. Why? Because the lead you breathe in, along with the dust or vapors, is a poison, a poison that stays in your body.

Instead of removing all the old paint, chip off only what's loose. Then use liquid sandpaper to feather the remaining paint to a smooth surface. Prime any unpainted areas, and repaint with latex or oil/alkyd.

How to tell if a building has lead-based paint— In 1972, the U.S. government passed laws restricting lead content in paint to less than one-half of one percent. Most manufacturers have stopped using lead altogether. But how do you tell whether an old building was painted with lead-based paint? One way is by the age of the building. Any pre-1972 building is a suspect.

The other clue to look for is *fume staining.* This is a brown discoloration that bleeds through the top coat. Look at Figure 8-5 for an example. This white-painted siding looked dull and patchy. The home was over 30 years old, and that gave the clue: Lead, and other minerals, from the lead-based paint were leaching to the surface. The solution was to bleach the entire surface, prime it, and then repaint.

Fume staining from lead-based paint
Figure 8-5

Spray Painting

Any dry powder in the air can cause an explosion, if there's enough of it and a flame or spark to ignite it. Even flour dust or sugar dust will explode under the right conditions. Spray painting creates the right conditions. This is true of nearly all paint spraying. And it's one of the reasons that the UFC requires an approved paint booth or painting room. These booths have explosion-proof lighting and ventilation fans. Most have an air scrubber or filter system. (A scrubber is a water-spray system through which the contaminated or dust-laden air passes.)

To reduce the chance of an explosion, follow these simple steps:

- Keep the area well ventilated.
- Keep the spray wet, not dry and powdery.

• Don't spray paint in a very hot (over 100 degrees F) enclosed room.

• Finally, make sure that all flames and spark-generating devices are turned off.

Common-Sense Safety Tips

Here are a few more common-sense tips. Follow them, and you'll have a long and trouble-free career as a painter. Ignore them at your peril.

• Collect all rags used for applying or for cleaning up paints or other coatings. Soak them in water to prevent spontaneous combustion.

• Be sure your power cords and tools are in good condition.

• When using electric or gas-operated equipment, keep it away from combustibles.

• Gas-operated equipment (an airless or a waterblaster, for example) has a muffler. This muffler gets hot enough to set grass and weeds on fire, if it touches them. Also be sure such equipment has a spark arrestor. If you're painting outside in dry air, keep a fire extinguisher or water hose handy.

• Don't strip paint with a torch: use chemicals instead. I've seen homes go up in flames hours after the torch was turned off. The torch had started an unseen fire inside the walls. This is especially likely on wood lap siding.

Most of today's paints are synthetics, containing man-made chemicals. Many of these chemicals are dangerous if misused. Your body is a remarkable instrument: it can expel tiny amounts of chemicals and poisons. But since you're a painter, you're exposed to much more than tiny amounts. You must protect your health by taking proper precautions, wearing protective clothing, and using safety devices. Wear gloves, eye protectors, a dust mask, steel-tipped shoes and full body coverings. It'll cost a little more and take a little extra time, but you can't afford *not* to do it.

chapter nine

Choosing Paint

A can of paint is a can of paint.

I've heard that said many times, but it's just not true. Paint manufacturers have made a mistake. They haven't made it clear that paint is chemistry — modern, complex, high-tech chemistry. Choose the wrong paint and you'll do a poor paint job, no matter how skilled you are with a brush, roller or spray rig. But it's easy to avoid costly and time-consuming reworks. Select the right paints and coatings the first time.

Why Learn About Paint Chemistry?

Problems don't usually appear in the first 90 days after painting. They show up months — sometimes years — later. You say, "So what, then it's someone else's problem." Wrong. It's your problem. Even if the paint fails 18 to 48 months after your warranty period, you'll be blamed. Paint should last 7 to 10 years, and properly selected and applied paint *will.* But choose the wrong products, and you're in trouble. It's your reputation on the line — the reputation that builds your business by bringing referrals and repeat customers.

This chapter covers some of the chemistry of paints. Take the time to read it for one very important reason: The knowledge will help you make money. Here's how:

1) If you know the problems associated with paints and surfaces, and understand the solutions, it's easier to sell yourself and the job.

2) You'll eliminate most of your in-warranty reworks.

3) Over the years, you'll build a reputation for knowing your job and doing quality work. This, in itself, can earn you hundreds of thousands of dollars in increased sales.

4) As your reputation as a professional grows, you need less of that expensive advertising. Think about how saving just 10% of your advertising budget will affect your profit margin.

5) A bad reputation will hurt you. Everyone remembers the painter who applied a coating that peeled or blistered. People will spread the word. How much will that cost you? What's your excuse?

6) Contractor license tests in many states require knowledge of the chemistry of paints.

I know you aren't a chemist. But you expect to be a professional painter and earn a decent living. Nearly every problem shown in this book could have been prevented, if the painters knew their job. For example, the community I live in has 870 homes. Half of them are under five years old, but one-third need to be repainted already.

The paint contractors who originally painted them are all out of business. No one will rehire them. Several other painters have gone out of business in the last few years too, because of costly paint failures. One manufacturer also went out of business. Their products weren't formulated for the climate.

"But," you say, "why should I have to know the products? I'll just ask the paint store salespeople. They're the ones who should know." You're right, they should know. But do they? And do they know your job conditions? Most are clerks, not paint experts or painters. No, the final responsibility rests with only one person, you. Your knowledge of the products you apply will save the day.

Why Paint in the First Place?

The answer is simple. Most people say you paint to add color, but that's not the whole reason. You paint because it's the least expensive way to protect the surface. Even if you paint everything white, you've still accomplished the basic purpose: protection.

The two big errors in paint selection are choosing paints by color and cost. Don't do it. Choose paints by need. Here are the needs I'm talking about:

- How much protection is needed?
- How long must the protection last?
- What are the environmental conditions?

Worry about color and price only after these needs are met. It doesn't matter that the paint only comes in black, and costs $100 a gallon. If it's called for, use it. Using the wrong paint will leave the surface unprotected and speed deterioration — resulting in a loss that makes the cost of the paint look like peanuts. So the question becomes: Is protecting the item worth the cost to protect it? That's up to the owner. But it's your job to make sure the owner understands that it's protection he's buying — or not buying.

Of course, color does have its advantages. And, since you're painting anyway, you may as well use some color. Color sets the mood, adds satisfaction, and makes the old look new. Color turns a cold room warm, a warm room cool. Color makes the mood happy, restful, rambunctious. Color adds elegance, design, and brightness. (More about all this in Chapter 10.)

But here's my point: Poor paint selection is a major cause of paint failures. You just can't pick any old can of paint off the shelf, slap it on, and expect it to perform properly — even if the color is perfect. The rest of this chapter will show you what to look for in selecting the right coating for the job.

Paint Quality

Paint prices are based on supply and demand, brand name, solids content, carrier type, and factory additives. Some of the price also provides the profit for manufacturer, wholesaler, and retailer.

Inexpensive paint is fine if it's good paint — but cheap paint is cheap paint. Those $4 and $5 a gallon paints generally have a low solids content. This translates to poor coverage, poor durability, and short life. Is saving $4 or $5 a gallon worth the time and effort of doing the job twice?

Buy brand names, but don't be afraid to try some off-brands. I've found some to be quite good. And I've found some major brands to be so-so, and very over-priced. Most of the paint I buy costs $12 to $18 a gallon in 1987 prices. It's probably the best paint value. If you're not in a hurry, wait a while. Most paints go on sale two or three times a year, so you can stock up then.

Remember, you're about to spend hours — maybe weeks — painting. Buy paint that's right for the job.

What You're Paying For

Fortunately, reputable paint manufacturers recognize all the problems I've listed. They've taken steps to prevent, or at least lessen, these problems. Chemists select formulations and additives that will overcome potential problems. Most provide an attractive and reasonably durable product.

Today's quality paints have additives to fight dirt, grease, and mildew, additives to make them penetrate and adhere better, additives that make them flatten out so brush and roller marks aren't obvious, additives to speed drying, additives to help prevent drips and splatters, and additives to reduce the effects of sun, rain, heat, and cold. Does paint made to be sold for $5 or $6 a gallon have these additives? What do you think?

Quality Paint

Quality paint isn't necessarily so expensive that it prices you out of the job. It just has to meet job requirements and last long enough to justify the expense of application. Also, a quality paint must:

- Flow freely for easy application, yet not so freely that it flows right off the surface
- Flow at a rate that ensures proper leveling
- Flow properly at a wide range of temperatures
- Have a reasonably short drying time to minimize the pickup of dirt, dust, and insects
- Have a long enough drying time to assure leveling
- Have enough bond to stay on the surface
- When dry, be somewhat resistant to chemicals and water
- Be hard enough to protect the surface it covers
- Be soft and rubbery enough to withstand abrasion
- Stay flexible through variations and extremes of temperature
- Not destroy the surface through any chemical reactions
- Be reasonably safe for normal handling
- Have a suitable solvent for cleanup
- Withstand the effects of weather
- Withstand the effects of external energy fields such as electromagnetic radiation.

Most paints are transparent to electromagnetic radiation in the nonvisible range, such as radio and TV signals. Lead-based or metallic-pigmented paints are not. Don't use these paints on a surface that you want this radiation to pass through, a microwave antenna enclosure, for instance.

Every paint, though, can't be formulated to overcome all problems. It would be much too expensive. What we have, then, are specialty paints. These are paints with just enough of selected additives to do the intended job. Note that I said *intended job.* It's up to you to choose the right paint for the application. Using an interior paint on the exterior of a house is courting disaster, for instance. And using exterior paint on an interior is a waste of money.

Paint Longevity

How long does paint last? I guess this is one of the most common questions about paint. The answer depends on some variables: Interior paints are subjected to smoke, waxes, grease, soaps, and chemical and cooking fumes. Typical physical abuse includes hitting, bumping, and cleaning.

Exterior paints face most of the same hazards as interior paint and also must withstand the weather. This includes abrasion by wind-blown rain, dust, and dirt.

Now add the normal oxidation and chemical reactions that occur in both interior and exterior paints. Consider the possibility of poor or improper paint selection, or poor surface preparation, and you have real problems.

Yet despite all this, most paints have an average useful life span of between seven and twelve years. The right paint, properly applied, should last that long. The exception is marine paints, which have a useful life of two and a half to three years.

Longevity of Plastic Coatings

Seven to twelve years is a good expectation for most exterior and some interior applications. Plastic based paints are designed to meet some special requirement such as extra strength. A plastic's actual life-span should be almost double the figures above, if the coating was applied correctly.

Plastic coatings lose about one-half their resilience in seven to twelve years. As they age, they become less resistant to weather and damage. They become more brittle, and eventually crack. Use of U/V stabilizers, chemical additives, waxes, and wax-cleaner-restorers delays this action, but won't prevent it. Aging is normal. But your clients should know about the problem and plan for eventual replacement. (When vinyl house siding came out, no one knew about this problem. The product was new, and no longevity testing had been done.) So think of a plastic coating as paint that will need to be replaced periodically.

Today's paints and coatings go through extensive longevity testing. Though they don't publicize it, I'll bet most manufacturers could tell you how long a coating will last in a given exposure. Most people expect to repaint in seven to ten years, anyway. And the government doesn't require that the information be made public, since paint failure isn't a safety hazard.

Still, most painting is done before the paint fails completely. People repaint because they're tired of the color. Or because the paint hasn't been maintained. Sometimes repainting is needed because the wrong paint was used. Or because surface prep wasn't done, or was done wrong, and the coating has been destroyed.

Paint Porosity

Once applied and dry, the paint film has holes between the molecules. These holes can fill with oil, grease, dirt, and bacteria or fungi. Denser paints have fewer holes, and less chance of becoming contaminated.

This is why you use dense gloss or semi-gloss enamels for kitchen and bath areas, rather than a more porous flat paint. The denser the paint, the less water will penetrate it. Dense paints are good vapor barriers for these areas.

Paint Expansion and Contraction

All materials expand and contract with temperature and moisture changes. To stay firmly attached, the paint film you apply should expand and contract at the same rate and the same percentage as the surface it covers. If you use a paint with the same *percentage* of expansion and contraction as the surface, but with a different *rate* of expansion and contraction, the paint will peel. The opposite is also true: If the rate is the same, but the percentage is different, the paint will peel.

Use the right paint for the surface— This is why manufacturers make paints for specific surfaces. They formulate the paint to have an average rate and percentage closely matching the average rate and percentage of the intended surface.

Look at Figure 9-1. This shows what happened on a door when the paint and the surface expand at different rates. The right side is somewhat protected from the weather; the left side isn't. At the right, the wood expanded, cracking the paint. Toward the center, the paint expanded, and started to peel. At the left, the paint and the wood expanded and contracted at different rates, and the paint is flaking. (The solution is to strip it completely, fill the cracks, sand it smooth, prime, and then topcoat. It's probably cheaper to replace the door.)

Epoxy paints— These work well on cement or cement block for two reasons. First, the percentage and rate of expansion and contraction of epoxy and of cement are nearly identical. Second, epoxy is tough and abrasion-resistant. Don't, however, use two-part epoxies on wood exteriors. Their rates and percentages of expansion and contraction are too different.

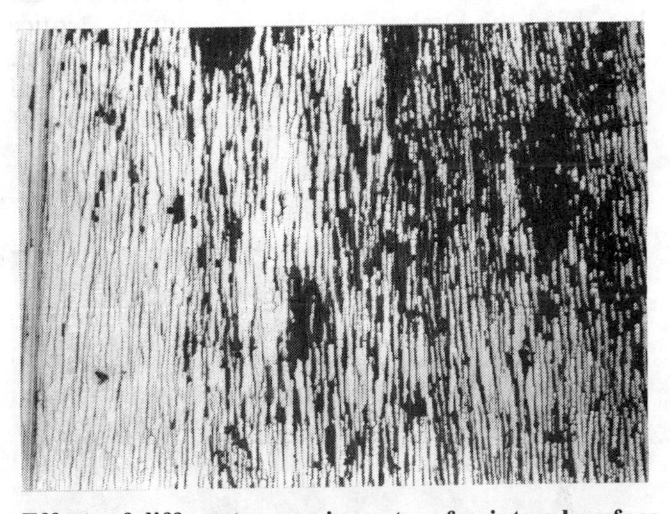

Effects of different expansion rates of paint and surface
Figure 9-1

Heat Absorption and Paint Color

When choosing paint, don't overlook the effect color has on heat absorption and reflection. Here's what you need to know: A white or light-colored surface reflects light, and therefore, heat. A dark color absorbs light and heat.

Absorbed heat causes paint and surface to expand at different rates— The paint color affects both the paint itself and the surface under it. If a dark paint allows heat to enter the surface material, and the material expands faster than the paint, the paint will separate from the surface. The opposite is also true: if heat causes the paint to expand faster than the surface, it will separate. In either case, the paint cracks, flakes, or peels.

Clear varnishes on hot, sunny surfaces cause the surface to soak up more heat. The surface expands. Eventually, the varnish separates. This is because clear varnishes dry hard and don't expand much. Instead, use a clear varnish containing a U/V inhibitor, or use a penetrating-oil-type varnish or coating in these areas.

Heat dries out the paint and surface— House trim is commonly painted a darker color than the siding. This dark color absorbs heat, and transmits it to the trim. The trim on the south side of a building will almost always crack, split, flake and peel faster than the trim on the north side. Of course, the water content of the lumber, the type of lumber, the paint formulation, and the color of the paint all affect the trim's (and the paint's) longevity.

Look at Figure 9-2. This dark brown fence is on the southeastern side of a building. The sun's heat has dried the lumber, causing it to shrink. Notice the vertical line where the post meets the panel. Also note that the knots are starting to open and crack. (The solution? Allow a full season or two of weathering, then repaint.)

Heat causes lumber to dry out
Figure 9-2

Paint and U/V Radiation

Ultraviolet (U/V) radiation from the sun also destroys paints. Ultraviolet radiation tends to intensify damage caused by heat. In very hot or sunny areas, use exterior paints with U/V inhibitors.

Look at Figure 9-3. Although there are several problems here, the main culprit is long-term weathering. This building is in an area of Northern California that has several months of temperatures over 90 degrees. U/V radiation has literally caused the paint to dissolve into the air. Notice that the sun-protected areas at the right and the top of the door show much less damage than the unprotected areas.

Paint Oxidation

Some paints are labeled *photo-sensitive* or *photochemically sensitive*. These paints contain pigments that will either oxidize or change color. Never use them on exteriors, or on interior surfaces subject to sunlight, such as window sills.

Chalking paints— Some paints are formulated to oxidize. They form a powder that's easily washed off with water. The surface dirt is washed away, along with the oxides, during a rain storm or a hosing-off. These paints are often called *chalking paint.* They are for exterior use only. Don't use them for interiors. They'll stain furniture, clothing, hands, and anything else that rubs against them.

There are other problems with this type of paint. First, the oxidizing process consumes the paint film. Eventually there's none left. Second, it stains. Never use oxidizing paint above a surface you don't want to have stained. And never use it on dormers, or on roof protrusions like vent pipes or a chimneys. It will stain the roofing.

Figure 9-4 shows what I mean. The stucco below the siding is stained the same color as the siding. The rain has washed the chalk onto the stucco below. (Solution? Bleach and repaint the stucco. And either repaint the siding with a non-chalking paint, or wash it down more often.)

Most paints will chalk a little. But a non-chalking paint that shows severe chalking indicates an abnormal paint failure that needs investigation.

Effects of U/V radiation on paint
Figure 9-3

Chalking paint stains stucco wall
Figure 9-4

Paint Fading

Once a paint is dry, each particle of pigment is surrounded with a binder that holds the film onto the surface. As this binder is worn off — by weathering, U/V radiation, or abrasion — it no longer surrounds the pigment. Now the pigment is free to be blown from the surface by wind or washed away by rain. As the amount of pigment decreases, the color *fades*. The paint loses its hiding power and the underlying surface color begins to show.

Look at Figure 9-5. At the left is a newly-painted surface with the pigments locked in. The binder slowly gets worn down and the pigments start to loosen. Eventually, the entire coating is consumed and no longer protects the surface.

Sometimes the binder, or another additive in the paint, oxidizes, resulting in a whitish surface. This is called *apparent fade*. The surface looks faded, but actually isn't. A wash with oil or water corrects the problem. A self-cleaning paint will show apparent fade if there's been no rain for a long time. A good wash-down is all that's needed to bring back the color and newness.

An expertly-applied, professional-looking paint job can look like a million dollars and be worth nothing. And a sloppy-looking job can look like nothing and be worth a lot. The difference is the choice of the proper paint. That's one that meets, or exceeds, the requirements of the job.

All cxterior paint should have color-fast pigments, be U/V protective, be nonphotochemically reactive, and have good resistance to weather.

Basic Paint Formulations

The basic theory of paint formulations is simple: Suspend a pigment in a liquid vehicle. When this mixture is applied to a surface, the vehicle evaporates and the pigment stays on the surface to protect and color it. But this coating would soon come off if there were no binder (glue) to hold the pigment in place. We now have the three basic ingredients of all paint: pigment, vehicle, and binder.

One of the simplest paints is calcimine, made of water with zinc white and glue mixed in. It's used on plaster walls and ceilings. A similar mix is whitewash, used for building exteriors. It's water with lime or whiting, maybe zinc white, white glue, salt, and sometimes rice flour mixed in. It's used mainly for outbuildings on farms because the cost is low. It doesn't last very long, or give good protection. Figure 9-6 shows an old barn that has been coated with whitewash. I think the picture speaks for itself.

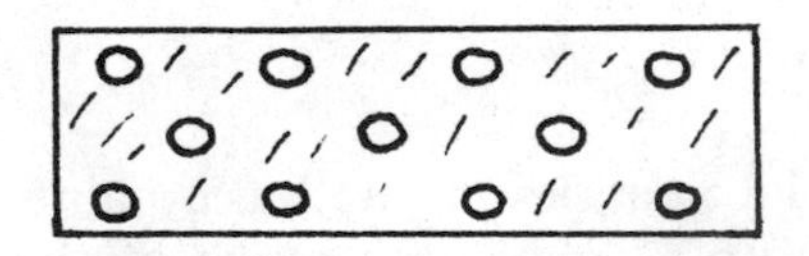

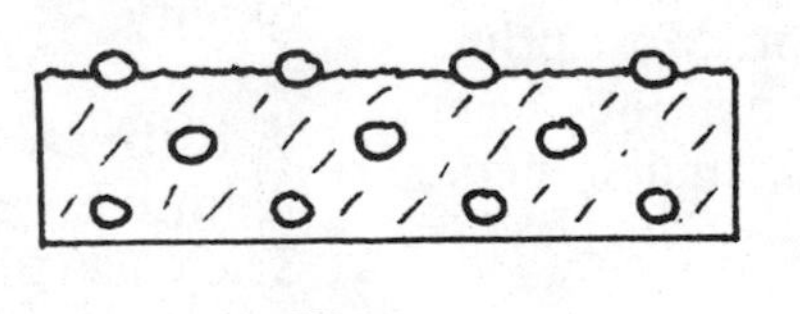

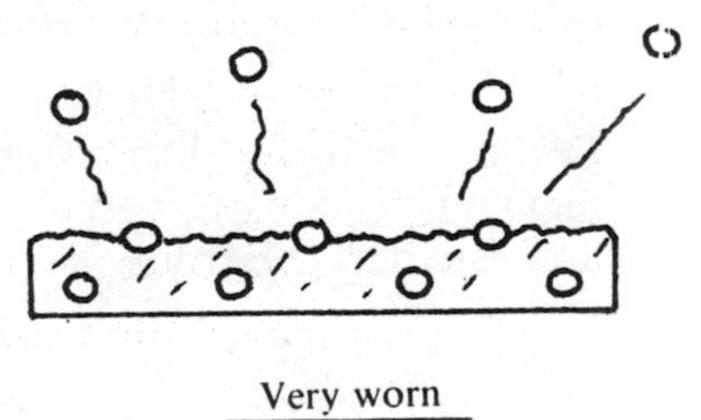

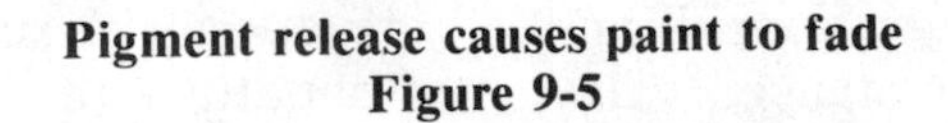

Pigment release causes paint to fade
Figure 9-5

Whitewash — a simple paint formulation
Figure 9-6

Pigments

All pigments must, first, have the same attributes as the basic paint. They also add special qualities, such as color. Not only color, though. There are many colorless pigments that add special qualities. Pigments, unlike additives, are inert insoluble particles suspended in the liquid. (Additives are soluble and react chemically with the vehicle or with the surface.) The pigment particles must be large enough to do their intended job, yet small enough to remain suspended in the vehicle. Pigments fall into two main groups: prime and inert.

Prime pigments— These supply color. Most prime pigments today are synthetic. Years ago they were natural, derived from plants, clays, and mineral ores.

Inert pigments— These pigments are used to meet special requirements. Many inert pigments are derived from clays and ores. Others are synthetic. Clay, mica and talc (a form of magnesium silicate) add strength and abrasion-resistance. Clay, calcium carbonate, mica, magnesium silicate, and talc increase durability and weathering qualities. Powders of aluminum and bronze add heat resistance. Red lead and zinc chromate prevent rust. Cobalt or manganese salts speed drying time. Fillers of ground shells from nuts or sea animals increase coverage.

Carbon black increases U/V resistance. It absorbs the ultraviolet light, protecting the paint base. Barium sulfate blocks U/V and is especially effective against X-ray radiation.

There are pigments to add electrical resistance, pigments to control fungus and mildew, and other pigments to control barnacles and algae. There are pigments available to provide just about any paint quality.

Additives

Additives give special qualities to the basic paint. They usually make up a small percentage of the total ingredients. They are liquids, or solids which dissolve in the vehicle.

Additives shouldn't cause the basic paint to lose any of its qualities, with one exception: chemical reaction with surface materials. Acids are additives that attack the surface, to clean it of oxidation and other contaminants that may prevent proper adhesion.

Additives are used to provide better adhesion, durability, chemical and water resistance, drying-time control, and spreadability. They also provide flame resistance, insect control, mildew and fungus control, algae and barnacle control, and increased gloss. Other additives aid in wetting or surface penetration, or aid the set-up of plastic bases.

Here's a sample list of some common additives. It's not complete, by any means.

- Some nitrogen compounds control micro-organisms; others make the paint more fire-resistant.

- Insecticides such as cinerin, dieldrin, Paris green, and most commonly, chlorpyrifos, are added to help keep flying and crawling insects off the paint surface when it's wet. They also help control insects when the paint is dry. Paris green is also a prime pigment, producing a shade of green. It's made of arsenic trioxide and copper acetate. As you might guess, it's highly poisonous, as are most additives.

- Pentachlorophenol is a bleaching and water-proofing pigment. It stains and kills plants. Use paints with this additive very carefully.

- Marine boat hull paint can contain oxides of copper, or tributyltin, to control algae and barnacles.

As you can see, there are many different pigments and additives for paints. Most are added in small quantities. All have a reason for being there. And a good many of them are poisonous.

Catalysts

The one group of additives I haven't mentioned so far is the catalysts. These initiate the chemical action of two-part paints such as epoxies, polyester, or fiberglass (polyester and glass fiber). Cobalt is also added sometimes to control curing or set-up times. A fiberglass catalyst called *sunlight catalyst* prevents the polyester from curing unless it receives U/V from the sun. It can't be used with color pigments, only with clear coats.

Catalysts don't increase the volume of the material, so don't figure them into your coverage calculation — except for epoxy catalysts.

Epoxy catalysts— An epoxy catalyst is actually a curing agent: It becomes part of the material by chemical action. The volumes of both the resin and the curing agent are used to determine coverage. Curing agents for epoxies vary, according to the epoxy formulation.

Both catalysts and curing agents work chemically by forming or starting a reaction that produces heat. Be careful not to use too much catalyst. It can prevent leveling of the paint (because of rapid drying), cause stress cracking, and in extreme cases, fire.

Polyester Resins and Fiberglass

Have you ever used polyester resin as a paint? It can be a good choice. It's waterproof, it doesn't collect algae or barnacles, and it's highly resistant to fungus, mildew, chemicals, and damage. These qualities make it an excellent choice for marine or wet-area use. Fiberglass, in cut or mat form, should always be used when physical strength and longevity are required.

Polyester resin can be colored, using colorant designed for polyester. It can be made flame-resistant. Certain chemical additions can vary the workability and strength. These additives lower the resin cost somewhat, but also lower the weather resistance. You can obtain a glass-smooth surface with polyester resin. Just lay a sheet of cellulose acetate, or cellophane, on the wet resin and smooth it. When the resin is cured, the cellophane strips off easily.

If you're not now using polyester, look into it. It can solve a lot of tough painting problems. One source is:

Gem-O-Lite Plastics Corporation
5525 Cahuenga Boulevard
North Hollywood, CA 91601

They also offer a book, *Polyester & Fiberglass,* written by Maurice Lannon, that has a lot of helpful information about fiberglassing. A note about fiberglass: If you ever tried to repair cracked, chipped, or broken fiberglass, you probably found that the new fiberglass didn't stick well. Next time, clean the old surface with acetone. Then try Shell Chemical Epon 815 for the patch. It's an epoxy, and will generally stick better than new polyester. Sanding the surface also helps.

Specific Paints and Coatings

Now, let's get down to business. As you've seen, paints and coatings are formulated for specific surfaces. As a professional painter, you should know what coating should be used on what surface and the characteristics of each coating. Here are the facts you should know.

Aerosols

I'll start with something that's considered a paint, but technically isn't. It's really an application method: the aerosol. Aerosol paints are extremely fine particles of pigment mixed with a gas. The gas is the propellant. It dries rapidly, leaving the pigment to cover the surface. Common propellants are hydrocarbons, nitrous oxide, and carbon dioxide. Except for carbon dioxide, these are toxic and flammable. Be careful to avoid open flames, overheating the cans, or puncturing them. They are explosive. Empty the cans fully before putting them in the trash.

Aerosol paints come in many colors and in many formulations for specific purposes. There are aerosols for appliances, toys, automotive use, ceilings, fabrics, and plastics. They're often softer than you'd like, and scratch easily.

An aerosol can covers about 40 square feet. They're useful for painting small or irregularly-shaped areas. Large areas are best done by conventional spraying. I don't recommend aerosols unless you're painting a one-time item or doing touch-up. You pay a very high price for the coverage you get. Most of the cost is in the can and nozzle system, not the paint.

Problems with aerosol nozzles— Nozzles are often a problem, especially after they've been used once. The instructions on most cans say to clean by turning the can upside-down and spraying. This works some of the time.

When a nozzle clogs — and it will — manufacturers suggest you use a fine sewing needle to clean out the hole. If not done carefully, this damages the opening, changing the spray pattern. Here's my solution. Shake the can for two minutes, then dip the nozzle in liquid sandpaper for thirty seconds. Remove and spray. You may need to do this more than once for badly-clogged or dried nozzles. But it generally works. Another solution is to replace the nozzle with one you know is good. Some smart person should package extra nozzles. Maybe at five for a dollar?

Aluminum-Fiber-Filled Coatings

These, and the newer aluminum-fiber-filled vinyl roof coatings, help control heat. They can reflect up to 75% of the sun's U/V radiation. This makes them ideal for mobile home roofs, exterior walls on farm and commercial buildings, and exteriors of chemical storage tanks. Remember, though, that any of these coatings that contain asphalt can be removed with gasoline and other petroleum products.

Asphalt

Technically, this isn't a paint. It's a surface coating that offers good protection, however. Asphalt is a petroleum-based product that can be applied with brush, roller, or spray to most surfaces. If you heat the asphalt it makes spreading easier and improves coverage. Commercial equipment for heating and spraying costs around $2,000.

Asphalt is mainly used for driveway and roof coating. It also waterproofs most woods, cements, and metals. I've used it on retaining walls, basement walls, planters, foundations, for auto undercoating, submerged lumber, and for fences. As a sound-deadening material, apply it to wall interiors, interior metal surfaces of appliances, and as an undercoating on sinks and tubs. But check your local building code if you intend to use asphalt for sound deadening on walls of normally occupied buildings.

Asphalt is toxic and flammable. It burns, so be careful what you apply it to. There are many different formulations, from very thin to very thick, and from no filler to metal and mineral fillers.

Asphaltum— This is the thinnest of the asphalts. It spreads easily. I use it on wood and fences. It's also good as a rust-preventive coat on metal flashing before the roofer applies shingles. Follow with a roof cement to assure proper water seal. (When using heavier-bodied asphalts on roofs, reinforce the liquid with special cloth roofing tape. This helps prevent cracking from the sun.)

Creosote— This coating is similar to asphalt and is used for similar purposes. It's a coal-tar derivative used for protecting wood from insects and rot. It looks like black tar. You'll see it on railroad ties, phone poles, fence posts, and other lumber that touches or is embedded in the earth. Never burn creosote-coated wood. The fumes are dangerous. Dispose of creosoted lumber at a legal dump or landfill. And check with local authorities before using it at all. Its use is controlled or banned in some areas.

Cement Paints

These are water-based paints to which pigments have been added. Some of the newer ones are epoxy-based. Cement paints are based on a mix of portland cement, calcium chloride, aluminum stearate, hydrated lime, and zinc sulfide or titanium dioxide. Up to 50% fine sand may be added for filler. Since this "paint" is basic in nature, you can use it to cover or line large diameter water-filled pipes. And because it's a base, it counters the rust (acid) action on iron.

You can also use cement paint as a filler for cement block. With epoxy additives, it acts as a water sealer for cement block walls, walks, and so on. It's applied in thick coatings (1/8" to 1/4"), usually with a trowel or spray equipment. You buy it in dry form and mix it with water for use.

Drying Oils

Danish oil, antique oil, tung oil, and linseed oil are all classified as drying oils. Many paints, varnishes, and lacquers have drying oil bases or additives. Drying oils come from various plants, such as flax seed (linseed oil), walnuts, hemp, poppy seeds, sunflower seeds, soybeans, and the oiticica and tung trees. They dry into very tough elastic films that protect wood surfaces. They all work on the same principle: They dry out and harden by absorbing oxygen from the surface and the air.

These penetrating-oil finishes are fine on woods that are unfilled and open-grained, such as walnut or oak. They don't work as well on tight-grained woods like pine and fir. Use them on furniture, cabinets, window sills, and moldings.

Danish and antique oils— These are fine finishes for furniture, providing a soft, warm glow and good protection. They usually deepen the color of the wood a bit. They penetrate deep into the wood fibers. When they harden, they bind the fibers together, yet allow the full color and grain to show. They resist water, alcohol, food and stains, as well as scratches and abrasions. Both contain a drying oil mixed with a plastic and a vehicle. The usual mix is 35% oil/resin (or oil/plastic) and 65% mineral spirits.

They're excellent finishes for both the beginner and the professional. Application is simple. Just rub on the surface with a clean cloth or rag. Most dry in four to eight hours. Prevent spontaneous combustion of the oil-soaked rags by washing or soaking them in water after use.

Fish oils— These are also drying oils that are fast being replaced with synthetics. Fish oil was used as a metal protector in primers and paints. It works like this: Metal rusts when exposed to air and moisture. Fish oil hardens by absorbing moisture and air. As the oil hardens, it prevents the penetration of more moisture and air, preventing rust.

Tung oil— Also called Japanese wood oil or China wood oil, this finish is quite expensive and is being replaced by the polyurethanes. The oil comes from the seed kernels of the tung tree. This tree grows mainly in China, but has been imported to the southeastern United States. Tung oil is an excellent drying agent. It resists acids, alkalis, water, stains, and alcohols.

The printing industry uses tung oil to bind inks used in printing on metal. Paint manufacturers use it as a vehicle or binder in many fine varnishes, waterproofing paints, lacquers, and furniture finishes.

At one time, the Japanese mixed the powdered bark and flowers of sumacs with tung oil to produce a very rich and protective finish. *Japanning* is the technique of using these mixtures to produce a brilliant, deep, surface coating. The items had to be kept in dark caves for weeks between coats, so they'd dry properly. Thank goodness for modern chemistry! Although I must admit that I've never seen a synthetic coating that rivals the crystal-clear, red-toned, hard-as-rock finish produced by those Japanese craftsmen.

Enamel

Real enamel is a mix of quartz, silica, borax, lead, feldspar, and mineral oxides. When heated to 1,500 degrees or more, it melts and fuses together, forming a glass-like coating that resists acids, corrosion, and water. That makes it an ideal finish for bath and kitchen fixtures, appliances, utensils, signs, and industrial equipment. It's also used in some jewelry and stained glass. Painters don't use real enamel. What you use is a natural or synthetic paint we just *call* enamel because the finish looks like real enamel.

Enamel Paint

These paints are based on varnish, polyurethane, or alkyd resin. The alkyd resin is a type of thermosetting plastic.

Some of the newer enamel paints are acrylic and can be thinned with water. Enamel hardens to a tough film, is weather resistant, and adheres to most surfaces.

Enamel paints come in many formulations, for both interior and exterior work. I don't really recommend them for exterior work. Although they form a tough, abrasion-resistant skin, they're susceptible to water damage from within. Enamel paints, though, are hard to beat for interior work such as trim and doors.

Many automotive enamels must be baked to harden properly. But I've used them as air-dry enamels with fair success. Automotive enamels come in colors and have a depth of shine that other formulations don't. You can't find Candy Apple Red in any other kind of paint.

Full-gloss enamel— The hardest paint to use and achieve one-coat coverage is white full-gloss enamel. Why? The less the pigment content, the higher the gloss. The gloss is the dried oil or synthetic binder, usually an oil/alkyd resin. Colored full-gloss enamel is the second-hardest paint to achieve one-coat coverage.

The disadvantage of using full-gloss enamels is that you must plan on at least two coats. In fact, it's often best to use a base coat of flat enamel, tinted to the color of the top coat. Flat paints have a higher pigment content that covers the surface dirt and any marks left behind after cleaning.

Full-gloss enamel has another disadvantage. It reflects so much light that it shows any little surface defect. Flat or semi-gloss enamels hide these defects better.

Fillers

These are formulated to fill holes or pores in wood before the top coat is applied. The exterior formulations are ideal for renewing cracked and split trim lumber. The two main forms are paste filler and filler paints.

Paste filler— To use this, first thin it to spreading consistency. Then rub, with the grain, into the surface. Let it dry a few minutes, then rub *across* the grain. Give it a final rub *with* the grain. This assures that all pores get filled. Allow the material to dry, sand lightly, and apply a wash or sealer coat.

Most fillers are stainable, but only to a limited extent. So don't, if you're thinking about staining it, try to level an entire surface with paste filler.

Filler paints— Unlike paste fillers, these should not be thinned. Use them just as they come from the can. Most go on over a primer, and under the top coat. Some, though, can be used as the top coat. Most can be tinted to the lighter pastel colors, but not to bright colors. They come in several degrees of thickness. Choose the one best suited to the application.

Fire-Retardant Paints

These contain nitrogen compounds that puff when exposed to flame. This puff, or blister, forms an insulating dead-air space between the flame and the surface being protected. Most fire-retardant paints are oil or oil/resin based. They are *not* fireproof.

Fungicides

These aren't strictly paints or coatings, but are used in conjunction with them. Pentachlorophenol is the most common one. It's a fungicide and disinfectant that is mixed with oil and resin. Its main use is as a wood preservative for shakes and shingles. It also bleaches darkened lumber back to a natural, light tan color. Be careful when using any product containing pentachlorophenol. It's poisonous, it kills plants, and it stains painted woodwork, brick, and concrete.

Heat-Resistant Paints

These use alkyd resin or silicone resin as a base. One type uses aluminum in a varnish. As the varnish burns off, the aluminum coating remains to protect the underlying surface.

Indoor Paints

If not latex, they're varnish, or alkyd resin with color pigments. Indoor floor paints are alkyd resin or polyurethane resin with mica, talc, and clay added for increased abrasion-resistance.

Latex Paints

These can be natural or synthetic emulsions, though synthetics make up the bulk of the products available. Latex is actually polyacrylic, or polyvinyl acetate, mixed with water and pigments. Latex is the best known and most used paint today. It cleans up easily with soap and water. It can be tinted to almost any color, and in flat, semi-gloss and full gloss mixtures. Note, however, that the semi-gloss and gloss paints are somewhat more difficult to apply than many oils and usually require two or more coats for proper coverage. Some of the full gloss and even some flat latex paint is very "slippery," making application difficult.

Latex paints have a tendency to "breathe" (allow moisture to pass through them), making them useful for painting concrete, stucco and wood. Coverage and durability is generally very good. Touch-up is usually easy within the first few weeks after application.

Masonry Paints

Masonry paints use polyvinyl acetate or acrylic emulsions. They're considered a latex-type paint, and form a tough skin that resists the alkali of the masonry. Unlike "cement" paints, masonry paints aren't cement based. They have a normal paint consistency. Masonry paints usually perform well as a moisture seal.

Metallic Paints

These can be made from gold, silver, bronze, or aluminum powders in an oil or oil/alkyd base. The metal powders float to the surface to give the metallic look.

Metal-Protecting Paints

These use oil, oil/alkyd, or fish oil as a base. Red lead or zinc chromate is added as the rust preventive.

Oil of Cedarwood

This is a specially refined oil from cedar. It's used for re-oiling cedar chests and closets. The oil puts back the natural aroma and protection of the cedar. I don't recommend it for exterior use since it costs about $20 a quart. This can get expensive.

Outdoor Paints

If not stains or latex-based, outdoor paints are generally formulated from natural or synthetic drying oils. The natural oils used are linseed and soybean oils. These paints usually contain iron oxides, zinc oxides, titanium dioxide, and other additives to resist insects, fungus and mildew.

Special-Effects Paints

Here are a couple more finishes you can offer to your customers: finishes your competition doesn't have. And that's what this book is all about: Helping you learn the products available, and how to use them, so you can make a better living in the painting business.

Cobweb finish— This finish mimics the cobwebs you find in the corners of your living room. It comes in red and black, in quarts or gallons, and is applied with standard spray equipment. It's good for walls, equipment, and cabinets. In fact, it's excellent for masking minor surface imperfections. Since it also hides dirt well, it's a good covering for garage or workshop walls.

Crackle lacquer— This paint is designed to crackle, or "alligator," when sprayed over a gloss tint base lacquer. It comes in black, blue, brown, clear, cream, green, red, yellow, and white. It's ideal for creating a production-fast antique look. But it probably shouldn't be used on walls except for special effects.

Splatter paints— These are specially designed for use on furniture and cabinets, and come in aerosol cans.

Here's a trick that's creates the look of pecan wood in much of today's furniture. It's called *distressing.* You can use black or brown enamel or brushing lacquer, instead of the special paint. First, paint the surface with a background color. Let it dry. Then, splatter one or more colors of paint onto it. To splatter, wet the brush with the paint, and sharply tap the brush against a wooden stick. The surface should be 18 to 24 inches away. When doing different colors, let each color dry before applying the next.

Many walls and floors in older homes were painted this way. It's an effective disguise for rough surfaces. It also works well in closets, cabinets, and utility-type rooms.

Wrinkle finish— This is a special finish, mainly used on equipment. It masks small imperfections well. The easiest way to do it is to buy wrinkle paint, in aerosol cans, from electronics hobby stores.

Another method is to first coat the surface with one coat of enamel. Then, when it's semi-dry, use lacquer for the second coat. The different drying rates of the paints will form the wrinkles.

A third method is to start by spraying droplets of enamel on the surface. When they're dry, respray with a smooth coat of enamel. This last method is used by many typewriter, computer, and equipment manufacturers.

Stains

These are dyes dissolved in either drying oil or water. There are pigmented stains also, but they're opaque, not transparent or translucent.

Varnish

This isn't a single product, but a large group of similar products. Varnish is the name given to a group of liquid products that, when dry, form a hard lustrous surface. Although usually clear, they may be tinted. The one natural varnish comes from a type of sumac found in Japan.

Varnishes are broken into two groups: spirit varnishes and oleoresins. Spirit varnish is a resin dissolved in a quick-drying solvent such as alcohol. Common spirit varnishes are shellac, Japan, and lacquer. Oleoresinous varnishes are a mixture of resins and drying oils in a base of turpentine or petroleum. A common varnish of this type is spar varnish, used to protect boats and exterior surfaces.

Varnishes are classed by the amount of oil, in gallons, that's mixed with 100 pounds of resin. This is called the *cut* or the *cut of the varnish.* For example, Number 30 tung varnish is 30 gallons of tung oil mixed with 100 pounds of resin. The lower the number, the higher the resin content. The higher the number, the less resin — and the higher the gloss.

Figure 9-7 gives useful information on these kinds of coatings. I don't like to be a borrower, but when someone's right let's give him credit. This chart comes from the book *Refinishing and*

Finishing, by Star Bronze Company. This 53-page booklet is good reading, and has tips and procedures for paint stripping and for reworking wood floors. (Zip-Guard is their name for their polyurethane wood coating.) The address is:

Star Bronze Company
Box 2206
Alliance, Ohio 44601-0206

Lacquer— This is another finish from Japan and China. Natural lacquer comes from a tree found in those countries. The sap from these trees is drained, strained, and then dried until it's syrup-like. You'll see this dark-brown lacquer on many oriental antiques.

Lacquer is a good protector: It's very hard, and it resists water, stains, and chemicals. But it doesn't resist U/V radiation very well, so don't use

	Alkyds & 1 can epoxy & varnish	Lacquers	Shellac	Polyurethane (Zip-Guard)
Recoat time	4 or more hours	1-2 hours	1 hour	2-3 hours
Surface condition after dry	Flexible	Hard	Hard	Hard and flexible
Shock resistant	Good	Average	Poor	Excellent
Scuff-abrasion mar resistant	Average	Poor	Poor	Excellent
Alcohol resistant	Average	Poor	Poor	Excellent
Household detergent & water resistant	Average	Average	Poor	Excellent
Odor	Low	Strong	Strong	Low
Waxing for protection	Recommended	Recommended	Recommended	Not necessary
Slip resistant	Average	Average	Poor	Excellent
Lap characteristics on 2nd coat	Excellent	Poor	Poor	Excellent
Stain effect on wood	Yes	No	No	No
Color retention of film	Darken with age	Little or no change	Darken with age	Little or no change
Brush cleaning	Paint thinner	Special solvents	Special solvents	Paint thinner
Application	Mop brush	Brush mop	Brush	Brush mop or spray
Application over old finishes	Generally	No	No	Yes
Application over filler and stain	Yes	No	Yes	Yes
Wear comparison	2nd	3rd	4th	1st

Courtesy: Star Bronze Company

Coatings comparison chart
Figure 9-7

it on areas like window sills. Most dry lacquer is nontoxic, though some commercial lacquers can be poisonous.

As with other fine finishes, we now have synthetic versions of lacquer. Today's lacquer is made from the cellulose in cotton. It's bathed in either acetic acid or nitric acid. This produces cellulose acetate or cellulose nitrate, respectively, both of which are highly flammable. The vehicle is either butyl alcohol or butyl acetate.

Lacquer is dangerous. A chief ingredient in dynamite is the same cellulose nitrate you find in lacquer.

Two other formulations of lacquer are resin mixed in turpentine, and lac (see shellac) mixed with ethyl alcohol. There are also acrylic-resin lacquers. These are usually used to protect polished metals such as brass and chrome.

Lacquer is used on cars, furniture, in the paper industry and in the commercial painting industry. Lacquer dries very quickly. That makes it easy to keep dust- and insect-free. All the fast-drying lacquers must be sprayed on. And, because they're so flammable, special rooms and spray booths must be used. In fact, many companies won't sell lacquer to anyone who can't show a state license and a county building permit for its use. Even the small aerosol cans are potential bombs.

There are some slow-drying brushing lacquers on the market, but they're difficult to use on a large surface.

Shellac— This coating is made from insects. The lac bugs from India and Burma give off a sticky, protective substance. It sticks to the branches of the soapberry and acacia trees the insects live in. The branches are removed from the trees and washed. The sticky material, now called seed lac, is melted, pulled into thin sheets, dried, and broken into flakes. The flakes, when dissolved in alcohol, produce orange shellac. If the flakes are bleached before being mixed with the alcohol, we have white shellac.

Shellac has been a very popular finish for cabinets and furniture, woodwork and trims. But it's being replaced by the synthetic varnishes because of its high cost and its poor resistance to alcohol and water. Many painters, though, still use shellac as a sealer or undercoat. It's excellent for sealing off a stained surface to keep the stain from coming through.

Spar varnish— This is a coating used for window sills, boats, and other exterior work. Some spar varnish is lacquer-based. Others are based on liquid plastic.

In general, liquid plastics are a blessing for the painter and the do-it-yourselfer. They're easily applied with either brush, roller, or spray. They dry to a super-tough film that resists water, alcohol, chemicals, heat, abrasion, and household cleaners. They're a little bit flexible. This keeps them from chipping, peeling, and flaking. Most make good wood sealers when diluted. Usually, they're thinned with one part mineral spirits to two parts varnish.

But, along with all these good points come some bad points. Drying time is very long. Most plastics need at least three coats, with a one-week drying time between each coat. And most require sanding between each coat. Though they can be used over most other coatings (varnish, paint, stain, and some lacquers), they can't be used over shellac. They'll cause lifting if used over shellac. On some lacquers, they may "blush" or turn white. It's best to do a sample spot before doing the whole item.

There are many formulations and each has a specific use. General-purpose polyurethane can be used for furniture, doors, trim, cabinets, and similar items. Special-purpose formulations should be chosen for exterior use, boat hulls and decks, bar tops, and floors. For exterior and window sill use, be sure you have polyurethane or spar varnish with a U/V inhibitor. On floors and stairs, apply two thin coats, followed by one heavy coat. This gives maximum durability.

Two-part plastics— For that deep, deep shine, and that "item-under-glass" look, use the two-part polymer or polyurethane products. One part is a catalyst that sets the plastic. Drying times are very long, so keep coated items in a dust-free and insect-free area until no longer tacky.

Two-part glass coat paints are polyester resin. They've been replaced, however, by the one-part polyurethanes for most applications.

Wood Green

This product is replacing creosote for most applications. It's used in lumber for house footings, or any lumber in contact with earth or cement. It's also used to make backer board (green board) for kitchens and baths where moisture may be a problem.

It's anti-rot, anti-disease, and anti-insect, when properly applied. Properly applied means applied by *injection,* not by brushing. Factory-injected lumber is best, but it's possible to apply wood green with a long soaking — 24 to 72 hours. It's best to buy it from the factory, since factory-injected lumber will last 30 years or more.

Wood green contains copper napthenate and, sometimes, arsenic. Both are deadly if absorbed into the body. Arsenic, especially, accumulates in the body until it kills you. So be careful. Wear gloves and protective clothing when handling either the liquid or the coated lumber. Never burn treated lumber — the fumes are poisonous. If burned in a fireplace, fumes will get into carpets, curtains, furniture, and clothing. Dispose of lumber treated with wood green at a legal dump or landfill.

Wood-Preservative Paints

These are made of oil with creosote, cuprite, Paris green (highly poisonous), or, most commonly, pentachlorophenol. They can also be petroleum-based hydrocarbons, such as asphalt and bitumen. Some wood protectors, outdoor paints, and primer paints contain balsam, an oil resin from plants. It's similar to turpentine, and is sometimes sold as terpene solvent.

Primers

Primers are important enough to deserve special mention. Primers are formulated to do what the top-coat paints can't do. Not using primer — or using the wrong primer for the job — is throwing money down the drain.

Most bad paint jobs are caused by the painter (usually a do-it-yourselfer) using the wrong primer. Or none at all. To see what I mean, look at Figure 9-8. The paint is flaking and the wood grain is showing through. No primer was used on this job. Also, note the brush marks: this was probably painted with a very old can of paint from which the solvent had evaporated. It could also have been painted with a very poor brush or in very high heat. (In any case, the solution is to strip to bare wood, prime, and paint again.)

Some paints require special primers. This means that both the primer and top coat contain chemicals which must react with each other to produce special top-coat properties. Here's the primer rule-of-thumb:

Use aerosol spray primers under aerosol spray paints, and use brush-on primers under brush-on paints. Why? Because aerosols use different solvents from those used in brush-on paints. Mixing the two can cause the top coat to lift or wrinkle.

Functions of Primers

Here's a list of the main reasons for using primers. Each primer serves some, or all, of the following functions:

Effects of not using primer
Figure 9-8

1) It hides the surface, so the top-coat color is clearer.

2) It acts as a moisture barrier, to prevent moisture from getting to the paint bond. This is especially important when painting metals or masonry.

3) It acts as a bonding agent between the surface and the top coat.

4) It acts as a chemical inhibitor between the surface and the top coat.

5) It limits the absorption of the top coat into a porous surface or subsurface. You use less paint, and get a smoother surface.

6) It can recondition old paint to retain its bond to the surface.

7) It can recondition old paint to bond with new paint.

8) It acts as a rust inhibitor.

Paint Warranties

What happens if you've done your prep work, used the right primer and top coat, but the paint job still fails? Well, maybe you can take advantage of the paint manufacturer's warranty.

Most paint manufacturers give a one year warranty. But, like most such contracts, it's limited to replacement of the same amount of paint (to cover the fault). It doesn't include labor.

There are no warranties for exact color or texture match. Warranties are void if improper preparation caused the failure. (And improper prep is the most likely cause.) There are no warranties that cover improper use or application. Read the label: It lists your rights under the warranty. It also lists whatever disclaimers apply.

I can only think of two instances where you might have a claim. First, if the paint turns to powder and falls off. Second, if acid rain causes blue paint to turn red. Actually, paint manufacturers do extensive testing for color retention, durability, longevity, bond strength, expansion/contraction, chemical and air pollution reactions, weatherability, and application methods. Manufacturing defects are extremely rare. In other words, you're on your own. Read and learn as much as you can about proper painting. Buy good quality paint, take your time, and do it right.

Color

Acceptability is the key to exterior color selection. You want to recommend colors that are generally considered appropriate for the intended use. For example, barn red is O.K. on a farm. But a barn-red home is inappropriate in a neighborhood of tan Tudor-style homes. That's an extreme example, but here's a similar problem I've faced. I've been asked by neighbors to repaint my own home, which is light tan. It's in a development of medium tan and brown homes. White, or any color close to white, is unacceptable. My house is the lightest on the street. In the opinion of at least one of my neighbors, that makes it stick out like a sore thumb — and break the harmony of the whole neighborhood.

I don't think the color of my home is so bad, though I may paint it a little darker tan when it's time to repaint. But another neighbor of mine painted his house pea green just to spite his busybody neighbors. He's now ostracized by some and is being sued by the owner's association.

That brings up an important point. Many apartments, co-ops, condos, and housing projects have covenants and conditions or association by-laws that permit only certain colors on exteriors. If your customer selects an exterior color that's very different from the color of other homes on the block, ask if there are any restrictions on choice of color. Many of the owners just don't know or don't remember the rules. If a project has gated or guarded entrances, you can bet there are rules for color selection. Check the color with the complex management. Do your customer a favor: Start asking questions if color selection may be an issue.

Residential Exteriors and Color

Even in areas without rigid rules, color acceptability depends on the color and style of adjacent properties. Neighborhood harmony and property values are at stake. If you're asked to use a color that's totally wrong for the neighborhood, you have two choices:

1) Suggest to the customer that their color may not be the best choice. Propose a more acceptable alternative. Use good judgment. The color shouldn't be your personal preference. It should be a matter of professional opinion. You don't have to like the color. It just has to be acceptable to the owner and his neighbors.

2) Just pass on that job.

My suggestion is to avoid the controversial. Making money on one job isn't enough. Don't earn a reputation as a trouble-maker. Helping one customer make a point isn't worth angering several dozen neighbors. You want to be invited back, not run out of town on a rail.

This is truly a judgment call on your part. We all like to express our individuality. Selecting a house color is one way to demonstrate how unique we are. But there's a fine line here. Self-expression that offends others is called bad manners. Stay clear of that type of action.

Exterior Color Schemes

Color schemes can be strikingly different and attractive. Others are strikingly different and downright obnoxious. Exterior paint color can either blend or contrast with the color of items you can't or don't want to paint: roof covering and rock, brick, or wood facing on the house. The choice is either blending or contrasting. A bad choice is called *clashing*. Let's see what makes colors blend or contrast well.

In the following discussion, *base color* refers to the main color of the home's exterior. It can be siding, shingles, stone, brick — whatever. Bright base colors are generally not acceptable. Dull colors and pastels usually are.

Rock work is neutral, so most colors combine well with it. Brick is a different matter, though. Green paint next to red brick may give a house a permanent Christmas feeling.

Clear coats— Some developments, mine included, do not allow clear coats which let the natural wood show through. Clear coatings need frequent maintenance. If they're not maintained, they give the building and the neighborhood a run-down look.

Exterior trim colors— Exterior trim colors also must be acceptable. Inside your house, you can have green walls with tan trim if you want. But tan trim on the exterior of a green home wouldn't be accepted in many areas.

Black trim is O.K. with almost any base color, but black isn't acceptable as a base color.

Trim in the neutral range (tan, brown, gray, black, and white) works well with most base colors.

Not all colors work well as trim on homes with neutral base colors. A tan or brown base color should have trim colors that are also tans or browns. A gray base color only looks good with white or black trim. A white base color looks good with black, some blues, reds, greens, and yellows. But white doesn't look good with purple, bright yellow, orange, or most pastels.

Dark colors on the roof and base make a building seem further away and closer to the ground. Here we use color as a tool. Properly chosen color can be a powerful tool. The next section shows some of the ways color affects how a home is perceived.

Creating Illusions with Trim Colors

Normally, you'd paint all the trim on a house the same color. But this isn't always the right thing to do, as you'll see. It depends on the shape of the building, and what you want the apparent shape to be. You can change the apparent shape by using contrasting color vertically, horizontally, or both vertically and horizontally.

For instance, if you want the building to stand out, paint the trim in darker, brighter colors. To make the building blend in with nature, paint all the building the same color, a color that closely matches the background.

To make a narrow building look wider, use a darker, bolder trim color horizontally, a light or base color vertically. To shrink the height of a two-story house, paint the bottom story in light colors, the top story in dark colors. Use light colors to make a building look larger, dark colors to make it look smaller. Just be sure to make any two-color combinations "break" at a natural break line, such as a corner, trim piece, or second story overlap.

Effects of poor trim color choice— Color can make a building appear larger or smaller, closer or farther away, shorter or taller than it actually is. Figures 10-1 and 10-2 show homes with trim that is improperly painted, in my opinion.

The house shown in Figure 10-1 is two stories, with almost no horizontal trim. It's all vertical lines, from the ground to the roof. It's painted light yellow, with dark brown trim. This makes it seem narrower and taller than it is. Perhaps a horizontal trim piece, or a two-tone paint job, would have been better.

Effect of too much dark vertical trim
Figure 10-1

The house in Figure 10-2 is a single story with lots of wood trim. It's painted tan with brown trim. The effect is confusion. The surface is broken into many parts by all that brown trim. A better solution would be to paint all trim between the corners and the doors the same color as the siding.

Effect of too much dark horizontal and vertical trim
Figure 10-2

Selecting the illusion—Sometimes, when I'm painting the exterior of a building, it just doesn't look right. So I make a scale drawing, make a dozen copies, and experiment with some colored pencils. After I choose the basic scheme or illusion, I paint a few scrap boards and put them against the background. Often a very slight change in the shade of a color can radically change the appearance. I've seen buildings go from cheap looking to very rich looking with just a small change of shade. Try this on your own home. It's remarkable what you can do with illusions.

Camouflage with color— Color can also be used as camouflage. Use the base color on areas you don't want people to notice: electrical boxes, gas meters, air conditioners, heat pumps, fences, and drainpipes. Look at Figure 10-3. The meter panel was painted the same color as the siding. It almost disappears. And the gutters are the same color as the trim for an almost perfect blend.

Caution: Metering equipment is often owned by the utility company, not the building owner. They may not be happy if you start painting their equipment. If in doubt, check.

Use color to accent areas you want people to notice. Entrance doors, windows, flower boxes, fences, and shutters are a few examples. Use darker, brighter colors or high contrast colors for these areas.

Camouflage with color
Figure 10-3

Color Fade Tendencies

Some paints change color as they fade, especially on exteriors. Sunlight, U/V, rays, weather and smog all have an effect. Figure 10-4 gives you an idea of what happens when various colors fade. Of course, this is general information only. Fade can't be predicted accurately. There are too many ways to formulate paint. If you must know how a paint will fade before you apply it, ask the manufacturer. Most manufacturers do extensive fade tests and will give you the information if you ask.

Industrial and Institutional Exteriors and Color

In this kind of painting, there's more latitude in color choice. Unlike residential work, you can try different colors and color schemes. Factories, for instance, create the illusion of large size by using two tones. The building is painted white, or some light base color. Then a wide horizontal stripe in a contrasting color is painted from corner to corner. For this effect, be sure to choose a color that doesn't look cheap. Strong browns and brown-oranges are good. Bright blue, green, or yellow are not. Of course, some business want to look cheap. Discount stores, ice cream shops, and so on — should be painted with bright, cheap colors. The idea here isn't physical illusion; it's psychological illusion.

Another psychological illusion is that of strength: use a bold, horizontal stripe. Create a forward, or progressive, illusion with a bold, diagonal stripe. Bottom left to top right is the best way to angle it.

Color can help you play tricks. First, decide the image your customer wants to convey. Then sketch the design with paints, colored pencils, or markers. You're the color expert. Experiment. Suggest what can and can't be done with color.

Interior Colors

There's much less problem with home interiors. The colors only have to be acceptable to the people who live with them. But sometimes your advice is called for to avoid an obviously bad color choice. This section should keep you from making the classic mistakes.

Decorator Colors

Some customers will select colors by name. "I want this room in Cinnabar; this one in Amethyst; and that one in Old Rose." What are you going to do? Show up with rust-red, purple, and dark pink? Figure 10-5 lists some fancy color names and identifies the real color of each. It will help you translate those "decorator colors." It isn't exact, since everyone has a different perception of color.

Color and Artificial Light

The type of light in a room changes the way color is perceived. All artificial light adds a color of its own. Figure 10-6 is a chart that lists the effects of different lamps on color. *Daylight, standard cool white, deluxe cool white, standard warm white,* and *deluxe warm white* are all fluorescent tube lamps. As you can see from the chart, careful selection of both paint color and type of lighting is important.

Color Reflection

The furnishings in a room — the carpet, drapes, chairs, tables and accessories — all affect the way a

Color	Fades to	Remarks
Black	Brown, purple, or green	Many black pigments are actually very dark shades of brown, purple, or green
Blue	Lighter blue, or purple	
Brown	Lighter brown, or tan	
Green	Lighter green, or yellow-green	
Orange	Lighter orange, or brown	
Purple	Lighter purple, or pink, or gray	
Red	Lighter red, or pink	
Yellow	Yellow-brown	
Clear coating	Yellow-brown	Total destruction to white, for many products
Light pastels	White or close to white	

Color fade tendencies
Figure 10-4

Color name	Color
Almond green	Yellowish-green
Amber	Orange-yellow
Amethyst	Purple
Azure	Clear sky blue
Blond	Lt. yellow-brown to dark gray-yellow
Chartreuse	Bright yellow-green
Cinnabar	Brownish-red
Coral pink	Yellowish-pink
Cornflower	Purplish-blue
Crimson	Strong purplish-red
Cyan	Blue, bluish
Ebony	Black
Emerald green	Bright green
Indigo	Dark blue
Ivory	Yellowish-white to pale yellow
Jacinthe	Orange
Jade green	Light bluish-green
Lavender	Pale purple
Lemon	Greenish yellow
Lilac	Purple
Magenta	Deep purplish-red
Maroon	Very dark red
Murrey	Purplish-black
Navy	Grayish-purplish-blue
Nile green	Pale yellow-green
Old gold	Dark yellow-brown
Old rose	Grayish-red
Olive	Yellow to yellow green
Peach	Yellowish-pink
Pea-green	Yellow-green

Color name	Color
Pearl	Bluish-medium-gray
Pearl gray	Yellowish to light gray
Powder blue	Pale blue
Primrose	Light yellow
Rose	Purplish-red
Royal blue	Purplish-blue
Royal purple	Dark reddish-purple
Russet	Reddish-brown
Sable	Grayish-yellowish-brown
Saffron	Orange to orange-yellow
Salmon pink	Orange-pink
Sap-green	Yellow-green
Sapphire	Deep purplish-blue
Scarlet	Dark reddish-orange
Slate black	Purplish-black
Slate blue	Grayish-blue
Sulphur yellow	Bright greenish-yellow
Tangerine	Reddish-orange
Tattletale gray	Grayish-white
Tawny	Brownish-orange to light brown
Teal blue	Dark greenish-blue
Terracotta	Brownish-orange
Turquoise	Light greenish-blue
Tyrian purple	Crimson
Ultramarine	Vivid blue
Umber	Yellowish-brown
Vandyke brown	Brown-black
Venetian red	Red
Viridian	Chrome green
Vermillion	Reddish orange to bright red

Decorator color name translations
Figure 10-5

wall or ceiling color looks. Deep, rich colors reflect more light off walls. They can change the appearance of the wall color entirely.

For example, carpets of deep reds, greens, blues, and yellows tend to change wall color to drab gray. On the other hand, carpets of light browns, tans, grays, and whites have little effect on wall color.

Color and Light Reflection

Light reflection is important to your customers. Your customers have to read, work and play in the rooms you paint. A poor choice of colors can cause problems. Even when painted white, interior rooms are often dark. If you use a paint with a low percentage of reflection on ceilings or walls, the room feels like a cave. The results are eye strain, high electric bills, and a room that feels cold.

Bright white isn't a good choice on walls in an industrial assembly room. Too much reflection from a wall results in poor workmanship, low productivity, eye strain, fatigue, and general boredom.

Figure 10-7 shows *percentage of reflection.* This is a measure of how much of the light striking a surface is reflected back from it. The percentage of reflection depends on the location and amount of light in the room, of course.

Balance the light reflection in a room— It takes a proper balance of color and reflectivity to make a good paint job. Figure 10-8 shows the right balance of reflection in a room to create the most comfort and the least eyestrain. Control reflection with the method of application, the type and color of paint, and the other materials used.

Code:		
	Good:	Looks same as if under natural light. Sunlight.
	Fair:	Slightly less vivid than good.
	Dull:	Not vivid, lacking luster or brilliance.
	Brown:	Light source has little blue light, color shows up brownish
	Yellow:	Light source has little blue light, color shows up yellowish.
	Soft:	Pinkish cast due to red light of lamp.

	Type of Lamp					
Color	**Incandescent**	**Daylight**	**Standard cool**	**Deluxe cool**	**Standard warm**	**Deluxe warm**
Maroon	Good	Dull	Dull	Dull	Dull	Fair
Red	Good	Fair	Dull	Dull	Fair	Good
Pink	Good	Fair	Fair	Fair	Fair	Good
Rust	Good	Dull	Fair	Fair	Fair	Fair
Orange	Good	Dull	Dull	Fair	Fair	Fair
Brown	Good	Dull	Fair	Good	Good	Fair
Tan	Good	Dull	Fair	Good	Good	Fair
Golden yellow	Good	Dull	Fair	Fair	Good	Fair
Yellow	Fair	Dull	Fair	Good	Good	Dull
Olive	Brown	Good	Fair	Fair	Fair	Brown
Chartreuse	Yellow	Good	Good	Good	Good	Yellow
Dark green	Dull	Good	Good	Good	Fair	Dull
Light green	Dull	Good	Good	Good	Fair	Dull
Peacock blue	Dull	Good	Good	Dull	Dull	Dull
Turquoise	Dull	Good	Fair	Dull	Dull	Dull
Royal blue	Dull	Good	Fair	Dull	Dull	Dull
Light blue	Dull	Good	Fair	Dull	Dull	Good
Purple	Dull	Good	Fair	Dull	Dull	Good
Lavender	Dull	Good	Good	Dull	Dull	Good
Magenta	Dull	Good	Good	Fair	Dull	Good
Gray	Dull	Good	Good	Fair	Soft	Soft

Courtesy: Department of the Navy, Bureau of Yards and Docks, Design Manual; Architecture NAVDOCKS DM-1

Artificial light changes color appearance
Figure 10-6

Test Colors Before Painting

Very few painters do this. Too bad. It would prevent many a botched job. As you've seen, shade of paint changes through the day as light levels and types of light change. Sunlight is very different from incandescent light. Fluorescent light is different from both. There can also be a lot of light reflected from carpets, furniture, drapes, even trees outside.

To test colors, paint a large sample of the color on a piece of board (or on the wall). Have the customer check the color from different areas in the room and at several times of day. Yes, it takes time and effort — but it can prevent costly repainting. I've seen blue walls turn purple under changed lighting conditions. And of course, the customer blamed the painter.

Problems with Color Choices

You'll have customers with poor taste. You'll have customers who don't understand the way perception affects color. And you'll have customers who are color blind. Color blindness is a physical handicap. The eye just doesn't see some color differences. Usually it involves just reds and greens, but a rare form causes a person to see only shades of gray.

You're the expert. If your customer makes a bad color choice, don't be afraid to suggest that another choice might be better. Suggest an alternative. You don't want to hear, "This color looks terrible! Why didn't you tell me before you painted?" You may get paid for the job, anyway. But an unhappy customer won't tell anyone that

Color	Percent of reflection
Marine Corps green	4
Deep Navy gray	7
Passive green	7
Passive maroon	7
Fire red	7
Medium Navy gray	14
Deep green	14
Spruce green	15
International orange	15
Radiation purple	15
Clear blue	19
Terra cotta	20
Medium tan	21
Vivid orange	23
Medium green	25
Bright green	26
Pearl gray	46
Light blue	50
Highlight buff	55
Light green	55
Brilliant yellow	58
Sun tan	60
Peach	64
Soft yellow	75
Light ivory	75
Off-white	75-90
Bright white	90-99

Percentage of reflection of colors
Figure 10-7

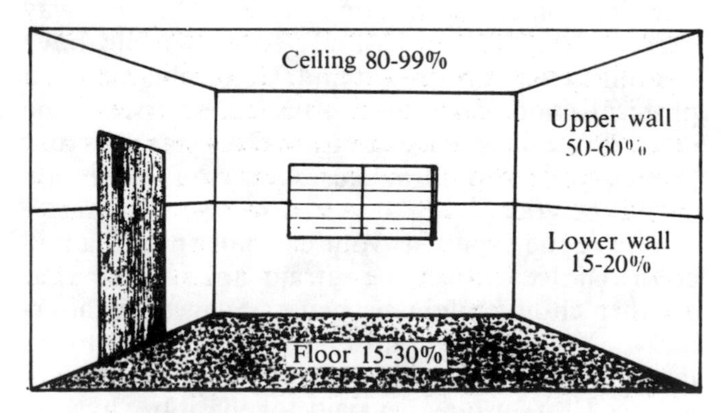

Balanced light reflection in a room
Figure 10-8

they chose the wrong color. It will be your fault: You didn't know your job. And anyone who shows the least interest will hear about it. Although this kind of advertising is free, it's not quite the kind of advertising you want.

Responsibility for Color Selection

I can hear you saying, "Hey, this stuff isn't my responsibility. I just apply what the customer told me to." Let me emphasize once again, you're the expert. You expect to be paid like a knowledgeable professional. For that fee, your customer expects you to keep them from making a mistake. They may have reasons for not following your suggestions. That's O.K. But at least you made an effort to point out the problem. No one's required to take your advice. Your job is to give it when appropriate.

But notice that I said "point out." Don't argue. Your customer makes the final choice. Just make your point. Let your customer take it from there. Say something like, "You know, Mr. Smith, I like this color, but it just might cause..." and here insert whatever the problem is. That's all that's needed. You've agreed with their taste and, at the same time, told them nicely that maybe the choice is wrong.

Remember, the final decision is theirs — they're paying the bill and have to live with the decision. Chances are they'll follow your advice, so you'd better be right. I've had customers come to me, months after a job was finished, to thank me for helping them make the right choice. Now *that's* good advertising.

Government and Industrial Color Code

Sooner or later, you'll want to bid factory, institutional, and possibly government jobs. They can be some of the most profitable work in the painting business. I've included several charts to show you some of the colors specified by various agencies.

The Interstate Commerce Commission (ICC) specifies lettering and background colors for labels identifying certain items. Figure 10-9 is the ICC code. Figure 10-10 lists the National Safety Council color code for identifying various dangerous and safe materials. They are usually the color of the container or piping on mobile equipment and tanks. Or they may be used as label colors only. Check with your customer for proper application.

Used to identify	Background color	Lettering color
Poisons, explosives, tear gas, other poisonous gases	White	Red
Compressed gases	Green	Black
Flammable liquids and fireworks	Red	Black
Acids	White	Black
Inflammable solids and oxidizing materials	Yellow	Black

Interstate Commerce Commission label color code
Figure 10-9

Used to identify	Color
Valuable materials, not to be disposed of	Purple
Fire-protection equipment	Red
Protective materials. Antitoxins, antidotes, medicines	Blue
Safe materials, those that are normally safe to handle	Green, black, white, gray

National Safety Council color code
Figure 10-10

Figure 10-11 lists the standard industrial piping color code. Figure 10-12 lists the military piping and fluid valve code. The military uses a lot of dull colors: grays, greens and browns. For most items, one of these is the base color. The valves are painted the primary color: red, blue, yellow or so on. Pipes are generally striped lengthwise or like spaced wrapping.

Figure 10-13 shows the standard industrial color codes for equipment and fixtures in factories.

Customers will usually specify colors to use. Unfortunately, your customer may not understand the rules set down by the various government agencies. Take the initiative of painting 500 feet of oxygen pipeline orange in a hospital and you'll probably have to take the initiative in chipping orange paint off 500 feet of pipe — at your expense.

Federal Color Standards

Many manufacturers offer color chips that will help you confirm your customer's color choice with the appropriate agency. Most government agencies require you to keep a 3-inch by 5-inch color chip in the job folder at all times. You can get color chips in *Federal Standard #595a*. It has a full set of chips for federal property. Get a copy of it from:

Federal Supply Service
Standardization Division
Specifications and Standards Branch
Washington, D.C. 20402

You can get additional information from:

Superintendent of Documents
U.S. Government Printing Office
Washington, D.C. 20402

Ask for the Index of Federal Specifications and Standards. The Index is updated monthly and sold on a subscription basis.

You can get single copies of most specifications from the General Services Administration regional offices. They're located in Boston, New York, Washington, Atlanta, Chicago, Kansas City, Ft. Worth, Denver, Los Angeles, San Francisco, and Seattle. Look for listings in the phone directory under U.S. Government.

Color	Meaning
Yellow, orange	Danger
Green, gray, white, or black	Safe
Blue	Protected, as antidotes
Purple	Safe, valuable

Industrial piping color code
Figure 10-11

Color and Estimating

The price of a paint job and the color you use are related. Several factors are involved. I can hear you say, "What difference does it make whether I paint something brown or yellow? A gallon of paint is a gallon of paint." That's not always so. Some colors cover a surface in one coat, but many won't. It's a big mistake to assume one-coat coverage when two are needed.

Some Colors Need More Than One Coat

White, most bright colors, and most semi-gloss or full-gloss colors need an undercoat, or two or more

Color	Meaning	Color	Meaning
Red	Fuel	Orange, blue	Pneumatic
Red, gray	Rocket fuel	Yellow	Lubrication
Green, gray	Oxidizers	Blue	Coolant
Red, gray-red	Water injection	Orange, green	Inert
Blue, yellow	Hydraulic	Green	Oxygen
Brown, gray	Air conditioning	Brown	Fire protection
Gray	De-icing	Orange	Compressed gas
Yellow, green	Rocket catalyst	Brown, orange	Electrical conduit
Yellow, orange	Monopropellant	White	All others

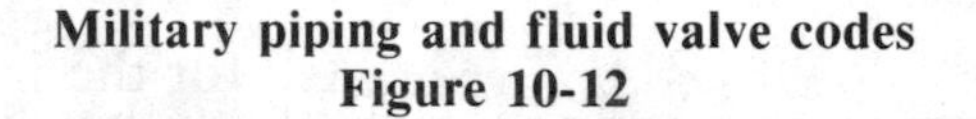

Military piping and fluid valve codes
Figure 10-12

coats, to hide the surface. Be careful. Know which colors may need more than one coat, and plan your estimates accordingly. Here are some other situations where you may need more than one coat:

- When painting light colors over dark surfaces.

- When making a complete change from one color group to another. A color group is the related shades of particular color — browns, reds, oranges; yellows, greens, blues; black, grays, whites, for example.

- When painting over bright colors.

Color	Used for
Gray	Fixed mechanical equipment
Buff	Working area walls, backgrounds
Red	Standpipes, hose connections, fire equipment
Yellow and red	Fresh water hydrants, yellow with red tops
Bright yellow	Mobile shop machinery, as loaders and forklifts
Dark yellow	Large equipment, as trucks, bumpers with black stripes
Blue	Electrical boxes
Orange	Inside electrical boxes
Yellow-orange	Buses
White and green	First aid equipment, supplies

Industrial color code
Figure 10-13

Purchasing Color

Cost depends on how you buy color. If you are just learning the business, let the paint store do your color mixing. Yes, it costs a couple dollars more per gallon. But your paint dealer has the know-how, the colorants, and the equipment.

Once you're established and know more about paints and colorants, do your own color blending. It saves money. Here's an example. Paint has a 35% to 42% markup. If a gallon of paint costs your supplier $10.00, he'll retail it for about $14.00. If you get a 10% trade discount, you'll pay $12.60. (The best discount you can get is probably 15% off list. That $10 gallon would go for $11.90 at a 15% discount.) On mixed colors the supplier also charges you for a few ounces of colorant. For the material, the markup, and his labor you'll pay about 50 cents an ounce. For a three-ounce mix, your cost is back up to $14.10.

"But," you say, "my supplier tints my paint for free." You're wrong. Colorants cost him money. You can be sure this cost is included somewhere in his price.

Save money by coloring paint yourself— Now, if you buy *tint base,* no color added, and you buy in quantity, the supplier will usually give you the $11.90 price. A quart of universal colorant costs about $10.00. That's 31 cents per ounce. So this gallon of paint now costs you $11.90 plus 93 cents for colorant, or $12.83. You saved $1.37. Big deal, you may say. But it adds up fast when you're using 10 gallons of paint a day, 5 days a week, 50 weeks per year. That's 10 times 5 times 50 times $1.37, or $3,425. Now do you believe me? Color *is* related to profits.

Mixing Your Own Colors

To mix your own colors, you'll have to buy colorants. Figure 10-14 lists the universal colorants you should stock. This list is from the company I deal with, Mohawk Products. Other companies may give the colors different names. Colorants are available in 4-ounce and 8-ounce tubes, half-pints, full pints, quarts, and gallons. As you can see, it's a long list. And it requires a fair-sized cash outlay. But stocking these colorants will pay for itself when you're dealing in volume. Remember also that you'll have to buy measuring and mixing equipment.

Here are a couple of rules to remember when actually mixing the paint:

1) Never add more than eight ounces of colorant to a gallon of paint base.

2) *Stir* the colorant into the base until it's the shade you want, then shake. Many paints, especially latex, shouldn't be shaken for more than three minutes. Shaking longer than that aerates the paint. Then you've got some more problems. Aeration causes bubbles which adhere to brushes, rollers, and the surface you're painting. As the bubbles burst on the surface, they leave pits or voids.

A note of caution: Some manufacturers' color chip cards are printed with color mix formulas. But don't think you can copy the colors just by following the formulas. For example, "1U" doesn't necessarily mean 1 ounce of umber. It could easily mean 1/8 ounce of cobalt blue. Companies code their color mix formula and each company's code system can be different.

Black, drop	Red, liberty
Black, lamp	Red, venetian
Blue, cobalt	Red, vermillion
Blue, Prussian	Sienna, burnt
Blue, ultramarine	Sienna, raw
Brown, Van Dyke	Umber, burnt
Green, dark	Umber, raw
Green, medium	White, flake
Green, light	Yellow, lemon
Ochre, French yellow	Yellow, medium
Pink, rose	Yellow, orange
Red, bulletin	

Universal colorants
Figure 10-14

Colorants for Stains and Lacquers

These aren't the same as universal colorants. Don't try to use universal colorants in stain and lacquer. Many colors are available. Mohawk Products, for example, lists 64 colors of just one variety of stain. It's a dry powder aniline and must be mixed with a solvent for use.

Colorants for Cement Paints

Cement and water-soluble dry mix paints require colorants formulated specially for these types of paint. Don't use universal colorants. Cement paint colorants are formulated so they won't be destroyed by the alkali in cement.

Not All Paints Can Be Colored

Tinting paint to a different color isn't always possible. Suppose you want to tint white paint to a dark color and select a can of white paint that already has as much pigment as it can hold. Any more pigment and it won't bond to the surface. Adding more pigment unbalances the paint. The ratio of pigment to binder changes. This causes several things to happen:

- The feel of the paint changes.
- The drying time changes.
- The paint appears dull or flat.
- Chalking occurs.
- The paint's life span is shortened.
- Apparent fade occurs.

Use a *tint base* whenever you want to mix your own colors. It's made to have the right amount of pigment *after* you add the colorant.

Coloring white paint— If you absolutely *must* tint a white paint, don't use universal colorants. Instead, use a premixed paint of the same type and brand. Just mix the two together. You'll get a color change, without sacrificing the qualities of the paint.

Coloring primers and metallic paints— I've also seen painters try to color primers and metallic paints. It's poor practice. Primers are formulated with the maximum amount of pigment, so you'll have the same problems described above for coloring white paint. Not only that, the acids in the primer may react unfavorably with the colorants.

Coloring metallic paints can cause all of the above problems, plus the possible formation of gas, which will blister the paint.

In general, if the can says it can be colored, then color it. If the label doesn't specifically say it can be colored, don't. You'll save your customer — and yourself — a lot of problems.

Getting an Exact Color

I can see it now. Some painters who've read the paragraphs above will run out, buy the entire selection of colorants, and be color tint experts in a few short weeks — experts who will try to match any color on any surface. *Don't do it!* You'll only get yourself in trouble. It takes years of experience to be able to match colors exactly. In fact, an exact match is nearly impossible. A close match is all you can hope for.

Remember the nature of color— Color is light. Specifically, it's reflected light. The same paint on the same surface can have a different sheen, and even appear to be a different color, when applied with different methods. Figure 10-15 shows how the application method affects the apparent color. The roller coat in the middle of the figure will look flattest. The brushing will be next. Since the spray coating is the most smooth and even, it will give the greatest reflection. Note that the "bounce" of each is different. The more scattered the bounce, the flatter the paint will appear. The surface will look 'shiny' when the reflected light is more regular and parallel. Other variables affect the color we perceive. I've mentioned some of them already in this chapter. Here's a list for you to consider.

- The source of the light (direct or reflected, natural or artificial)
- The angle at which the light hits the surface
- The angle from which we view the surface
- The surface material (smooth or rough)
- The colors of items near the surface
- The chemistry, gloss, and pigment content of the paint
- The consistency of the paint mix
- The age of the paint
- The manner and direction of application
- Our eyes' sensitivity to paint color and gloss

The best you and your customer can hope for is a close match – *a color that's acceptably close to the desired color.*

Color and Touch-Up Work

This brings us to a discussion of touch-up work. I define touch-up work as any paint job where part of a surface is being repainted to match another part. The trick is in trying to make the two paint jobs look as though they were one.

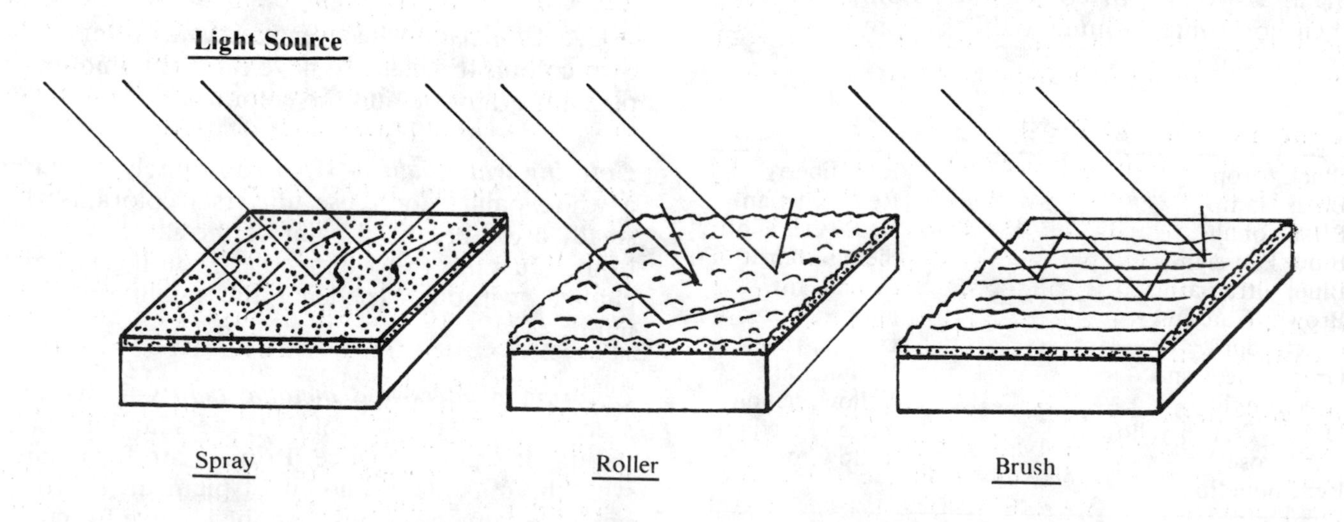

Application method changes apparent color
Figure 10-15

My advice is, don't try it. Too many things can go wrong. A perfect match is nearly impossible. Paint the whole surface or none of it. But if you *have* to do touch-up work, observe a few basic principles to eliminate most of the color variables.

When To Do Touch-Up Work

Some people wouldn't notice mismatched paint if you pointed it out to them. Others would. They're very sensitive to it. To them it's a sign that someone didn't care, a sign of a cheap job with shoddy workmanship, or a sign of poor planning. These people are sensitive to color and shade. Now, I ask you, who's going to hire you to do touch-up work. Someone who doesn't care about a mismatch or someone who's particular about color? The latter of course. They're critical of the prior paint job and will be critical of your work too unless it's that one-in-a-million perfect match.

So I'll repeat my first suggestion. Don't try to touch-up items. Leave them until the entire surface can be painted.

Of course, every rule has its exceptions. Here's one. If the area to paint is very small, say a small chip in a corner, you can do a successful touch-up. Small spots under 1/8-inch aren't noticeable. But anything over 1/2-inch almost always is. And anything in between depends on the size of the whole surface, the surface material, the gloss, the color and how far the surface is from normal eye level.

Here's the second exception. If the paint job isn't more than a few weeks old and you have left-over paint from the same batch, try it. Careful work will make an acceptable touch-up. Use the same type of tool (brush, roller, sprayer) that the original paint was applied with. And apply the new paint in the same direction. Feather the new paint outward to blend with the entire surface.

Touching Up Old Work

Is the original coating more than a couple of weeks old? Do you have to use a new batch of paint? Then paint the entire surface, from one natural break point to another. For example, paint a wall from corner to corner and from floor to ceiling. If the color match is close, it will probably blend with the other walls.

Painting only one wall— Often, only one wall needs to be painted. Instead of trying to match the color to the other walls, use a different but complementary color. The right color can give a room an entirely new feel — something many rooms need. If the room has three white walls, try one wall of blue, or tan, for example. Or paint the fourth wall white if the other walls are a pastel.

Use this trick on interiors only. It's just not accepted on exteriors. You'll seldom see three exterior walls on a house painted white and the fourth brown. Instead, change one wall by using a material other than paint. For instance, consider applying hardboard siding.

Properties of Color

Color not only affects our feelings and moods, it can also change the size, shape, or identity, of a room, a building, or an object. Using color, we can make an object appear closer or farther than it is, make items tend to disappear, direct vision to a certain spot, heat or cool an area, or provide comfortable light to a area.

Color and the Home Environment

We help control our home environment with color. A home in the desert will be cooler if its exterior is painted white and it has a light-colored roof. In colder climates, a dark roof and darker exterior will absorb heat, warming the interior when the sun shines.

A black-painted cement block wall stores the sun's heat for several hours. This is the principle of passive solar heating systems.

Warm rooms and cool rooms— The compass orientation of a house affects interior heating and cooling. Look at Figure 10-16. The rooms facing north get less light than other rooms. They'll be cooler and damper. Paint them in warm colors, like off-white toward brown, or pastels of yellow, red or orange. The south-facing rooms get lots of light and will be warmer and more dry. Paint them with cool colors: grays, blues or greens.

Sunlight changes room color— Sunlight varies through the day as the sun travels from east to west. It's cooler and harsher in the morning hours than in the afternoon hours. Again, look at Figure 10-16. Daylit rooms will change color and appearance through the day. Take this into account when choosing colors.

Color affects room comfort and heating and cooling requirements. That makes color selection an important decision. You charge a professional price for professional work. Part of your responsibility as a painting professional is helping your customer make good color decisions.

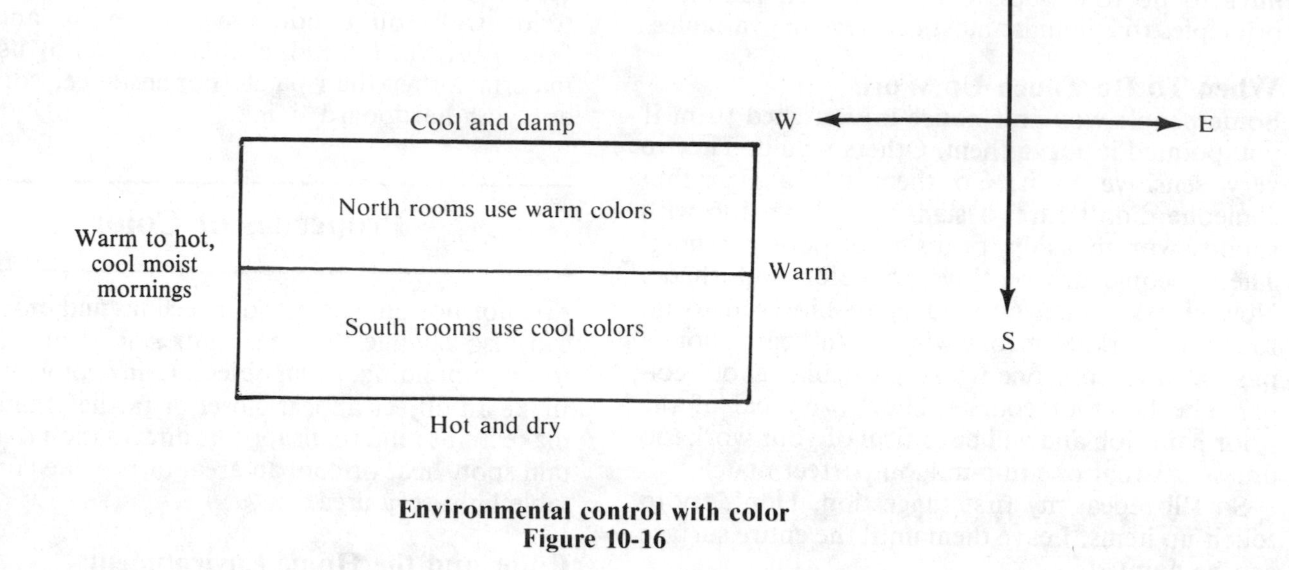

Environmental control with color
Figure 10-16

White Paints

When is white paint no longer considered white? That's a hard question. Any set of color chips has several "whites." Technically, there's only one white. In practice, any very light pastel is considered white. Why? Because it's hard to produce a white paint with enough pigment to hide properly. So, most manufacturers add some other color to improve the hiding ability. In the brown range, whites go nearly all the way to beige. In order of increasing amount of brown: white, soft white, eggshell, Navajo white, antique white.

Here's a related question. What's the difference between white and clear? The terms can be confusing. White is opaque; you can't see through it. Clear coatings let light pass right through to the underlying surface. The color of the surface below determines the color we see.

Sometimes "white" refers to clear, but this can be confusing. White shellac is an example. Here, "white" refers to shellac that's been bleached but still has some color. Clear shellac would be a better term.

Psychology of Color

No matter what names we give them, colors have very real psychological effects on us. We associate colors with our experiences. We give color a feeling.

Color Associations

Look around at our natural surroundings. We associate certain colors with the seasons or moods of nature. In spring we see cool green grass and new leaves; the sky is clear and refreshing. In summer, the grass is browner, and flowers are blooming. Fall brings the reds, yellows and browns of leaves turning. Winter brings white snow, gray skies, and bare black branches. Figure 10-17 lists some of the psychological color associations.

Attention-Getting Colors

Colors in nature are the light or medium shades of browns, greens, yellows, blues, and reds. Colors that don't occur in nature tend to attract attention. These are usually brighter colors. Some of the attention-getters are:

- Bright blue
- Bright green
- Bright red
- Bright yellow
- Orange

- Fluorescent green
- Fluorescent orange
- Pure white

These colors are ideal for packaging, displays, signs, and protective clothing. They also help create a party atmosphere. Use them with caution, though. An office, a room, or a house painted any of these colors will be overpowering.

Color	Feel	Mood or Association
Black	Cold	Forbidding, conservative, authority
Brown	Warm	Restful, earthy, businesslike
Gray	Cold	Gloomy, threatening
Silver	Cool	Metallic, rich
Gold	Warm	Rich, expensive
White	Cool	Clean, new, pure
Ivory, off-white	Warm	Rich, solid
Dark purple	Cold	Royal
Pink, light purple	Warm	Restful, love
Blue	Cool	Calm
Light blue	Cool	Restful, love
Green	Cool	Tranquil, new, safe
Bright green	Cool	Safe
Olive, some grays	Cool	Military, authority
Greenish-yellow	Cool	Fresh, new, safe
Yellow	Warm	Festive, cheerful
Bright yellow	Hot	Caution, attention
Orange	Warm	Festive, cheerful
Bright orange	Hot	Caution
Red	Hot	Danger, anger, warmth
Bright red	Hot	Danger, attention, sexy

Pyschological color associations
Figure 10-17

Maximum-visibility colors— Our eyes are more sensitive to some colors than to others. Therefore, we can see these colors at greater distances and angles. The military services use these colors on survival clothing, life rafts, buoys, markers, and emergency equipment. The colors are *bright yellowish-green, international orange* (adopted by both OSHA and the International Safety Council), and *high-visibility yellow* (OSHA color).

Maximum-Contrast Colors

Maximum contrast between two colors makes it easy to distinguish one color from the other, even in dim light or with poor eyesight. Almost any medium to dark color shows up against white. Not all colors show up against black or other dark colors. One combination used regularly in signs, decorations, and printing is black on white or white on black. The next most-used combination is blue on white. Remember that white lettering on a dark background appears smaller than it really is.

Metallic Color Associations

Metallic colors have a certain strength, solidity, and richness to them. Metal is thought of as being indestructible. Since the 1960's, car manufacturers have put metallic paints on their cars to lend a feeling of durability, solidity, and richness. Figure 10-18 lists some of the psychological associations for various metallic colors.

Color	Feel	Association
Aluminum	Cool	New, durable, modern
Brass	Warm	Rich, durable
Bronze	Warm	Rich, old, solid
Chrome	Cold	Hard, durable, sparkly
Gold	Warm	Rich, classy
Silver	Cool	Rich, mellow

Metallic color associations
Figure 10-18

Color Schemes to Avoid

There are also some color schemes you should avoid. You can, however, use these combinations in a room *if* one is the primary room color and the second is used in small amounts as an accent color. Try to avoid the color schemes listed in Figure 10-19.

Red and green	Blue and yellow
Red and orange	Blue and violet
Red and violet	Blue and green
Yellow and orange	Yellow and green

Color schemes to avoid
Figure 10-19

Color and Plants

Be careful when choosing colors for surfaces near plants and gardens, such as house fronts, retaining walls, fences, and planters. The paint must stand up to watering and chemicals used on the plants. It must also not harm the plants. White, for instance, may reflect too much light and cause plant leaves to burn. Reflected light will also tend to dry the soil quickly. On the other hand, white may be a good color in dark areas that don't get direct sun because the reflected light will be good for the plants.

Effects of Colors on Plant Growth

How well plants grow depends on the color of the light they receive. Plants grown indoors under cool-white or warm-white fluorescents develop lots of green foliage, but few flowers. If flowering is desired, add some reddish color to the surrounding area. Plants grown indoors under incandescent light will have pale green or yellow foliage. The surrounding area should be painted a blue or purplish color. Here's a list of the effect of various colors on plant growth.

- Yellow, green, and yellowish-green have no effect.
- Blue or violet causes foliage to be darker green.
- Red produces more, bigger, and brighter flowers.
- White produces growth balanced between foliage and flowers.
- Black slows growth.

I hope this chapter has made you a believer. Color has a major influence on man's perception of his environment. You select and apply color. Do it wisely. Follow the principles described in this chapter.

Surface Cleaning and Preparation

Before you can clean and recoat a surface, you'll often have to find out what the old coating is. Why? Because some paints and coatings aren't compatible. They react unfavorably with each other. For example, you can't successfully put lacquer over enamel. And if the old coating is stained, blistering or chalking, it's prudent to do some testing to find out the reason for the failure before recoating.

Testing the Old Surface

There are some fairly reliable tests you can use to determine the reason for a paint failure.

Blister Test

Blisters or bubbles are caused by moisture trapped behind the paint. The question is, where's the moisture coming from? This is an easy test: Just pop the bubble. If you see a bare, dry surface, it was painted while the surface was wet. If moisture is coming through the surface, eliminate the source of the moisture. If you see paint under the bubble, the problem was caused by someone painting over a wet coating. Scrape away the blister and repaint.

Another cause of blisters is oil. Oil keeps paint from adhering to the surface. Then air and moisture get in and expand the paint. Metals are especially likely to have oil problems. They must be thoroughly cleaned. See the section on cleaning galvanized metal later in this chapter.

Paint Adhesion Test

Press a strip of clear adhesive tape to the surface. Leave it for a few minutes, then pull it off with a jerk. If the paint comes off too, it hasn't adhered to the surface. The solution is to strip off the old paint and repaint.

Chalking Test

There are two ways to test for chalking. The first is the *black glove test.* It's shown in Figure 11-1. Rub the surface with a black rubber glove. If there's white or colored powder on the glove, the paint is chalking. In Figure 11-1, chalking paint made the surface look faded. A good wash with TSP (trisodium phosphate) brought back the bright, new finish.

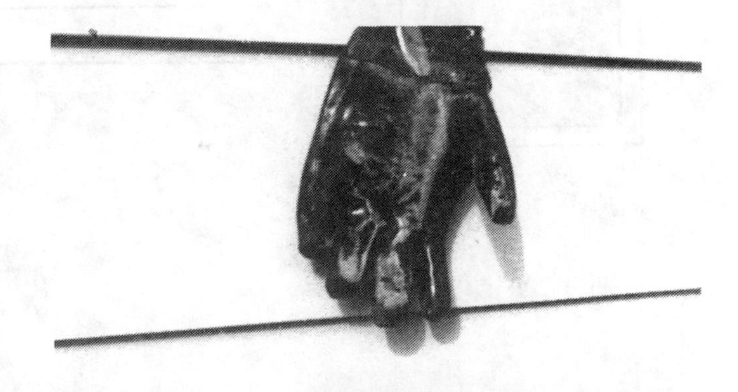

The black glove test for chalking
Figure 11-1

The second test uses clear adhesive tape. Press the tape lightly on the surface. Remove it gently. Press the same piece of tape on the surface again. If it doesn't stick, or sticks poorly, the surface is chalked. Clean it with the cleaner for aluminum siding, described later in this chapter.

Condensation Test

Here's a test for condensation on inside walls. Tape a one-square-foot piece of aluminum foil to the wall. Seal it tightly on all sides with tape, as in Figure 11-2. Leave it for two or three days, then remove it carefully. (The tape will pull loose paint from the wall.) Examine the foil.

Water on the wall side of the foil indicates seepage through the wall. Water on the room side of the foil indicates condensation. If the problem is condensation, the room's ventilation must be improved or the source of moisture reduced.

You may see both seepage and condensation. If so, fix the seepage problem first. It's probably moisture leaking through the roof or through a masonry wall. Then retest the room. Seepage could have been making the room air extra moist, causing the condensation problem too.

Lacquer Test

Test the area in question with 100% lacquer thinner. If the coating dissolves, it's lacquer.

Oil/Alkyd Paint Test

To test for oil/alkyd paint, use a solution of two tablespoons lye crystals dissolved in one cup of warm water. Apply this to the painted surface. Wait ten minutes. If it's oil/alkyd, the paint will bubble and blister. Latex paint won't be affected.

Shellac Test

Test the area in question with 100% denatured alcohol. If the coating dissolves, it's shellac. Remove shellac with a solution of half ammonia, half water.

Cleaning the Old Surface

Now you know what's covering the old surface. Can you start painting? Not quite yet. First you have to clean the surface. This is the first step in prep work. The rest of this chapter is a "recipe book" for getting rid of most stains you'll encounter.

But first, you need to know how to apply a neutralizing rinse. When you get to the formulas for stain removal, many will tell you to finish with a neutralizing rinse. These are the rinses they refer to.

Neutralizing Rinses

Water is the best rinse for removing most chemical

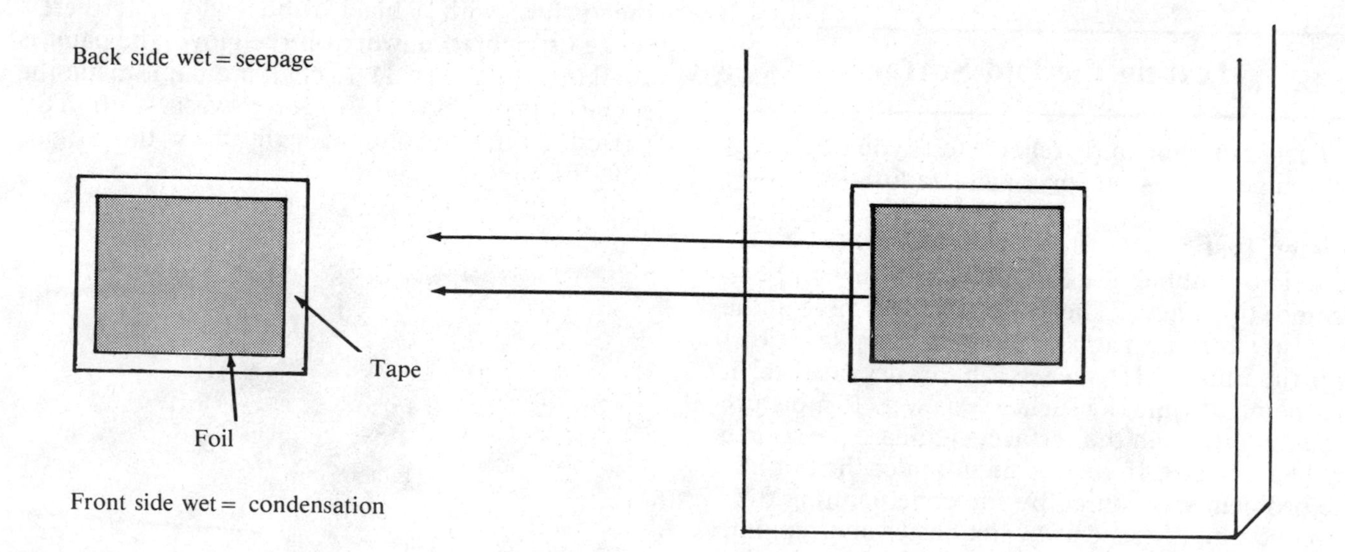

24-hour condensation test
Figure 11-2

residues. Sometimes, however, a water wash isn't enough. You want to stop the action of a caustic, an acid, or bleach. This is when you need a chemical rinse. The following formulas are called for in the sections on removing stains, and on cleaning large areas. You'll likely use them often.

Bleach neutralizer, type A— One cup white vinegar, mixed with one quart clear water.

Bleach neutralizer, type B— One cup borax, mixed with one quart clear water.

Muriatic acid neutralizer— Use a solution of 10% ammonia and 90% water or a 10% potassium hydroxide and 90% water solution (lye).

Tannic acid neutralizer— Tannic acid (also found in tea) is an acid stain from cedar and redwood. Use one quart of denatured alcohol, mixed with one quart clear water.

Welding flux neutralizer, type A— Six ounces of sodium carbonate (washing soda), mixed with one gallon water.

Welding flux neutralizer, type B— Six ounces of sodium bicarbonate (baking soda), mixed with one gallon water.

Recipes for Removing Stains

Many times you'll have to paint over a surface that's been marred by children, pets, or just plain carelessness. But painting over stains is a waste of time. They'll just bleed right through the new top coat. Save yourself time and energy by removing the stain before you paint.

This section tells how to remove just about any stain you'll ever encounter. Sometimes several stains are treated the same way. They're listed after the first mention of the treatment.

Adhesive tapes and contact papers— These are hard to remove because the top surface usually can't be dissolved. Figure 11-3 shows a typical example. The tape has to be peeled off. A thin razor knife is handy for this. The adhesive that remains when the paper is removed usually comes off with xylene or toluene. If peeling takes too long, score surface and soak it with xylene or toluene to speed peeling.

Remove adhesive tape and residue
Figure 11-3

Animal urine— Use one tablespoon white vinegar, mixed in one cup lukewarm water. Follow with a soap-and-water wash, then rinse with clear water. This also works on tea stains.

Asphalt, asphaltum, tar— Use either mineral spirits or paint thinner.

Ballpoint pen ink— Use trichloroethylene if you have it. Otherwise, try a solution of half household bleach and half water. Follow with bleach neutralizer, type A. This works on all inks.

Beer— Wash with soap and water. Then treat like animal urine. Mustard takes the same treatment.

Bleach— Use type A or type B bleach neutralizer.

Blood— Scrape off the excess, then wash with soap and water. Follow with a solution of one tablespoon clear, non-sudsing ammonia mixed with one cup clear water. Then treat like animal urine. Treat all these the same way:

Dye, water-based
Egg
Fruit
Ice cream
Milk
Vomit

Butter— Scrape off excess, then treat like ballpoint pen ink.

Candle wax— Try trichloroethylene, or xylene, or a solution of half alcohol and half water.

Candy— Remove excess, then treat like bleach.

Catsup— Use soap and water.

Chewing gum— Treat like candle wax. Heavy deposits of gum can be chipped off if you freeze them first. Use ice cubes, or commercial *Freeze Mist* spray. It's available in electronics hobby stores. The same method works on lipstick.

Chocolate— Use soap and water, then treat like animal urine. This also works on cocktails and coffee.

Cosmetics— First try soap and water. If that doesn't work, treat like candle wax. If it's still there, treat like blood. It all depends on the material the cosmetic manufacturer used as a base.

Dirt stains— Wash with one tablespoon household detergent in one gallon of water. Rinse with clear water.

Furniture polish— Furniture polishes are like cosmetics; each manufacturer has a different formula. If a silicone wax (such as Johnson's) was used, apply a heavy coat of the same wax. Wipe it off after it has cut through the older wax and dirt. Follow with a solution of half alcohol and half water. Finally, wipe with mineral spirits.

For a non-silicone wax, use the alcohol-water solution only, then wipe with mineral spirits. If that hasn't worked, treat like blood and animal urine. Trichloroethylene and xylene may also work. Use the same methods on shoe polish.

Glue, water-based— Wash with soap and water, or treat like blood.

Gravy— Wash with soap and water.

Grease or oil— If it's a silicone-type grease, treat like candle wax. Toluene or mineral spirits will work on most non-silicone greases. For extremely heavy, caked-on or burnt-on grease (as on a barbecue) use a solution of one cup lye crystals in one gallon water. Be careful! This solution is very caustic. It also dissolves oil-based paints.

There is an alternative you can use. Try automotive "tar and bug" remover, followed by a degreaser. This works well on most cooking oil splatters.

Household-type cements and glues— If you know the type, use the solvent recommended for that glue to soften it. Then wipe off the excess. Follow with mineral spirits or xylene.

Mud, dirt, clay— Wash with soap and water.

Nail polish— Chip off excess; follow with acetone or mineral spirits.

Perfume— Treat like candle wax, then follow the steps for animal urine.

Rust stains— Mix four ounces oxalic acid crystals in one gallon water. Brush on. Allow to sit for three or four days, then rinse.

Soft drinks— Wash with soap and water. It that doesn't work, try the steps for blood and animal urine.

Sour milk— This is casein — the same material used to make many paints and glues. Removal is difficult, at best. Try a solution of one cup lye crystals in one gallon of water. Scrub with steel wool and the lye solution, then rinse well. Be sure to wear protective clothing (eye protectors and gloves) when using lye.

Tar— Use trichloroethylene or xylene. Mineral spirits will also work on most tar stains.

If any of these stains remain after the treatment recommended, a solution of half household bleach, half water should do the trick. Always allow this solution time to work. It usually takes several hours. Follow with a bleach neutralizer.

Commercial stain removers— Of course, you can always try some of the commercially-available stain removers. For tough cases, you can try one of the "anti-vandal" sprays or top coatings that remove or prevent the adhesion of spray paints and waxes.

Here are a few of the commercial products, with addresses of the manufacturers.

Fantastik, from
Texize Chemicals,
Division of Morton-Norwich Products
Greenville, South Carolina 29602

Oakite, from
Oakite Products
50 Valley Road
Berkeley Heights, New Jersey 07922

Anti-Vandal Spray, from
Crown Industrial Products
Hebron, Illinois 60034

Vandax Spray, from
Great Lakes Pollution Control
Box 248
Dexter, Michigan 48130

DWR-11, from
KRC Research Corporation
315 North Washington Avenue
Morristown, New Jersey 08057

No matter how you remove the stains, don't forget to spot-prime or seal the surface before applying the top coat. Later in this chapter you'll find recipes for sealers.

Cleaners for Large Areas

Your prep work will often involve cleaning large areas. Here's a list of cleaners that should help you.

Acoustical ceiling cleaner— Mix one quart of household bleach with one gallon of water in which one tablespoon of TSP is dissolved. Blot, or better yet, spray this solution onto the ceiling. Wait six to eight hours. If it worked, neutralize with either bleach neutralizer type A or type B. If there are still some stains, repeat the process.

Wait 24 hours before trying to paint. And paint only if you must, using nonbridging paint. Nonbridging paint is formulated so it doesn't form a solid film surface. It doesn't "bridge" or fill in the sound-absorbing holes in acoustic ceilings. Also use nonbridging paint on popcorn ceilings to preserve the acoustic properties.

Aluminum siding cleaner—Dissolve two tablespoons of TSP in one gallon of clear water. Wet down the surface first, then clean with a sponge or a sponge mop. Be careful not to scratch the surface with the metal edges of the mop. Rinse with clear water. And work from the bottom up to prevent the cleaner from streaking the surface.

Asbestos shingle or panel cleaner— Use a solution of 10% muriatic acid and 90% water.

Brick, rock, and cement cleaner— Mildly-stained surfaces can usually be cleaned with one cup of muriatic acid mixed with one gallon of clear water. Follow with a clear-water rinse. Heavily-stained surfaces may need a half-muriatic acid, half-water solution. On extremely stained surfaces, use 100% muriatic acid.

Muriatic acid etches cement and brick, so the two stronger solutions require a neutralizer rinse to stop the etching action. Use either of the muriatic acid neutralizer solutions. It's not enough to just rinse with water. Any acid that has seeped into the surface will leach out and destroy your paint job.

A stiff-bristled brush is handy for applying muriatic acid. Always wear rubber gloves and eye protection. And avoid breathing the fumes. This stuff is actually hydrochloric acid and will injure lung tissue.

Efflorescence cleaner/remover— Use the same solutions as for brick, rock, and cement.

Efflorescence on cultured stone— Mix one part white household vinegar with five parts water. Use a bristle brush, not a wire brush. Wire will scratch the stone. Rinse with clear water.

Flux cleaner/remover for welded metals— Mix four ounces of commercially-available 85% phosphoric acid with one gallon of water. Scrub with a stiff-bristled brush, or one you've cut the bristles on. Wear rubber gloves and eye protection. Follow with a flux neutralizer solution.

Galvanized metal (also stainless steel) cleaner— Mix 1/4 cup kitchen detergent in one gallon of water. Rinse well with clear water. Finally, wipe down with mineral spirits. What you're trying to do is remove manufacturing oils from the metal. A wash of 100% denatured alcohol also works, but take care. Alcohol is flammable.

Gas stain removal— Gas stains are chemical stains caused by airborne gasses from sewers and many industrial plants. The paint turns dark. In many instances it turns a metallic gray, resembling graphite. Use a solution of 5% muriatic acid and 95% water, or a 3% hydrogen peroxide solution (right out of the bottle), or 10% bleach and 90% water. Rinse with clear water and repaint.

Glass cleaner— Mix one pint of non-sudsing ammonia with one quart of clear water. Don't use this mixture on glass in direct sunlight. It evaporates too fast and will streak.

Fume stain cleaner— A fume stain is the brown discoloration that bleeds through the top coat of paint. It comes from paint below the top coat that contains lead. Use a solution of one pint household bleach and one gallon of water to remove the stain. Apply with a stiff brush or a sponge. Allow several hours for it to work, then follow with bleach neutralizer type A or type B. Finally, rinse with clear water.

Here's a safety note: Don't sand lead-based paint. The dust created by sanding contains lead and is hazardous to your health. Follow the above procedure, followed by a sealer and a top coat.

Linoleum floor cleaner— Yes, linoleum — but not vinyl — can be painted with latex floor and deck paints. To prepare linoleum, first clean off dirt, wax, and polish with a solution of TSP, then wipe with mineral spirits. Sand lightly.

Mildew cleaner/remover— Use a solution of one quart household bleach, mixed with three quarts clear water in which one tablespoon of TSP has been dissolved. Blot or spray on the surface. Always wear rubber gloves and eye protection. Also, cover everything not to be bleached. The mildew should disappear within one hour. (For extremely heavy mildew, use a solution of half bleach and half water.) Follow with bleach neutralizer type A or type B, and finally rinse with clear water.

Plastic and Plexiglas cleaner— Use mild soap and water. Don't use abrasives or abrasive cleaners. Use soft rags and turn or rinse them often to avoid an accumulation of grit. Naptha and kerosene also work very well.

Rusted metal cleaner— Brush or sand off as much of the rust as you can, then coat with phosphoric acid (Naval Jelly). Rinse well with clear water and dry immediately. Then prime immediately to prevent rust from starting again. Finish with a top coat.

You can also use commercial rust removers, available at plumbing supply outlets. They contain sodium bisulfate or sodium hydrosulfate, or both. They're very caustic. If you use them, wear protective clothing, rubber gloves, and eye protection.

Stucco cleaner— Wash with one cup household laundry detergent mixed with one gallon of water. Rinse well.

Special Preparation Requirements

Some materials must be specially prepared to receive the top coat, even after they've been cleaned. Here's how:

Ceramic tile and glass— Sandblast lightly, or etch chemically with hydrofluoric acid to abraid the surface.

Painted surfaces— Wipe them down with liquid sanding solution. This roughens the surface, adding *tooth* in the language of experienced painters. Hand-sanding also does the job, but it's more work.

Polyethylene, polypropylene, polyvinyl chloride (PVC)— Clean with soap and water, followed with a clear water rinse. Dry, then use methyl ethyl ketone (MEK), available commercially, to clean and etch the surface.

Stainless steel— Give it a pickling bath. This is a 20% nitric acid/water solution. It removes the surface contamination of sulfur, selenium, and lead found in many stainless steels.

Whitewash remover— This won't fully remove old whitewash, but it does get it ready for a new coat of whitewash. Use 50% hot water and 50% vinegar, or 10% hydrochloric acid and 90% water as a prewash. Rinse it with clear water and apply new whitewash while the surface is still damp.

Recipes for Sealers

Here are some useful recipes for sealers. Use them to seal in stains, and as an aid in sanding. They make sanding easier by hardening any small fibers on the surface, making them easier to sand off. Also use sealers to prevent sap-laden woods and knots from bleeding. Knots are hard and smooth and contain a lot of resin. Figure 11-4 shows what happens when they aren't properly treated before painting. A sealer coat must be used under the paint.

Sealer coats are, by nature, thin coats. Apply them thinly. A build-up of sealer can cause the topcoat to chip, peel, or blister.

Here are the sealers to use under the top coats listed:

Sealer for nonsynthetic, oil-type varnish— Mix one part pure turpentine with one part of the finish varnish.

Sealer for varnish, shellac, lacquer, and most paints— For a filler sealer, mix one part four-pound cut shellac with four parts denatured alcohol. For a stain sealer, mix one part four-pound cut shellac with eight parts denatured alcohol.

Note that some lacquers may not dry properly or may "check" over a too-heavy coating or mix of sealer. It's much better to use a two-pound cut shellac with eight parts denatured alcohol for this sealer.

Sealer for urethane-type coatings— Make the sealer from the urethane being used. Mix one part urethane to one part mineral spirits.

Sand and seal knots before painting
Figure 11-4

Cleaning Brushes and Equipment

Clean latex paints with soap and water. Clean oils and alkyds with mineral spirits. Hardened oil or latex must be removed with either liquid sandpaper or paint remover (methyl chloride).

Remove epoxy resin from brushes with diacetone alcohol *before it's cured.* After it's cured, the only solution is to buy a new brush.

Remove polyester resin with acetone or nail-polish remover — before it has cured.

Concrete and Mortar Mixes

Lastly, here are some formulas for mixing your own concretes and mortars if your job includes patching concrete or masonry.

Concrete mix for foundations and retaining walls— Use one part cement, four parts gravel, and three parts sand. Add water to make a workable consistency.

Concrete mix for sidewalls, steps, and drives— Use one part cement, two parts sand, three parts gravel and enough water to make it workable.

Mortar mix Number 1— Use one part portland cement, one part lime, and 4¼ parts masonry sand.

Mortar mix Number 2— Use one part masonry cement, and three parts masonry sand.

In summary, here's the wisdom of this chapter in a nutshell: *Clean it before you paint it.* I hope I've given you some useful pointers to make that cleaning easier. In the next chapter, we'll look at dozens of problems you're bound to face sooner or later, with their causes and cures.

chapter twelve

Common Paint Problems: Prevention and Cure

If you've been reading and digesting the material in the earlier chapters, you understand the basics of paint and paint failures. Now let's get down to the specifics: looking at the problems you'll encounter, how to cure them, and how to keep them from occurring again.

You've heard painters say: "This so-and-so paint is no good! I just don't like that brand." The truth is, relatively little substandard paint is sold and very few paint failures are caused by the paint itself. All of the larger paint manufacturers are very good at manufacturing and testing their products. Nearly all reasonably priced paint will perform as intended *if* it's applied correctly. And that's the rub. Nearly all paint failures can be traced to the applicator. Either the prep work or application was faulty, or the painter chose the wrong material for the particular job.

I'll grant that some shoddy paint is being sold. Usually it's sold as the "house brand" of some cutrate discount house or chain store. Their business is selling at the lowest price possible to those who will buy only what they see as a bargain. Unfortunately, many of these "bargain paints" end up costing double or triple what a good medium-priced paint would have cost. They may need more than one coat, or even worse, they have to be stripped off so a good quality paint can be applied. This takes labor, and labor costs money. As a professional painter, you've no business using substandard materials. The materials on a paint job usually only cost from 15% to 25% of the job; the remainder is labor. It's not worth taking a chance on paint failure for such a small savings.

Bargain paint fails either because it's poorly formulated, has a low pigment content, or was overthinned at the factory. Thinner costs less than oils, plastics, and pigments. That makes it a temptation for some to fill the bucket with thinner.

Choosing a Paint with a Feel You Like

I've found some good quality brand-name paints that I disliked. Not because they didn't do the job or because they failed in service — they were good paints. It was just the *feel* of the paint that I disliked. With one brand, it felt like I was trying to paint with butter or grease. It was very difficult to control the brushing or rolling, even though the wall surface was properly prepped. Was it bad paint? No, just a different formulation and "feel." It's a personal judgment. To someone else, that paint may have been just right.

Here's my suggestion to the beginner, or even to the old pro: Try several different brands of paint. See which ones feel right to you and perform the way you like a paint to perform. You may like different kinds of paints from different manufacturers. Just because you like the latex from manufacturer X doesn't necessarily mean you'll like that manufacturer's enamel. Try different brands. You might even find a good paint that costs less, or paint that's sold by a dealer who's willing to give you more attractive discounts.

Handling Problems

Occasionally you'll come across a can of paint that's been frozen or improperly mixed. These aren't manufacturing problems. They're handling problems.

As a general rule, don't blame the paint for paint failure. It's pretty unlikely. Check for proper surface preparation, proper application, and proper selection of material before accusing the manufacturer. It can save you a red face.

Reasons for Paint Failure

Read this section carefully. Make yourself familiar with the most common problems. It can help you do the job right the first time. I can't identify the reason for every paint failure. There are just too many variables. But the list here covers most of the problems you're likely to see in a career of painting.

I've selected problems and solutions to help you do the job right. But you're the final authority in determining if the solutions are correct for you. You have to know your products, surfaces, and application methods to do it right the first time. Most of the knowledge a painter needs comes only from doing and experimenting.

In general, paint fails for the following reasons:

Sunlight— Ultraviolet radiation bleaches colors and attacks the paint-to-surface bonds. The heat generated causes expansion, which leads to cracking and peeling. Figure 12-1 shows an example.

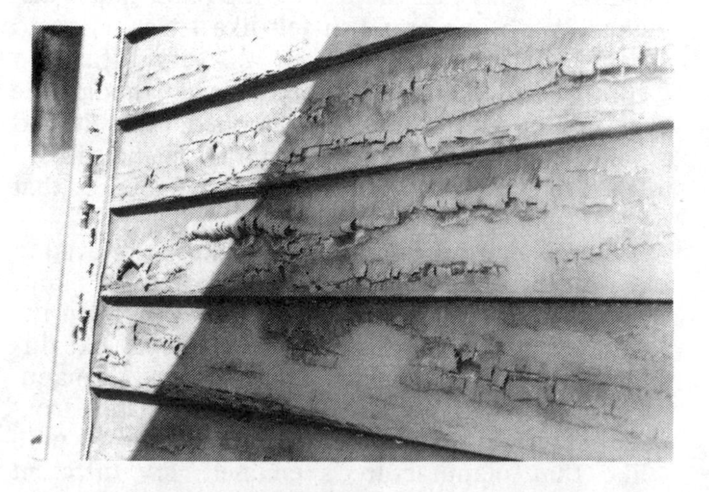

"Curling" paint caused by sun and water
Figure 12-1

Water— Rain brings acids and contamination to the paint bond, loosening the paint. Moisture from condensation, water seepage and leaks destroys the subsurface material and the bond between the subsurface and the paint. It also dissolves minerals, bases and acids and brings them to the paint surface.

Figure 12-2 shows a kitchen window with a condensation problem. The moisture is attacking the paint and the wood itself. It will take vents to solve this problem.

Moisture attacks paint and wood
Figure 12-2

Salts— Salts from the subsurface material and chemical reactions attack the paint-to-subsurface bond and loosen it. And salt used to melt snow and ice can cause the same damage. In Figure 12-3, paint is peeling from the brick for two reasons. The salt used on the sidewalk splashed onto the wall, and people use the wall as a footrest while waiting for the bus that stops here.

Chemicals— Chemicals are everywhere, in the paint, subsurface, air and cleaners. They attack the paint film from within, under, and above to destroy it.

Cold— Contraction of the subsurface and the paint at unequal rates causes cracking and peeling.

Paint build-up— Application of layer after layer of paint will lead to failure between the layers. This results in cracking, peeling and sometimes blistering.

Paint damage caused by salt and abrasion
Figure 12-3

Cleaning— Abrasive cleaners wear the paint surface film, leading to poor coverage and surface attack by contamination. Many of these cleaners contain bleach, perfumes and other chemicals that can damage the paint.

Accidents— Paint is abraded from the surface, causing chipping and scratching.

With all these in mind, we get to the ultimate problem that has caused almost every single paint failure listed: *Poor surface cleaning and preparation.* Take your time and do the job right the first time!

Paint Application Problems

The first group of common painting problems has to do with application methods. You're the one who's applying the paint to the surface and it's up to you to do it right — or clean up after someone else who didn't take the time to do a good job in the first place.

Here's the format I'll use for all of the problems. First I'll give a description of what the condition looks like, then the probable cause, the ways to prevent it, and ways to correct it. You'll find that prevention is almost always easier than the cures. I've arranged the causes in descending order of probability. The preventions and cures are arranged in the order in which you'll do them. I've organized the problems into six sections to make it easier to find a specific problem and its cure. Let's get started with the first section, paint application problems.

Bubbles and Craters

Description
Small bubbles, including some that have burst to form craters, appear on the surface of the painted item. The bubbles are most common on clear coatings such as urethanes. Craters are more likely when you're rolling out latex.

Cause

A) Air is getting trapped in the coating from one of the following:

1) The material is too thick

2) Faulty surface prep of the wood or material

3) Cracks or joints in the material or item

4) Air is entrapped in the brush or roller cover

5) Overagitation of the coating

6) Stirring or shaking a full-gloss type coating

7) Shaking a stir-only type coating

8) Applying coating at elevated temperatures

B) Trying to fast-dry a coating using external heat can cause it.

Prevention

A) Thin the coating material so it flows easier and trapped air can escape before skinning forms. And don't forget this trick: When thinning paint, always pour off about a pint per gallon of the paint into another container before you start thinning. Then, if you overthin the paint, you can put some of this extra paint back until you reach the consistency you're looking for.

B) Thin the coating to the consistency of the wash coat for a first coat on porous surfaces. This helps drive out trapped air.

C) To remove excess paint from the brush, allow the paint to drip off the brush, or dab it on the lip of the can. Don't drag the brush along the lip of the can. That usually takes too much paint off the brush, and causes paint drips on the outside of the can.

D) Don't overmix or machine-mix coatings that are marked "stir only." Don't machine-mix latex paints for over three minutes. If it happens, let the paint sit several hours so air can escape.

E) "Tic-tac-toe" the surface. First brush with the grain, then across the grain, and tip-off with the grain.

Cure

A) During application and before setup, burst any bubbles with the tip of the brush, a pin or a toothpick.

B) After partial setup, try a wash coat. The solvent may dissolve the surface film, leveling the bubbles.

C) After setup, allow the surface to dry for 24 to 48 hours, spot sand, feather, and touch up or recoat.

Poor Coverage

Description

The subsurface color shows through the newly painted surface, giving an uneven blotchy look or a slight discoloration.

Cause

A) The most common cause is overstretching the paint, trying to cover 500 square feet with a 400 square foot can. The pigment is simply too thin to provide proper coverage.

B) Overthinning the paint has the same effect.

C) Using poor quality brushes or rollers that can't lay down a good coating will leave a splotchy wall.

D) If you don't mix the paint properly, some of the pigment stays in the can, not on the wall.

E) Poor quality paint that has low pigment content can cause it, even if it's applied correctly. Figure 12-4 shows a building stained with an inexpensive stain that was low in pigment. An otherwise nice job becomes an eyesore.

F) Don't use pure white or tint base without adding colorant. The coverage may be correct. But remember this: White has to reflect all colors of light. White pigment spread too thin will allow subsurface colors to show through. Adding a little color takes care of the problem.

G) Even bright color paints can be stretched too thin. Bright colors require far less pigment to get the desired color. But pigment stretched too thin is still pigment stretched too thin.

Stained with stain too low in pigment
Figure 12-4

H) Semi-gloss or full gloss paints may not cover well because less pigment is used in the manufacturing process.

I) A poorly cleaned and primed subsurface may allow dirt, oils, grease and stains to bleed through.

Prevention
A) Use a good grade of paint and equipment.

B) Use correct amounts of paint and thinners for the job.

C) Plan on more than one coat for bright colors, whites, semi-glosses, or full gloss paints.

D) Use a suitable primer, then add a flat coat of paint to which you've added the final color. This flat coat provides the hiding power for the semi-gloss, full gloss or bright colored final top coat.

Cure
A) The normal cure for poor coverage is to recoat with the correct application of good quality paint.

B) If the problem is one of dirt or stains, these must be removed. This will usually require that the surface be completely stripped of all new paint, properly prepared, and repainted. You may be able to cover these areas with another coat. But somewhere down the line other problems will probably surface, including bleedthrough, cracks, blisters, or peeling.

Ropes, Pulls, Lap Marks or Roller Stipple

Description
The paint dries with obvious indentations from the brush or roller. The surface doesn't level to a smooth, mark-free finish.

Cause
A) Usually this is due to poor painting technique: too little material, excessive rolling or brushing, or poor quality brush or roller.

B) Painting in the hot sun or on a hot surface can

cause the paint to dry before it has a chance to level.

C) At temperatures below 50 degrees F, paint, especially enamel, can become too thick to perform well.

D) The wrong primer or sealer can cause the problem. Paint is being sucked into the surface, leaving an unbalanced surface film.

E) Using the wrong thinner or too much thinner can result in the same thing.

F) Excessive drafts or air movement can cause the solvents to evaporate too quickly.

G) Allowing insufficient drying time between coats can result in the new coat dissolving the prior coat.

Prevention

A) Use good quality paint and equipment.

B) Prime or seal the surface.

C) Paint at more moderate temperatures.

D) Apply a full thickness of material at the correct coverage rate.

E) Use good painting procedures.

Cure

A) If paint is still wet, *stop*, remove what you've painted, find the cause, correct and continue.

B) If paint has set up, allow to dry, sand smooth, and repaint.

C) If the paint has dried and there's too much of it, strip to a bare surface and recoat as though it were new work.

Voids, Skips, Holidays

Description

A newly painted surface shows voids in the paint, perhaps in only one or two places, or scattered over the entire surface. The voids are usually small areas 1 or 2 inches in diameter.

Cause

A) Painting in a poorly lighted room where you can't see what you're doing.

B) The roller passed over low spots on the wall surface.

C) A worn or poor-quality roller didn't carry enough paint to cover the surface.

D) If the nap of the roller is too short, it won't carry enough paint.

E) The surface wasn't properly cleaned, deglossed or degreased.

F) Poor quality paint or poorly mixed paint may be the culprit.

Prevention

A) Work in a well-lighted area.

B) Use proper rolling or painting techniques.

C) Use a roller with a nap at least 1/2-inch long.

D) Clean and degloss or degrease the surface before painting.

E) Use good quality paint and mix it properly.

Cure

A) Working in a well-lighted area, inspect the surface before paint has set. It's easy to touch up any voids while the paint is wet.

B) After paint has dried, but within a few hours of application, touch up voids using a semidry brush or roller. Feather wet paint into the surrounding area.

C) After paint has dried more than a few hours, wait 24 to 48 hours, clean the affected area, sand lightly, and apply touch-up paint.

D) After paint has dried several weeks or months, clean, degloss or degrease the entire surface and repaint.

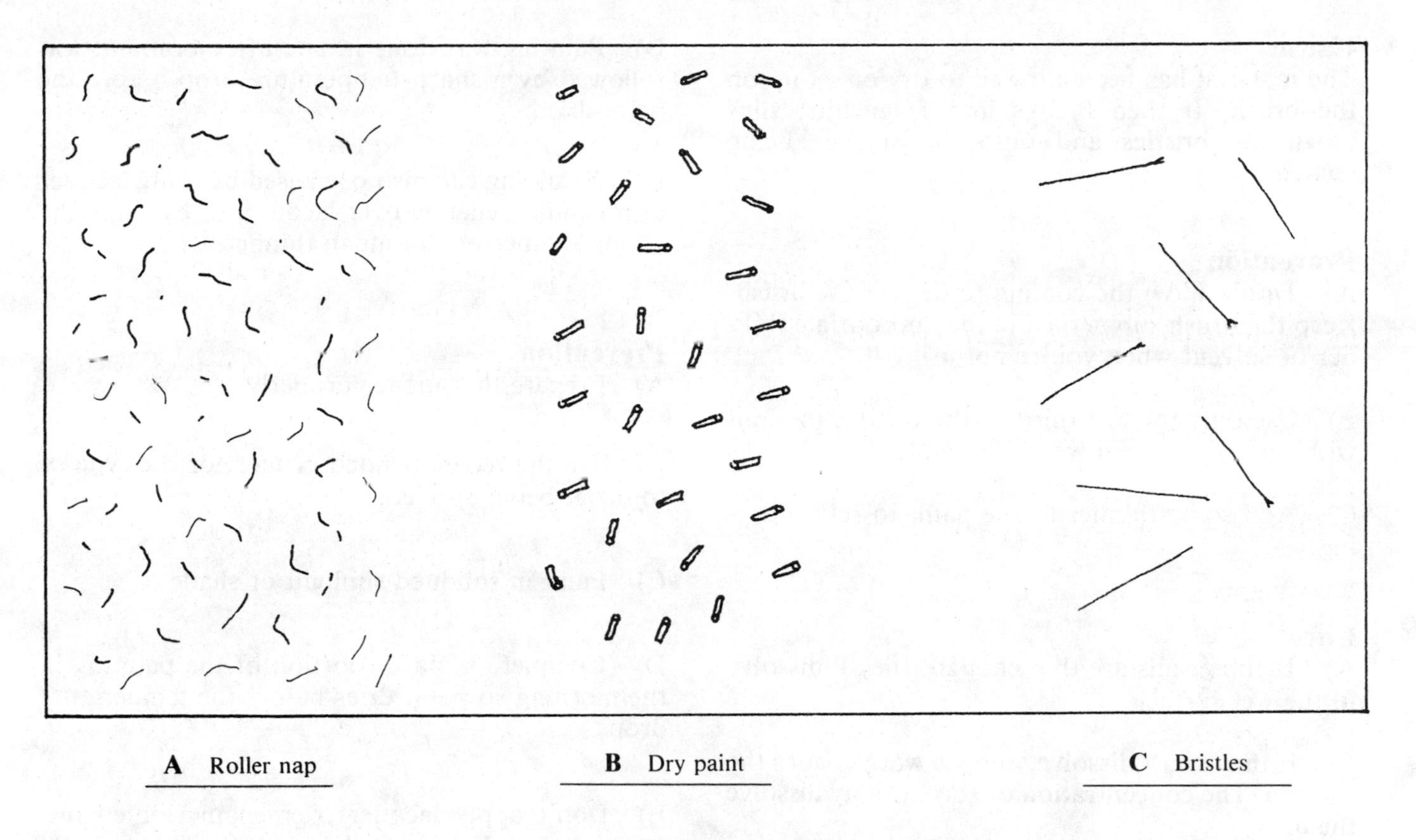

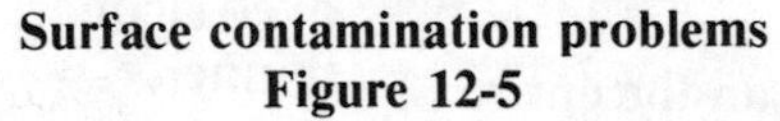
Surface contamination problems
Figure 12-5

Small Threads or Lint on Newly Painted Surface

Description
An otherwise decent paint job starts to develop small threadlike imperfections, which worsen as you continue painting. Figure 12-5 shows the three possible kinds of surface contamination from rollers or brushes.

Cause
A) A poor quality roller is being used. The mat or nap isn't properly bonded to the core material.

B) The applicator or roller isn't the correct type for the paint being used. The solvent or vehicle in the paint is dissolving the glue binder in the applicator, releasing the fibers of the mat or nap.

Prevention
Use the proper type and grade of applicator for the paint being used. If the label says to use it with water-based paints only, believe it. The paint accessory field is highly competitive. The manufacturers generally use the lowest cost material that's suitable for the job being done. Don't expect your brush or roller to do a job it wasn't designed for.

Cure
A) If you see the problem before the paint has started to dry, remove the threads with a small brush, needle or toothpick. Continue with a new applicator of the proper type.

B) If the problem is severe, wash the entire area with lint-free rags and appropriate thinner, allow to dry, and recoat using a proper applicator.

C) If paint has set up, allow it to dry hard. Sand or strip the affected area and recoat.

Tube-Shaped Particles on Paint Surface

Description
Small tube-shaped particles begin to show up in the surface as the paint is applied. It's generally most noticeable when using clear urethane or clear brushing lacquer. The middle section of Figure 12-5 shows the dry paint "tubes."

Cause

The material has been allowed to dry or set up on the brush. It then breaks into fragments, slips down the bristles and onto the surface being coated.

Prevention

A) Don't allow the coating to dry on the brush. Keep the brush submerged in the appropriate thinner or solvent when you're not using it.

B) Use only the last third of the bristles to paint with, and keep them wet with paint.

C) Add some thinner to the paint to retard drying.

Cure

A) If tube walls are thin enough, they'll dissolve in the wet surface.

B) If they don't dissolve, apply a wash coat to the surface. The concentration of solvent may dissolve them.

C) If neither A nor B works, then clean the entire surface down to the bare material or the last good coating. Start over with a new brush.

D) If the tubes have dried *in* the surface, try sanding. If that's not satisfactory, the surface will have to be stripped and recoated.

Wrinkling

Description

The surface takes on wrinkles or cracks within a few minutes to hours after being painted. The subsurface had a prior coat of paint or primer. This condition is clearly visible in the bottom board in Figure 12-6.

Cause

A) The surface wasn't properly prepared, wasn't deglossed, or wasn't primed.

B) The wrong primer was used, or the primer wasn't thoroughly dry.

C) Painting was done in full sunlight and the surface dried before the full film thickness dried.

D) Painting was done in late afternoon and was followed by a sharp temperature drop before the paint dried.

E) Wrinkling can also be caused by using lacquer over enamel, enamel over lacquer, or by using the wrong thinner or too much thinner.

Prevention

A) Prepare the surface properly.

B) Use the recommended primer for the type of top coat being applied.

C) Paint in subdued sunlight or shade.

D) Complete a major portion of the painting in the morning so paint dries before the temperature drops.

E) Don't apply lacquer over enamel or enamel over lacquer. Use the right amount of the correct thinner.

Cure

Remove all new paint and prep the subsurface properly. Repaint in the morning hours and out of hot sunlight.

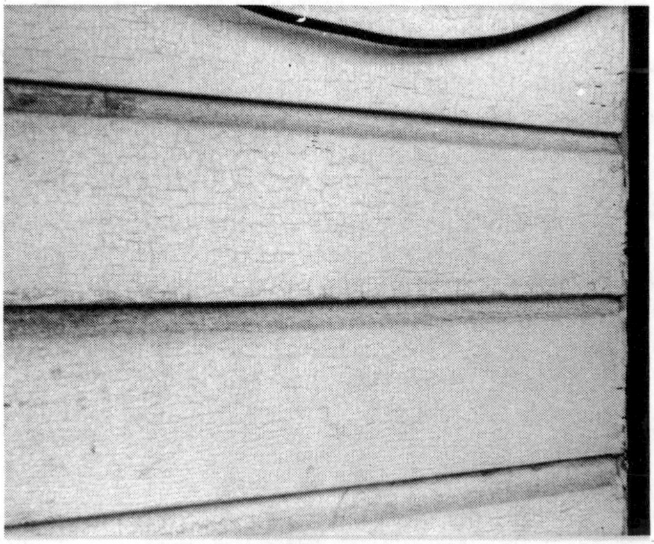

Wrinkling and alligatoring
Figure 12-6

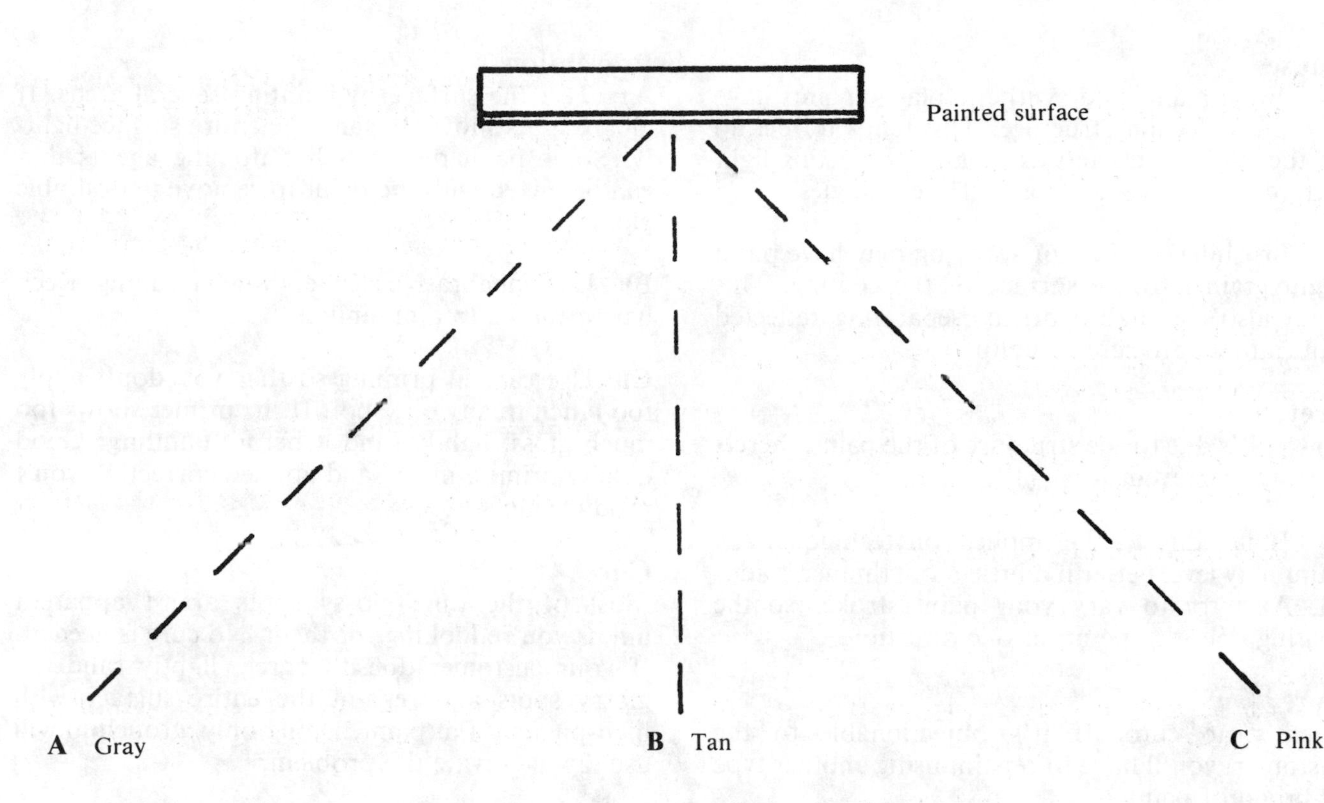

Different colors seen from different viewing angles
Figure 12-7

Color Change Problems

This category of paint problems includes color changes due to a wide variety of causes, ranging from irregularities in the molecular structure of the paint to leaks in the roof. Let's look at them one at a time, beginning with the paint that changes colors as it dries.

Wet Paint Changes Color During Drying

Description
Paint goes on one color and starts to change to another color.

Cause
It's normal for some latex paints and some stains to change color as they dry. I've seen paint go from brown to dark brown to green and then finally to the intended color, light brown.

Prevention
There's no way to prevent it. To make sure the final color will be right, test a section before doing the entire surface. Wait a few days and check the color against the color chip.

Cure
There's no cure if color change is a normal part of the drying process. But if it dries to a color that isn't what the chip shows, call the supplier for an adjustment. The manufacturer may or may not accept responsibility for paint that doesn't match a color chip. Most manufacturer's don't guarantee a good match. Your supplier may be more eager to resolve the problem if you're a good, regular customer. If the problem has happened more than once, then it's time to change brands.

Color of Entire Surface Changes with Viewing Angle

Description
The paint color appears to change color when viewed from different angles. You may see a rainbow effect as you view it from left to right or right to left. This effect is most common with exterior-type latex paints. Figure 12-7 shows how one wall can look three different colors, depending on the viewing angle.

Cause

A) Vinyl paints and synthetic plastic paints have a chain molecular structure. This tends to line up on the surface and acts as a prism, reflecting light differently when seen from different angles.

B) Brushing, rolling, or spraying may have put a slight grain into the surface of the coating. This grain also acts like a prism, separating reflected light into a spectrum of color.

Prevention

A) If it's due to the structure of the paint, there's no way to prevent it.

B) If it's due to the application technique, the paint may level better if a little more thinner is added. Also try to vary your paint strokes so the coating doesn't line up in one direction.

Cure

There's no cure. If it's objectionable to the customer, you'll have to repaint using another type or brand of paint.

Gloss Spots on Surface When Painted with Flat Paint

Description

The entire surface may be glossy when viewed at an angle or there may be glossy spots, usually 1 to 3 inches in diameter.

Cause

A) Surface gloss is normal at certain viewing angles with some types of paint. Gloss will be greater if you're coating a very hard, smooth surface. You'll see the reflection of subsurface as much or more than the gloss of the paint itself. Many texture coatings will look glossy because the raised texture is very flat and hard. This is true of Mediterranean knock-down, with its rather large flattened areas.

B) If the surface is glossy only in spots, it may be caused by abrasion from the cleaning you did. Rubbed areas are harder, thus shinier.

C) Glossy spots may be caused by very hard areas of the subsurface, by overpriming, by using low-pigment paint, or by improper mixing of the binder, vehicle and pigment.

Prevention

A) Test the surface by painting several areas. If glossy spots show up, sand the entire surface lightly. Some paint dealers sell flattening agents that can be mixed with the paint to remove undesirable gloss.

B) Use nonabrasive cleaners when cleaning. Keep hard rubbing to a minimum.

C) Use care in priming so that you don't apply too much to any one spot. If the primer shows too much gloss, lightly sand it before painting. Good quality primer, mixed and applied correctly, won't produce glossy spots.

Cure

Most of the time glossy spots aren't apparent unless you're looking for them. No cure is needed. If your customer doesn't agree, lightly sand the glossy spots and repaint the entire surface with high-pigment flat paint. Doing only a touchup will usually aggravate the problem.

Fading on Cement or Plaster

Description

Paint appears to have faded or turned whitish on the entire surface or in isolated spots, usually one to two years after last being painted.

Cause

A) Fading is caused by ultraviolet light from sunlight.

B) Photosensitive color pigment in the paint can cause fading.

C) Lime burn from excessive alkali in the subsurface is a possible cause.

D) Moisture coming through the surface can also cause fading.

E) On an improperly primed surface, the vehicle may be absorbed into the surface, leaving unbonded pigment on the surface (chalking).

Prevention

A) Use proper exterior paints, stains and pigments.

B) Seal or prime with the correct primer.

Chemicals leaching from brick cause stains
Figure 12-8

C) Acid wash the surface to neutralize alkali.

D) Use only nonchalking paints for colorant additions.

Cure

A) First, try wiping the surface with a wet rag. If the color comes back to normal, the problem isn't fading but chalking. Follow the procedure I've recommended for chalking.

B) If the problem is actually fading, remove the paint from the affected area, apply the proper prevention technique, then recoat with the right paint and colorant.

Green Stain on Gray Brick

Description

Buff or gray brick or tile takes on a green or greenish tint.

Cause

Molybdenum or vanadium compounds are leaching from clay in the brick or tile. This condition is shown in Figure 12-8.

Prevention

There's no way to prevent it, other than keeping moisture or water from entering the surface

Cure

Dampen the surface with clear water, then wash with a solution of one cup of sodium hydroxide crystals or lye to ten cups of water. Rinse with plenty of clear water when the green color has disappeared. Don't wash with an acid solution. The stain will turn brown and set permanently.

Black or Dark Gray Wood Siding or Shingles

Description

Siding or shingles that are uncoated or coated with Clear Wood Finish (CWF) have turned black or

Wood siding turns dark in protected areas
Figure 12-9

dark gray. It's most apparent on areas that are somewhat protected from rain, such as under eaves or window sills. Figure 12-9 shows the condition.

Cause

A) This is a natural effect of weather and sunlight on a surface that hasn't been coated or has lost its coating.

B) Mold or mildew can cause it.

C) Leaching of the resins, oils, and water soluble chemicals in the wood can cause it.

Prevention

A) Allow the new wood to weather, opening the surface and leaching out the oils, resins, and chemicals.

B) Then try water blasting, drying, and a coat of a CWF-type product. CWF or *Clear Wood Finish* is a trade name for a product sold by The Flood Company. It's available in most paint stores and hardware stores selling paint.

Cure

A) Water blast or scrub the surface, then recoat with a CWF-type product.

B) If stains are very dark, try a bleaching solution followed with a neutralizing rinse, a clear water rinse, drying, and a recoat with a CWF-type product.

Premature or Spotty Graying of Clear-Coated Wood

Description

The entire surface or large areas of exterior siding or wood have turned light to medium gray. A CWF-type product had been used for prior coating.

Cause

The CWF-type product was applied in too thin a coating, or was applied improperly with an airless spray system. It can also occur if the wood still had mill finish or wasn't properly weathered at the time of application.

Prevention

A) CWF-type products should be brushed or rolled on to the surface until it can absorb no more. If possible, soaking works best. Board stock or shingles should be soaked for twenty minutes, then racked for drying.

B) Allow the surface time to weather to an even light tone before application of CWF-type product.

C) When using an airless spray system, be sure to wet the entire surface evenly.

Cure

A) Wash and dry the surface, then reapply a CWF-type product until the surface can no longer absorb more of the coating.

B) If the surface is very glossy and doesn't accept the CWF-type product, the mill glaze still hasn't weathered sufficiently. Either sand these areas lightly before applying the coating or wait until the surface has weathered before trying to apply a CWF-type product.

Whitish or Brownish Stains on Exterior Walls

Description

You'll see whitish or brownish stains on the lower portions of exterior walls. It looks like ultraviolet damage, fading or bleeding, but the upper wall areas aren't affected. Notice the light stain at the bottom of the wall in the house shown in Figure 12-10.

Water stains on exterior walls
Figure 12-10

Cause
This is caused by water, either rain or sprinklers. A brown stain is usually dirt splash. A light stain is from dissolved mineral salts in the water.

Prevention
A) Don't water the sides of buildings.

B) Install proper roof overhangs and gutters.

Cure
A) Normally all that's required is a good washing with TSP solution.

B) If the condition is allowed to continue for any length of time, the surface coating will become pigmented and change color permanently. Remember, many paint pigments are manufactured from ores and minerals. Try a bleaching solution after the TSP wash. If the stain isn't removed, strip the paint to the bare surface and finish as new work or apply a sealer coat before top coating. Make sure the source of water is removed or the problem will soon return.

Streaks Below Metal

Description
Dark streaks or stains appear under metal that's in contact with wood — usually under metal flashing, gutters, or spigots.

Cause
Chemicals in the wood are reacting to residue washed off the metal. The result is stained wood. The process is activated by moisture.

Prevention
A) Paint or prime metal flashing before wood siding is installed. Use aluminum or brass nails and fasteners in cedar and redwood.

B) Caulk seams with an acrylic or butyl rubber caulk. This helps keep water away from the wood surface and prevents activating the chemicals in wood.

Cure

A) Brush or power wash (water blast) the surface. If stains remain, rewash with a bleaching solution. Instructions for removing rust stains are in Chapter 11.

B) After washing, caulk all seams with butyl rubber or acrylic caulk.

C) After washing and caulking, you may want to recoat with a CWF-type product or the coating presently on the surface.

Yellowed White

Description

The entire surface that has been painted white begins to yellow within a few months of application.

Cause

A) This is probably the result of using the wrong paint, such as using an indoor paint on an exterior surface.

B) Moisture is leaching out chemicals from prior coats or the surface. This is most common when using latex paint.

C) There's no top coat, only white primer was used.

Prevention

A) Use the proper type and quality of paint for the application.

B) Find and eliminate the source of moisture before painting.

C) Remove old paint coatings before new paint is applied.

D) Never leave a primer without a top coat.

Cure

A) Reprime and top coat, using proper primer and paint for the type of application.

B) Find and eliminate the cause of moisture, then reprime and paint.

C) Strip to the bare surface and start over.

D) If there's primer only and it's less than 90 days old, a top coat should be enough to do the job. But first give the primer a good detergent cleaning, rinse with clear water and allow to dry.

Yellow Spots on a Ceiling

Description

The ceiling and upper wall area have 1/4 to 1/2-inch circular yellow spots, which may have formed waxy or hard droplets. These are usually found in laundry, bath, and kitchen areas.

Cause

Soap and grease spots have formed on the surface. Hot water vapor from washing has carried the soap's oils and fats to the ceiling. As the steam condenses on the cooler surface, it forms droplets which leave behind the solid oils when dry.

Prevention

Improve ventilation. Install a vent fan that will remove water vapor whenever hot water is being used for washing.

Cure

Wash the area with a TSP solution, then rinse well with clear water. Painting may not be required. If painting is needed, follow the normal procedures.

Yellow or Brown Bands on Acoustic Ceiling or Drywall

Description

Yellowish or grayish or brownish bands appear on acoustic ceilings or walls shortly after new construction. The bands will usually be at 4-foot intervals.

Cause

A) Moisture is seeping through the drywall joints because the seams weren't properly sealed with tape and joint compound. This is most likely in ceilings.

B) The joint compound wasn't given sufficient time to fully dry before being painted.

Prevention

A) Drywall seams in the ceiling should have the recommended three coats of joint compound, and be properly primed.

B) Each layer of joint compound should dry completely before primer or paint is applied.

Cure

A) Wait long enough for all moisture to escape. This will take several weeks because moisture must now pass through the top coating. Then apply a sealer and primer to the affected joints, and re-paint.

B) If you're painting over acoustic ceiling material, use a nonbridging acoustic ceiling paint to preserve the acoustic properties. Spraying is recommended, but a special sponge-type acoustic roller is available at professional paint stores. The roller cover has slots cut every 3/8''. These trap and hold any loose acoustic particles.

Water Rings or Stains on Acoustic Ceilings

Description

Medium to large irregular gray or brown spots are visible on an acoustic ceiling. Figure 12-11 shows a typical example.

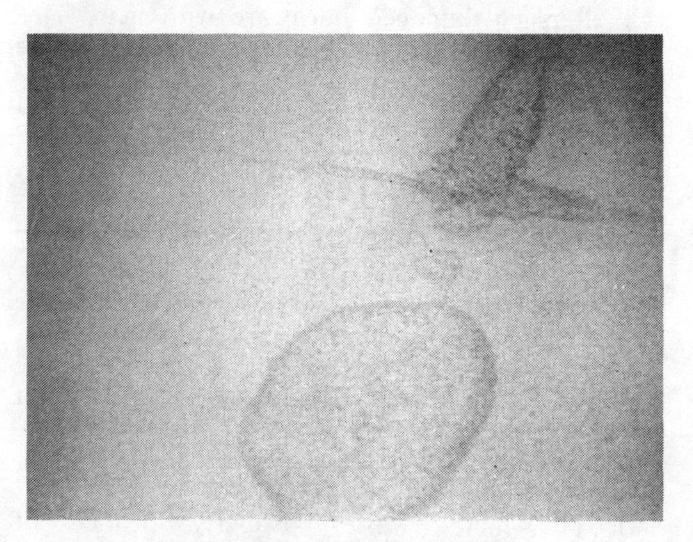

Water stains on acoustic ceiling
Figure 12-11

Cause

Water seeps in behind the ceiling, leaching out chemicals used to bind the acoustic particles together. It's probably caused by a leak somewhere in the roof or around pipes that extend through the roof. A less likely cause is heavy condensation around the pipe due to poor venting of interior moisture.

Prevention

A) Find and fix the roof leak.

B) Don't use acoustic tile in rooms where moisture is present: kitchen, bath, laundry.

C) Install and use a vent fan.

D) Don't apply acoustic material to bare wallboard. Use an oil-base primer first. This acts as a vapor barrier.

Cure

A) Locate and correct the cause of the water damage. Clean the area with commercial acoustic ceiling cleaner or the mixture described in Chapter 11.

B) If the problem still remains, let the room dry out for 24 hours. Then paint the affected area with titantiated shellac or thinned oil base sealer. Titantiated shellac is shellac with titanic oxide added as a white pigment. Don't confuse it with ''white shellac,'' a bleached shellac.

I recommend a spray application: aerosol is acceptable. After the sealer has dried, recoat with nonbridging acoustic ceiling paint.

Paint Peeling on Metal

The third category of paint problems includes both interior and exterior metals. First we'll cover the interior problems, starting with door knobs and hardware.

Worn or Peeling Coating on Door Knobs or Hardware

Description

The protective coating on door knobs, draw pulls, and other household hardware has worn off, exposing bare metal.

Paint peeling from exterior metal
Figure 12-12

Cause
It may be caused by general abrasion from usage, from abrasive cleaners used for cleaning, or by chemicals in the cleaners or on hands.

Prevention
There's no way to prevent normal wear. But you can recommend that the owners avoid using cleaners with abrasives or harsh chemicals.

Cure
A) Test for lacquer. If the coating is lacquer, remove the coating with lacquer thinner. If it's not lacquer, it's probably a urethane or enamel. Remove the remainder of the coating with paint stripper. Do not sand.

B) If the metal is brass, clean it with a commercial brass cleaner to bring back the shine. Rinse in clear water, towel dry, and top coat with spray lacquer or urethane. If it's aluminum, chrome, or stainless steel, clean to a shine with a mild abrasive cleaner that contains no wax or silicone additives. A paste of water and baking soda will sometimes work. Rinse well in clear water, towel dry, and coat with spray lacquer or urethane.

C) Replace damaged hardware with new. This may be the fastest, easiest and least expensive remedy.

Paint Peeling from Exterior Metal

Description
The paint forms large flakes and peels from the surface. It's most common on galvanized gutters, downspouts or flashings, usually from one to five years after application. Figure 12-12 shows paint peeling from exterior metal.

Cause
A) The metal wasn't properly cleaned of manufacturing oils, waxes, greases, or solvents. Fingerprints from handling can also be suspect.

B) Shiny metal surfaces, like chrome, provide poor adhesion for most coatings.

C) Metal was wet when painted.

D) It wasn't primed, or the wrong type of primer was used.

E) The wrong type paint was used.

Prevention
A) Clean the metal with the recommended cleaner before application of primers and top coats.

B) Use a suitable primer for the metal being coated. This insures a good bond between the surface, primer and top coat.

C) Use paint that's recommended for the application.

Cure
A) Scrape, chip, or otherwise strip the remaining paint from the surface. I recommend stripping the entire surface, since all of the old coating will eventually peel off if the priming was done wrong.

B) Prime with a suitable primer. Let it dry for 8 to 24 hours.

C) Recoat, using the correct type of paint.

Plaster, Wallboard and Cement Problems

The fourth category, and one of the most common, includes deteriorating plaster, uncured plaster, peeling on cement, drywall seams, corner chipping and nail popping. Some are simple to fix. Others take some detective work to find the cause. Once the cause is found, make sure it's eliminated before you repaint. Otherwise you'll probably find yourself doing the job over again in a few months.

Paint or Plaster Lifts When Brushed or Rolled

Description
During painting, sections of the plaster or drywall crumble or otherwise lift off the surface. The condition is common in older buildings and in areas subject to moisture, such as exterior walls. The damaged section is usually from 1 to 3 feet in diameter.

Cause
A) The most likely cause is severe water damage. We'll cover water damage in detail in the next chapter.

B) Condensation can also cause water damage.

C) Water-repellant plaster or drywall wasn't used in an area where moisture was expected, such as laundry, bath or behind sinks.

D) Wall has no vapor barrier or an improperly installed vapor barrier.

E) No sealer or the wrong sealer was used on surfaces subject to moisture.

Prevention
A) Use moisture-resistant drywall (greenboard) in potentially wet areas.

B) Use a vapor barrier on all exterior walls. The barrier should be on the inside of the wall, next to the living area.

C) Use an oil-based paint in normally wet areas.

D) Keep water off the exterior wall by repairing the roof at the overhang or installing rain gutters.

Cure
Remove the affected area of drywall or plaster and inspect for the cause of the damage. Correct the problem, then replace the wall material and finish as new. Don't bother to repair and refinish the wall surface until the source of water is found and corrected. There's no sense doing the job twice.

Alligatoring on Plaster Surfaces

Description
Paint on new plaster has irregular horizontal and vertical cracks that resemble the skin of an alligator. The top boards in Figure 12-6 show alligatoring.

Cause
The plaster has shrunk excessively because it was painted before being fully cured. Possibly im-

proper ingredients were used or the plaster wasn't mixed correctly. The moisture content of the plaster may be too low because the new paint blocked entry of external moisture. Maybe the moisture evaporated through the paint, causing the plaster to dry too soon. The paint film is cracking because the plaster below is cracking.

Prevention
Mix and cure the plaster properly before painting. Allow enough drying time and check for cracking. If there is any cracking, repair with a nonshrinking plaster or spackle before painting.

Cure
A) Either strip the surface to bare plaster and repair the cracks by rubbing with nonshrinking plaster or spackle, or score the entire surface and replaster. Cure properly, then apply oil-base primer and top coat.

B) If the problem is excessive, consider applying some other siding material over the plaster. See Chapter 19.

Paint Peels from Plaster (New Construction)

Description
Paint begins to peel or flake shortly after being applied to a newly plastered wall surface.

Cause
The plaster had a high moisture content when it was painted. The moisture is separating the paint from the surface and may be leaching lime and calcium salts to the surface. Lime is powdery and very alkaline; it prevents adhesion and attacks chemicals in the paint.

Prevention
Allow enough time for the plaster to dry and fully cure before painting.

Cure
Strip off the paint. Give the plaster time to dry and fully cure. Use a mild acid rinse to remove salts and powders, if necessary. The recipe is in Chapter 11. When fully dry, prime with an oil-base primer, dry and top coat.

Paint Peels from Plaster (Old Construction)

Description
Old paint is peeling or fairly new paint over old paint starts to peel or flake from wall surfaces.

Causes
A) One possible cause is water damage to the plaster.

B) Perhaps the wall had been previously coated with sizing that wasn't removed prior to repainting. This happens when wallpaper is removed and the wall isn't properly cleaned.

C) If the wall had been previously coated with a calcimine solution, moisture acting on the calcimine can leave a chalky surface that the paint can't adhere to.

Prevention
A) Find and eliminate the source of water. Repair the damaged area before attempting to repaint.

B) Remove all the old sizing or calcimine. Then seal old plaster with oil-base primer before repainting.

Cure
A) Remove all paint down to bare plaster by sanding or using a chemical stripper. Then prime the surface with an oil-base primer. When fully dry, apply a top coat.

B) Replace the plaster with wallboard or some other material. See the suggestions in Chapter 19.

Flaking or Powdering on Cement or Block

Description
On cement or cement block that has recently been painted, the paint begins to flake, powder, and discolor. It usually happens within a year after a new paint job.

Cause
A) Interior paint was used on an exterior surface.

B) Prior to the last painting, the surface was cleaned with acid, probably muriatic or hydrochloric, and rinsed with water only.

Prevention

A) Use the proper grade and type of paint for the application.

B) After cleaning with acids, a neutralizing rinse *must* be applied. The acid will remain in the surface pores and remain active for months, attacking the new paint from within. See Chapter 11 for neutralizing rinses.

Cure

Sandblast the entire surface down to bare cement. Then prime and topcoat with the proper paint for application. If you know or suspect that acid was used, apply a neutralizing rinse, followed by a clear water rinse. Let the surface dry for 24 to 48 hours before priming and painting with the correct materials.

Large Dull Spots on Newly Painted Plaster or Drywall

Description

Dull spots, usually 1 to 2 feet in diameter, appear on the surface of newly-painted plaster or drywall after the paint dries. The spots might appear chalky.

Cause

A) The plaster or drywall wasn't properly sealed. The paint vehicle sank into the wall, leaving only pigment on the surface. Pigment alone tends to dry dull and powdery.

B) The paint had too much thinner.

C) A brush or roller saturated with thinner or solvent was used for painting.

D) The wall was painted with improperly-mixed paint. The vehicle and pigments separated.

Prevention

A) Prime the area with a good grade of sealer and primer or polyvinyl acetate (PVA). These are available at professional paint stores.

B) Follow the paint manufacturer's recommendations on thinning.

C) After cleaning brushes or rollers, remove excess cleaner by squeezing or by rolling out on a piece of clean cardboard.

D) Mix the paint thoroughly before applying.

E) As you're applying the paint, check back to see if the paint is drying evenly. Immediately recoat any spots that are drying too fast.

Cures

A) Repaint using a full-strength unthinned top coat.

B) Allow to dry, spot prime, and then repaint the entire surface.

Corner Chipping

Description

Paint chips off the corners of interior walls, generally at waist to chest height, exposing the bare metal bead below. Look at Figure 12-13.

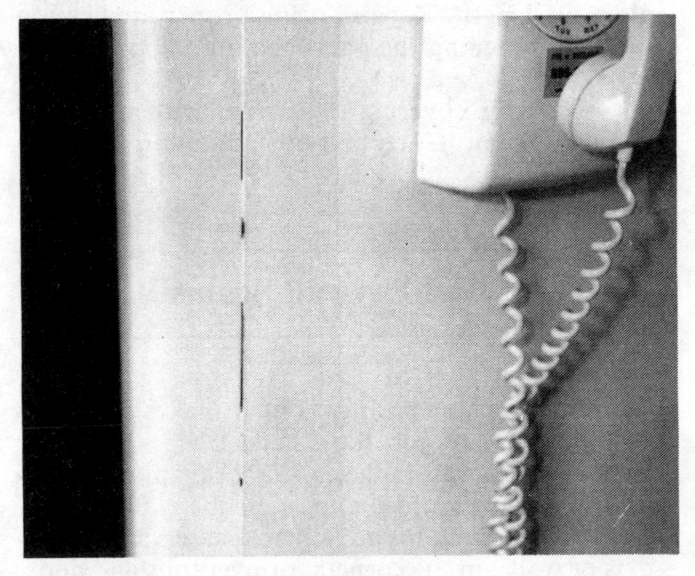

Corner chipping
Figure 12-13

Causes

A) Physical damage is caused by people or objects bumping into or rubbing against the corner.

B) Too thick or too many coats of paint were applied to the metal corner bead.

C) Poor brushing or roller techniques were used in painting corners, allowing paint to build up there.

D) Drywall joint compound was applied over the corner bead.

Prevention
A) Apply a thin or near-dry paint coating to corner beads.

B) When brushing or rolling paint to a corner, don't allow paint to build up. Use a semidry brush or roller to remove excess paint.

C) Use the proper technique when spackling on corner metal protectors.

D) Clean the corner bead by lightly sanding before painting.

Cure
A) Inform occupants that the damage was done by people or objects bumping into the corner. It will happen again if they're not more careful.

B) Sand the corner bead down to bare metal along its length. Repaint using proper application methods, keeping the paint coating thin.

C) Consider covering the corner with plastic or wood molding if one or both adjoining walls are paneled.

Bad Drywall Seams

Description
This isn't a paint problem but it makes an otherwise good paint job look bad. Long vertical or horizontal bumps, bulges, or cracks, approximately 2 to 4 inches wide, appear in the wall surface. They're generally found at 4-foot intervals on ceilings or walls, in the corners, or over tubs and doorways. See Figure 12-14.

Cause
A) Improper application of joint compound over drywall tape is the most common cause.

B) Butting and taping of untapered edges of one or both pieces of drywall can also cause it.

C) Warped framing lumber will normally result in vertical bumps spaced 16 inches apart.

Improperly taped drywall seam
Figure 12-14

Prevention
A) Proper taping and finishing of drywall seams is the answer. The drywall joint compound should be feathered out 14 to 18 inches from the seam.

B) Use dry, straight framing lumber.

Cures
A) Sand the entire area until smooth or slightly indented. Recoat with joint compound so it extends out 14 to 18 inches on each side of the seam. Then let dry for 24 hours. Sand smooth, prime and apply a top coat.

B) If the problem is severe on an entire wall surface, a new layer of drywall or an alternate covering may be a better solution. See the recommended alternate wall coverings in Chapter 19.

Nail Popping

Description
Small round spots of plaster, approximately 1/2-inch wide, separate or protrude from the wall surface. They may occur alone or there may be several in a vertical row. Nail heads may or may not be showing. It usually occurs weeks to months after drywall application. Figure 12-15 shows a nail pop with the plaster still in place.

Nail popping
Figure 12-15

Causes

A) Framing lumber contracts or expands away from the flat plane of the wallboard surface, forcing nails outward.

B) Flooring or other type of nails were used instead of ring-shank drywall nails. The nails used don't have sufficient grip.

C) Drywall nails were too short for proper grip.

D) Nails were hammered in too hard, breaking the wallboard paper or powdering the wallboard core.

E) The nails used weren't primed or were rusted before application of the joint compound.

F) Joint compound wasn't forced into nail dimples, leaving a hollow area between the nail and the top surface.

Prevention

A) Use only straight, kiln-dried framing lumber.

B) If the carpenters have to use green or S-dry lumber, let it dry before applying drywall. Replace lumber that warps.

C) Use only rust-free, ring-shank drywall nails.

D) Use proper drywall hanging and jointing techniques. Double-nailing the drywall will minimize nail pops.

Cure

Remove the plaster spot if it hasn't already fallen out. Set the nail with a punch. You could also pull the nail out, but that would probably cause more damage. For extra holding power, insert a second nail about 2 inches above the problem nail. Sand both nail heads and spot prime. Respackle and retexture if necessary. Let the repair dry overnight. Spot prime the repair and let it dry for several hours. Then touch up the top coat or repaint the entire wall.

Plastic and Vinyl Problems

This section covers two related subjects. The first is the common problems you'll have when using paints that have a high plastic content. Second, I'll discuss the things that can go wrong when paint is applied over plastics.

Crazing or Stress Cracking

Description

A very fine hairlike network of cracks appears on the painted surface. The cracks are random sizes and shapes and go in random directions. The surface was painted with a high plastic content paint or the paint is on a plastic surface. Figure 12-16 shows what it looks like.

Causes

A) If the problem is in the coating itself, the cause is usually ultraviolet light.

B) On coatings over plastics, the cause is probably incompatibility of the coating and the

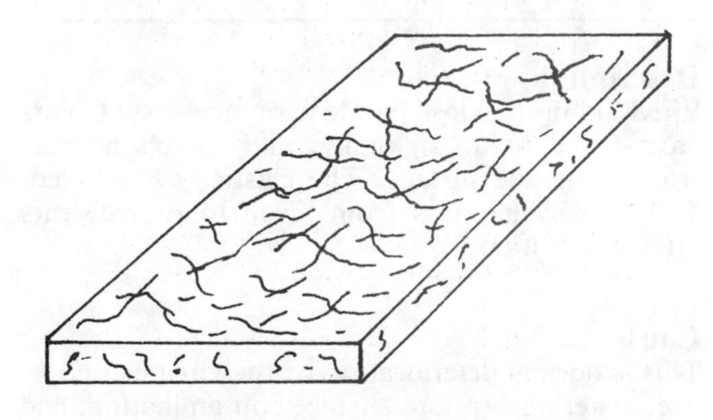

Crazing or stress cracking
Figure 12-16

plastic. The coating contains organic solvents or caustics or acids that have attacked the plastic.

C) It can also be caused by cleaning with a solvent that contains organic solvents or acids, or a cleaner that is highly caustic.

D) Another possible cause is expansion and contraction of the paint film due to temperature changes.

E) On two-part coatings like epoxies and polyesters, the problem can be the result of using a "hot mix" — adding too much catalyst in an attempt to reduce the setting time.

Prevention

A) To exterior coatings with a low pigment content, such as clear coatings, add a U/V inhibitor. Your paint dealer can recommend the inhibitor for your application.

B) Select the correct paint for the plastic being coated.

C) Don't use organic solvents (alcohol, ketone, esters, aromatic or chlorinated solvents) or harsh caustic cleaners (containing lye or high sodium content phosphates or bases) on plastics or high plastic content paints.

Cures

A) If the cracks are small, another coat of paint may fill them in. But the problem is likely to recur.

B) Strip the coating and start over, using the proper paint for the subsurface or a paint that has a U/V inhibitor added.

Vinyl Siding Deterioration

Description

Vinyl siding has lost its gloss or newness. Colors have faded or are splotchy, and there's hairline cracking in the surface. The elasticity has failed. This usually happens from seven to twelve years after installation.

Cause

This is normal deterioration of the vinyl caused by age, water absorption, surface contamination, and U/V radiation from sunlight.

Prevention

A) Either consider vinyl siding a surface coating material that will require periodic replacement, or don't install it in the first place.

B) A nonabrasive, nonchemical paste wax will protect the siding and make it last longer.

Cure

A) Clean the surface with mild soap and water to see if color and gloss are restored.

B) Try cleaning with a commercial vinyl-restoring paste. This will restore the surface for a while.

C) Clean the siding with an aromatic hydrocarbon solvent or xylene. These will attack the surface, removing oils, contaminants, and any remaining gloss. Paint with an acrylic latex paint. A light sanding with a medium grit paper will improve adhesion. Vinyl acetate latex paint will etch the surface and create a good bond. But don't use vinyl acetate latex until you've tried it on a small area to see if the final finish is acceptable.

Here's a word of caution. Painting vinyl siding is a temporary solution. It can add a few years of extra life to the surface, at most. Expect to replace the siding in a few years. It's almost impossible to repaint vinyl siding again because the vinyl will have deteriorated even more by then.

D) Replace the siding with new vinyl siding or an alternate material.

Wood-Related Paint Problems

Finally, we come to the largest category of paint problems — the problems that come up when wood is painted. Let's take a little stroll though diamond drops, past fish eyes, arriving finally at wood rot.

Poor Paint Adhesion to New Wood

Description

The paint isn't spreading very well or covering easily on a new wood surface. It feels like you're painting over wax or grease.

Paint flaking from lumber used for cement forming
Figure 12-17

Cause
Some woods have extremely tight fibers and pores. Others have a high oil or resin content. This is especially true of some imported specialty woods like marine grade plywood from Panama.

Prevention
There isn't any, except to select a different type of wood. Some woods just seem to repell paint.

Cure
A) Scuff sand the surface before continuing to paint.

B) Try a surface wash of lye solution: four ounces of lye dissolved in one gallon of water. Coat the surface of the wood and let it stand for 20 minutes. Then rinse and dry it. Be very careful. This solution is extremely caustic. Wear rubber gloves and eye protection.

Paint Flakes, Leaving White or Gray Powder Under It

Description
The paint flakes off from one to five months after a new paint job. White to gray powder is exposed where the flakes have fallen off. Look at the bottom member of the fence in Figure 12-17.

Cause
The most probable cause is that this lumber had been used for concrete forming before being reused and then painted. It's O.K. to reuse concrete forming lumber if it's thoroughly cleaned of all cement residue. Hosing it off isn't enough. That drives cement into the wood fibers. With age, the cement leaches to the surface and breaks the surface-to-paint bond.

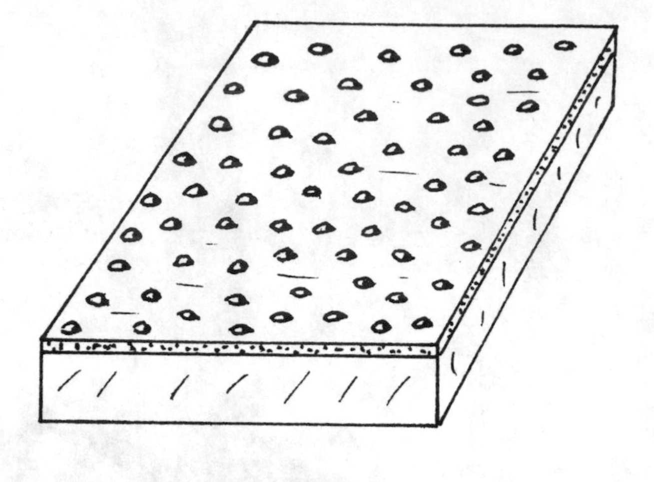

A Chemical blisters

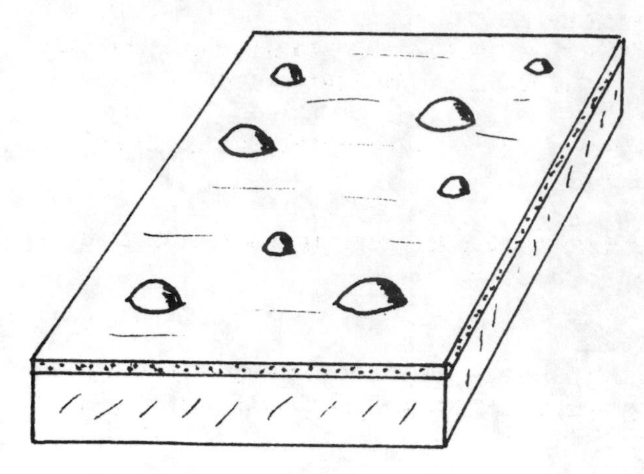

B Heat and water blisters

Blistering on painted surfaces
Figure 12-18

Prevention
Avoid painting any lumber that has been used for concrete forming. If you must use it, sand it first until all traces of cement are removed.

Cure
A) Scrape off all loose paint. Then water blast or sandblast the surface until the surface is in like-new, bare condition. Then prime and paint.

B) In some cases it will be easier to just replace the lumber.

Paint Peeling from Caulking

Description
Paint is cracking or peeling off caulking material used to caulk windows, doors, metal, or metal flashing.

Cause
A nonpaintable caulk was used. Many caulking products contain chemicals that repell paint. Silicon base caulk is one of these. There are, however, some paintable silicon caulks on the market, as well as precolored silicon caulk.

Prevention
Don't paint over silicon base caulk (unless the label claims the product can be painted). In fact, you shouldn't use any caulk that doesn't claim to be paintable. Butyl rubber or acrylic are usually good choices.

Cures
A) Strip off the old paint and caulking. Start over with a paintable caulk, and then repaint.

B) Strip off the old paint and caulk. Prime and paint the surface. Then seal with a silicon or other caulk.

Blisters, Chemical

Description
Fine, small to medium size blisters or bubbles form on all walls of a newly repainted surface. You break a few blisters and don't find any moisture inside. The undersides of the blisters are glossy. This happened two to three days after repainting. See Figure 12-18.

Cause
A) The new paint isn't chemically compatible with the old coating. The reaction between the two formed gas bubbles that created the blisters.

B) The old surface was contaminated by cleaners, smog, or other impurities.

Prevention

A) Use paint that's known to be compatible. See Chapter 11 for the testing procedure.

B) Properly clean and prep the old surface before repainting.

C) Paint a small test section and wait three days before continuing.

Cure

Remove all paint and sand the subsurface. Prime using a zinc-free primer. Then coat with a quality grade of paint.

Alligatoring

Description

The paint film has severe cracks both horizontally and vertically, forming a pattern that resembles the skin of an alligator. The crack pattern is very irregular. Alligatoring can occur on any painted surface within hours or days of painting.

Causes

A) Applying a paint coat that's too heavy.

B) Painting over a soft or non-dry base coat.

C) Painting over a wet surface.

D) Painting in full sunlight. The film surface dried too fast.

E) Poor surface prep, cleaning or priming.

Prevention

A) Prepare and prime the surface properly.

B) Apply the right paint thickness.

C) Don't paint a surface when it's exposed to full, bright sunlight.

D) Allow sufficient drying time between coats.

Cure

Remove all of the new finish. Clean, sand, and prime. Wait 8 to 24 hours. Then repaint in subdued sunlight or shade. Reread the section of this chapter on wrinkling.

Checking

Description

Cracks in the shape of small, regular squares appear throughout the painted surface. They're about 1/4-inch square and fairly regular in shape. You may be able to see old paint of different colors in the cracks. This type of checking occurs in older buildings that have been painted many times. Figure 12-19 shows the difference between checking and alligatoring.

Cause

Paint build-up is the cause. There are just too many layers of paint on the surface. Each layer expands and contracts at a different rate, cracking or checking the surface coat.

Prevention

Remove all old layers of paint down to the wood before repainting. If you suspect there are too many old coats of paint on a surface, scrape it with a knife or scraper and count the layers. If there are more than three layers, strip the surface before painting.

Cure

The cure is the same as the prevention — strip before painting.

Cracking

Description

A paint coating can crack with or across the grain of the wood. It's similar to checking or alligatoring but you can see unpainted wood in the cracks. It's usually larger than checking but smaller than alligatoring. Cracks across the surface are much straighter than those in alligatoring. Figure 12-19 shows the difference in outward appearance. Figure 12-20 shows the actual cracking in the lumber or plywood.

Cause

This isn't really a paint problem but a subsurface or wood problem. The wood has dried out and is splitting along the grain. Cracks across the grain are from expansion of the wood. This problem is most common on plywood sidings and lumber that's not kiln dried. It usually occurs from a few weeks to months after painting new construction.

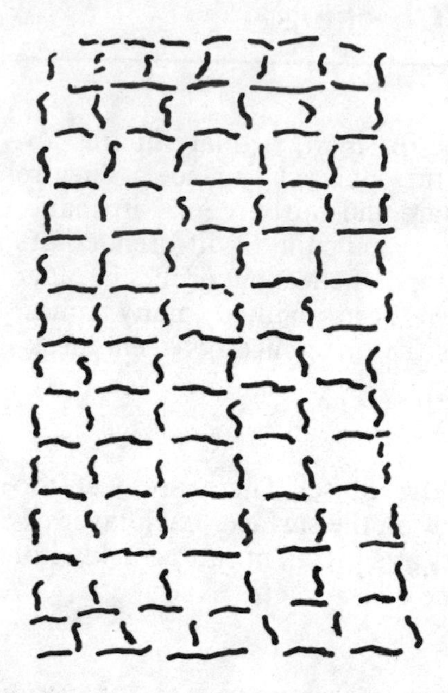

Checking

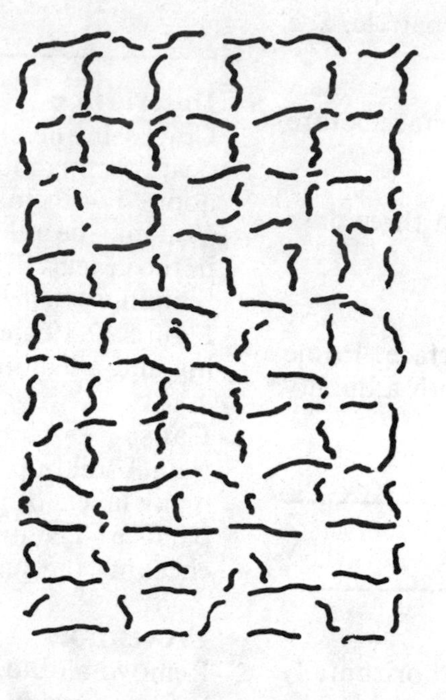

Alligatoring

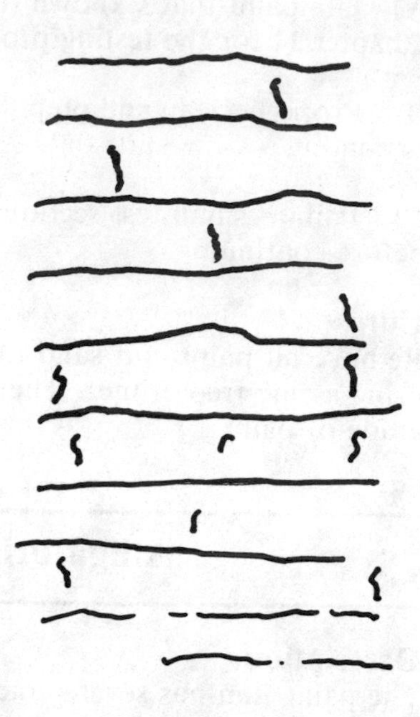

Cracking

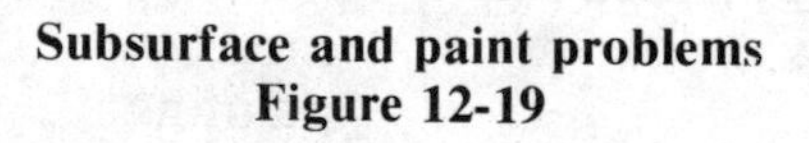

Subsurface and paint problems
Figure 12-19

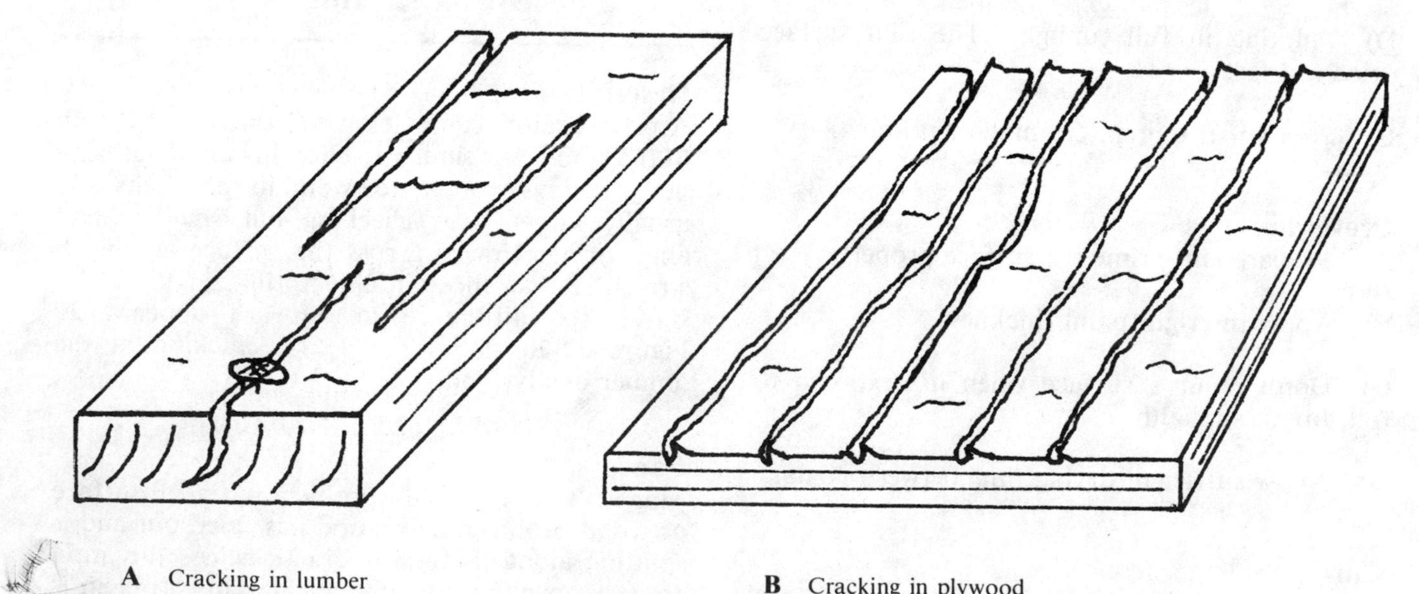

A Cracking in lumber

B Cracking in plywood

Grain cracking
Figure 12-20

Prevention
A) Use kiln-dried lumber. Use nails sized to do the job and no larger. Your building code lists proper nail sizes. Most cracks in timber start where the timber has been nailed. Predrilling of nail holes, or flattening the points of the nails so they cut rather than separate the wood fibers, can help prevent this problem. And don't drive nails less than 3 inches from the edge or end cut of lumber.

B) Allow ample time for the lumber to weather or dry out before painting. This may not be practical in new construction, but it's the best way to prevent cracking.

C) Use light colored paints. The cracks will still be there but won't be as obvious.

Cure
A) In plywood, sand off any cracks where the top ply has curled outward. Apply a filler to the cracks. In lumber such as 2 x 4's, fill the cracks with a glue that's not water based. Then either nail or clamp, or do both, until the glue has dried. After the cracks are repaired, prime, dry, and apply another top coat.

B) If the problem is too severe to repair, remove and replace the wood with new, properly dried wood. Prime, dry, and apply the top coat.

Diamond Drops

Description
Small glistening droplets of sap appear on the painted surface, days to weeks after painting. They may be white or clear, sticky or hard. They usually appear in areas subjected to direct sunlight.

Cause
A) Wet or unseasoned lumber was used.

B) The wood wasn't sealed or primed, or a poor quality sealer and primer was used.

C) Hot sunlight on lumber with a high sap content can bring sap to the surface.

Prevention
A) Use kiln-dried lumber.

B) Select low-resin lumber such as redwood for trim if possible. Pine and fir are a poor choice.

C) Allow lumber to cure for several weeks. Then clean the surface with turpentine or mineral spirits before painting.

D) Apply a good grade of the correct sealer and primer according to directions on the can.

Cure
A) Clean the surface with turpentine or mineral spirits, followed by a light sanding to remove the hardened sap.

B) Seal or prime using a good grade of sealer or primer. Then apply a top coat.

C) Remove and replace the trim with non-resinous lumber if the problem is too severe to repair.

Fish Eyes or Cat's Eyes

Description
Small spots on the wet painted surface that have a rainbow ring around them, usually 1/4 to 1/2 inch in diameter. The surface may dry with small spots or rings throughout.

Cause
A) Oil or grease on the surface or in the paint repels the paint, forming the spots. Light reflects in a circular rainbow pattern around the spots.

B) There could be an internal problem with the spray equipment that's allowing oil to seep into the paint.

Prevention
A) Clean all dirt, oil, grease, and other contamination off the surface before painting.

B) Test all spray equipment on a piece of scrap wood before attempting to paint the surface. Keep your equipment clean and in good repair.

Cure
A) While the surface is still wet, try to roll or brush out the spots. The problem will usually disappear as the paint solvent disperses the contamination.

B) Allow the area to dry, sand the spots, and then touch up. Repaint the entire surface if necessary.

Cracking, Flaking, and Peeling of Clear Gloss External Coating

Description

A clear gloss finish applied to exterior wood surfaces becomes brittle and begins to crack, flake, or peel, usually within two to three years of application.

Cause

If the problem develops seven to ten years after application, it's normal aging, weathering, and the effect of U/V radiation from sunlight. If it develops in less than three years, it's probably because the wrong material was applied in the first place. The coating wasn't elastic enough to withstand the expansion and contraction of the subsurface. When the problem appears in less than one year, the surface prep was probably faulty. The lumber wasn't aged or weathered and surface oils and contamination weren't removed.

Prevention

A) Weather the surface by allowing several months of exposure to wind, rain and sunlight. This dries the wood, opens or closes the pores, and removes the mill shine. If time is a problem, you can artificially weather wood by wetting it and letting it dry in the sun several times a day for a week or two.

After it's weathered, power wash the surface before application of a varnish.

B) Use only materials that are formulated for use on external surfaces.

C) Don't use varnish for large exterior surfaces. Use stain or paint of a CWF-type material instead.

Cure

In all cases, the old varnish must be removed. Chemical stripper, sanding, or sandblasting will be required. Clean the surface of all stains and residue and then recoat using the right material.

Clear Coating Worn Off on Cabinetry

Description

The surface coating around drawer pulls on cabinets lacks gloss, appears to be a different color, or is obviously worn off. This problem occurs with kitchen and bath cabinets that are finished with a factory spray coating of lacquer or urethane. A factory spray finish has little surface thickness and has a satin appearance.

Cause

It's probably due to normal wear of the surface from daily use. The surface coating is destroyed by water, oils, greases, or other contaminants on hands or by cleaning with chemical or abrasive-type cleaners.

Prevention

There's no way to prevent normal wear, except possibly using clear plastic guards around handles and knobs. Most people won't use these because they don't like the way they look.

Cure

A) Test for the type of coating. Is it lacquer, shellac, or some other material? If it's lacquer, what color is the exposed wood? If the exposed wood is a natural color, you're probably dealing with a lacquer stain. If you can't get the same type and color of lacquer stain, you'll have to strip the entire cabinet and recoat.

But here's a word of caution. In stripping or sanding cabinet woods, make sure you *are* working on wood. Many cabinets are made from a core of lumber or chip flake board that's covered with vinyl, paper or a very thin wood veneer. Sanding and some chemical stripping can destroy the surface. Then you have to replace the entire cabinet. Consider this alternative. Cover the worn cabinet surface with a wood veneer that you apply with contact glue or a heat-activated glue. Then finish it as though it were new wood.

B) If the exposed wood is close to the same color as the rest of the cabinet, you're dealing with clear lacquer over stain. Clean the area with xylene or alcohol to remove all dirt, wax and grease, being careful not to remove the stain. Then a clear coat of spray lacquer should bring the surface back to normal. You can also try using lacquer thinner to soften the nearby lacquer and spread this into the bare spot.

C) If the finish is shellac, clean with mineral spirits and apply new shellac to the area. Use white shellac over light colored woods, orange shellac over darker woods. You can usually tell a shellac coating because it's thicker and you can probably see some brush marks. Very few modern cabinet shops use shellac today. Some do-it-yourselfers and hard core "old timers" still use it.

D) If it's neither shellac or lacquer, you're probably working with a varnish, most probably urethane. Clean with xylene or alcohol. Then carefully touch up with new urethane — or strip the entire surface and paint as though it were new wood.

E) If the problem is confined to a small area, replacing the hardware may solve the problem. You can buy hardware with large escutcheons that may cover the bare spot. Measure the area you need to cover and shop for hardware that'll do the job.

F) If the problem is very severe, consider painting the cabinets with an opaque gloss or semi-gloss enamel.

G) Replacing all the cabinetry is one solution. But it's not worth the expense unless it's part of a complete remodeling project.

Worn Clear Coat on Wood Floors

Description
The clear coat on wood floors is no longer glossy. It's chipped, cracked, peeling, or shows signs of abrasion. The floor in Figure 12-21 is painted rather than clear-coated, but the effect is the same.

Worn coat on wood floor
Figure 12-21

Cause
A) If the coating is over five years old, it's probably just normal wear. This will happen in high traffic areas. The floor under furniture may still look almost new.

B) If the problem extends over all the floor, it's probably due to poor surface preparation or use of the wrong material.

Prevention
A) Use vinyl or carpet runners in areas of high traffic.

B) Keep floors clean and well waxed.

C) Use only those materials formulated for high traffic floor use.

D) Before applying any coating, sand the entire surface so it's free of greases, oils, waxes, and any other surface contamination.

Cure
A) To test for wrong type of coating, allow some water and some alcohol to stand on the coating for thirty minutes. If the coating turns white, it wasn't the right material to start with.

B) In all cases, the floor will require stripping to bare wood by chemicals or sanding. For sanding, use a floor disc sander for most of the floor and an edger along the walls. Don't let the sander work in one spot too long. That will create an oversanded spot that's hard to remove. Make the first sweep of the sander across the planks to smooth out the high spots. Then go back along the planks to catch the low spots and do the final smoothing. If the wood is discolored, use a bleach solution to bring it back to the original color.

Stain if desired, then apply a sealer coat. Allow it to dry for at least several days. Then scuff sand by hand and remove all dust. Make sure the top coat is formulated for floors. Let it dry for several days before anyone walks on it. The surface will be weak for up to a month. After a month, apply a coat of good paste wax to protect the surface.

C) Usually it'll be faster, cheaper and easier to lay some other floor cover such as carpet or vinyl than to restore the wood finish. This is especially true with parquet floors. The grain pattern makes it very difficult to sand properly once it's installed.

Wood Graining of Plywood Siding

Description
Newly painted plywood siding appears to have large unpainted areas or areas that show excessive wood grain.

Fiber destruction in a wood deck
Figure 12-22

Cause

A) The soft wood was splintered during factory sanding. New paint swells these fibers so they become more visible.

B) Factory sanding was incomplete, leaving rough areas on the surface. This is common when the panel is made from veneer that has softwood meeting harder wood.

C) The surface wasn't properly primed before painting.

D) The top coat hasn't covered sufficiently.

Prevention

A) Buy only quality siding that you've inspected.

B) Use enough of the right type of primer and top coat.

C) Sand panels before painting to equalize surface texture across the panel. But be careful. This can aggravate the problem rather than solve it.

D) After primer has dried hard, examine the surface problem areas and spot sand and reprime as needed before applying the top coat.

Cure

A) If you notice the problem before the top coat is on, sand and reprime before applying the final coat.

B) If the problem's not apparent until after the top coat is applied, you'll have to sand and apply a new top coat. But be careful! I've found that these splinters are very sharp after being painted. I sand with a short length of 2 x 4 covered with number 100 grit sandpaper. It works well and keeps your fingers free of splinters.

Wood Fiber Destruction

Description

Wood has decomposed and becomes soft and pulpy. There's no evidence of wood rot. This is evident in front of the door in the deck in Figure 12-22.

Wood rot in an old wood gutter
Figure 12-23

Cause
The wood wasn't properly sealed against rain or moisture. Water entered the wood. When the fibers have swollen then dried, they become brittle and break easily under pressure. In cold climates, the water turns to ice and expands, breaking the fibers. Unlike mildew, this can occur in areas of direct sunlight. But once the wood is damaged, it's vulnerable to attack from mildew.

Prevention
A) Keep the surface coating in good condition so water doesn't penetrate the wood. Seal all sides of the wood. If the wood is elevated and in an area with good air flow, you can get by with sealing the top, sides and ends only.

B) Use redwood or chemically treated wood for areas subjected to standing water.

C) Use alternate materials for areas subjected to standing water.

D) Build in a slight slope to improve water drainage.

Cure
A) Dig out the damaged area, apply a bleach solution, neutralize, dry, and then fill with wood filler. When it's dry, prime and apply a top coat. For areas that don't carry traffic, your top coat might be fiberglass and polyester.

B) Replace the damaged lumber with redwood or chemically treated wood. Then seal it properly. Special deck lumber is available that's treated and has beveled edges to prevent splintering. Check with your local lumberyard.

C) Replace the entire surface with an alternate material, such as cement or stone.

Wood Rot

Description
The wood has decomposed, becoming soft and pulpy, and is turning black. You can see black wood in the old wood rain gutter over the picture window in Figure 12-23.

Cause

A fungus, usually mildew or mold, has entered the wood surface and is consuming the wood fibers. This is the largest single cause of wood destruction. The fungus can't live very long without moisture. That's why the problem is more common in cooler, damp, and shaded areas: the north side of a building, under eaves, and behind gutters.

Prevention

A) If the problem hasn't spread too far, clean the area with a solution of bleach and water (50/50). When it's thoroughly dry, fill the damaged area, then seal the wood well so moisture and fungus can't penetrate it. Use mildewcide in the new coating.

B) In severe cases, you'll have to replace the damaged wood, then seal it well before priming and painting to prevent a recurrence.

We've covered a lot of ground in this chapter. You couldn't possibly remember it all. Maybe it wasn't even worth reading this chapter more than lightly. But this book isn't intended to be read once and then discarded. It's intended to be your answer book for all painting problems — for many years and on many jobs.

The next time you come up against a paint failure, I hope you're thumbing through the pages of this chapter, finding the cause and taking the right steps to correct the problem. That's what an experienced professional painter would do. I hope the information in this chapter helps you do the same.

Water and Fire Damage

Many of the problems you read about in the last chapter had something to do with *water*. Nasty substance, water. Read on and I'll show you.

Water is one of the earth's most plentiful and most dangerous substances. Why is it dangerous? For starters, it's an excellent solvent. There are few materials that won't dissolve in water, given enough time.

And here's another danger. Water is made of two gases, hydrogen and oxygen — and it has an irritating habit of trying to separate back into these gases when in the presence of other elements. The oxygen attaches itself to the element and the hydrogen is released into the air. The element itself now becomes something different — its atomic structure has been changed. The best example, of course, is rust. Iron in the presence of water turns to iron oxide, otherwise known as rust.

But creating rust is just the beginning. Water that isn't used in oxidation makes an excellent vehicle for carrying the rust from where it is to where it probably shouldn't be. Because rust is red, the water takes on a red tint. Eventually the water evaporates and the rust remains as a very effective pigment. In fact, iron oxide is used as a pigment in many paints. Removing this pigment keeps hundreds of painters busy every day.

Types of Water Damage

We all know that water tends to seek its lowest level. In the process, it will penetrate any porous surface, dissolving what it can and depositing on the surface anything that's too large to pass through the pores of that surface. The result? Water staining. If the water contained rust and the pores of the material it traveled through were small enough to trap this rust, the surface becomes rust stained.

Water and Rust Stains

Figure 13-1 shows a good example of both water and rust staining. This happened on the northeast side of a store. Several problems are apparent, in addition to just plain neglect. The gutters empty onto the siding and have caused severe water damage to the paint and siding. Because the gutters are rusted, they've also stained the remaining paint. Iron nails were used and they're rusting. Some signs of mildew are evident over the doorway. The solution is to bleach the entire building, move the gutters to a proper drainage location, and replace the badly rusted downspouts. Then spot sand where required, prime and repaint.

A few months later, I got another shot of this building. Look at Figure 13-2. The painter did a fair paint job, but he didn't fix the defects. One

Water and rust stains are evident here
Figure 13-1

The same building after repainting
Figure 13-2

gutter is still located where it will empty onto the siding. The original problems will return.

Paint Peeling, Blistering and Lifting

When water saturates an easily dissolved material, such as wallboard, the material changes size and shape. Then, if the water meets a barrier like the underside of a paint coating, it will try to combine with, or oxidize, the paint. Oxides of paint are powders, and powders don't have much holding or bonding power. So the paint peels, blisters and lifts.

Here's another problem with water: It not only flows down but also sideways and sometimes even up. A stain on the living room wall may be coming from the garage roof, ten or more feet away. Water can cling to rafters, beams, and wires by surface tension and move horizontally or vertically for many feet.

Figure 13-3 shows an example. There's paint peeling from the lower section of the siding. The siding was water-sealed before installation. It was spray painted with a primer and a latex top coat. So what's causing the problem? There's a void in one layer of the ply in the edge of this sheet where it peeled. When the owner waters the lawn, water hits this section and is sucked up through the void, loosening the paint from beneath. The solution is caulking the void, stripping the loosened paint, then repriming and repainting the area.

Water can also travel up to cause damage
Figure 13-3

Here's my point: Before repainting, find out what's causing premature paint failure. It's a waste of time to repaint without eliminating the source of the damage. Check for the presence of water, even if water damage isn't obvious. Keep a flashlight in your tool kit for probing into dark corners and attics.

Looking for Water Damage

When should you suspect water damage? Here's what I recommend: Look for water damage every time you repaint anything. Very few buildings escape at least minor damage from water. Here are some obvious danger signals:

- Normally hard surfaces, such as wood, plaster or drywall, are soft, spongy, or discolored.
- Paint coatings have begun to crack, blister, or peel.
- Siding or other wood is in direct contact with the ground.
- The building is located in a flood area, near a lake, river, or canal.
- Roof and roof flashings are in poor condition.
- Poor drainage or shallow roof slopes cause ice and snow buildup.
- Plants and trees are touching the building.
- Attic, laundry, kitchen, and bath areas don't have vent fans.
- Brownish rings are apparent on ceiling and wall surfaces.
- There are white rings or discoloration on lacquer or clear coated items.

What to do? If any of these conditions are present, stop. Don't bother to even open that paint can until the source of the water has been found and eliminated. Ask if the building owners are aware of the condition and what caused it. Maybe they know about the condition and have already made repairs. If not, they know about it now, and have the responsibility to make repairs. If you paint over these problems, then it's *your* responsibility. That can be costly, both to your pocketbook and your reputation.

Sources of Water or Leaks in Buildings

Let's look at the common sources for unwanted water. Some can be eliminated easily. If so, create a little goodwill by solving the problem yourself and at no cost to the owner. Others, like caulking, should just be part of your normal paint job. Still others are so difficult and time-consuming that you'll probably want to pass them up. But if you're a handyman as well as a painter, then by all means use this opportunity to generate some extra revenue. Just be sure you understand the problem and can make the proper repairs. Otherwise you may lose the entire job, including the painting.

There are many excellent construction books on the market. Some of them are published by the same company as this book, and can be ordered with the form at the back of this book. Buy and study a couple of good construction manuals. Knowing how a building is supposed to go together will help you in troubleshooting paint problems. It'll help even more if you decide to earn some extra dollars making repairs.

If the problem can't be repaired easily or if the owner won't pay for the repairs, you have two options: First, you can pass on the job and spend your time on something more profitable. Second, you can give the owner a note explaining the problem and advising that failure to have the repairs done will shorten the life expectancy of your paint job. Save a copy of your note. This will help protect you from future claims.

The water problem can originate on the roof, in the walls, or in the paint itself. We'll begin from the top, with the roof.

Roof-Related Water Sources

Overhang problems— There are two possibilities here. The roof overhang itself can be insufficient to shield the siding from water damage, or the shingles can have too little overhang, allowing the water to spill off behind the gutter.

Gutter problems— First, the gutters may be improperly installed with too little slope so water doesn't flow to a downspout.

Second, properly installed gutters may have rusted or rotted. Figure 13-4 is a good example of this. Water has seeped through the rotted wooden gutters and lifted the outer paint. These gutters must be kept coated with tar or asphalt on their inside surfaces, and they must not hold standing water. Check for proper drainage whenever you repaint. Standing water rots wood and rusts metal. The solution here is to replace with new gutters, preferably aluminum. If the condition is caught soon enough, you may be able to clean, prime, and repaint to retain the original architectural appearance.

Deteriorated wooden gutters cause paint damage
Figure 13-4

Gutters blocked by ice can cause damage
Figure 13-5

Third, the gutters and downspouts may be blocked, either from leaves or ice, or from roof material that's washed down into the gutter. The house shown in Figure 13-5 is in for some serious water damage. The snow has melted and refrozen several times, forming ice blockage in the gutters. This can cause damage to the roof, gutters, internal studs and possibly even the internal walls and ceilings.

Fourth, the shingle overhang may be too small. The water falls behind the gutters instead of into them. The shingles should overhang at least 1 inch.

And fifth, the gutter joints may be loose or not properly sealed, allowing water to seep through and into the fascia board.

Roof slope or drainage problems— In Figure 13-6, the shingle siding right above the lower roof is discolored because snow builds up against it. After it's restained, it needs a coat of waterproofing material.

Roof covering problems— Deteriorated roof coverings tend to leak. Even if the surface is still good, roofing nails may have popped through the shingles. Occasionally you'll find blisters on a flat roof. In a damp climate, look for tree branches or vines growing under shingles.

Flashing and seam problems— Worn, rusted, or improperly installed flashing will cause leaks sooner or later. Be alert for missing or damaged caulking or seals around chimneys, TV antennas, power lines, and other roof protrusions. On a flat roof, check for clogged drains. On a metal roof, look for split or improperly installed seams. Also check the gutter seams.

When you paint, make sure all seams are caulked. All flashing should have a prime coat, a top coat (preferably the same color as the roof) and a final coat of roof tar or asphalt. Seal flashing to the roof shingles with roof tar or asphalt.

Look at Figure 13-7. The metal roofing edge doesn't extend out 1 inch past the fascia as it should. It's in contact with the board, providing a direct trap for rain water. The result is severe dry rot in the fascia board.

Siding discolored by snow buildup
Figure 13-6

Dry rot caused by improperly installed metal flashing
Figure 13-7

Peeling paint caused by lack of vapor barrier
Figure 13-8

Equipment problems— Water may be coming from a leaky air conditioner or solar system on the roof. Is the float that controls the water level in an evaporative air cooler working properly? Is the condensate drain open and free-flowing?

Upper Wall Area Sources

If there's water damage on an upper wall area, look for a roof leak first. Check for each of the problems described in the section on roof problems. Then look for tree branches touching the exterior walls. If the source isn't on the outside, move inside to look for second floor leaks from bath tubs, pipes, and drains. Are there effective vents in the kitchen, laundry and bathrooms? Are they being used? Finally, examine the power and TV lead-in wires. Each wire should have a drip loop that sheds water to the ground before it can reach the structure.

Lower Wall Area Sources

Interior problems— After you've eliminated all of the roof and upper wall sources, check for broken pipes in the walls. Look for clogged or overflowing sinks, dishwashers, clothes washers or other water-using appliances. Are there plants or planters that could be the source of the water problem? Then examine all the doors and windows for old, missing or poorly applied caulking or putty.

Finally, think about vapor barriers and insulation. If there's no vapor barrier under a floor slab, moisture could be seeping up from the soil below. Was the ceiling and wall insulation installed with the vapor barrier on the wrong side or without a vapor barrier? Every wall needs a vapor barrier on the interior side of the insulation.

Look at Figure 13-8. See the paint peeling on the siding next to the downstairs window? This is an older home without proper vapor barrier in the walls. Water from the inside caused this damage. The solution is to install vent plugs in the outer wall siding. The room just behind the peeling should be painted with an oil-based paint that will keep the water vapor out of the wall. All old exterior paint should be stripped off and a coat of latex paint should be applied to the exterior walls.

There's one more interior source to check for: condensation. Remember the test for condensation back in Chapter 11? Of course, if you see something like the storm door in Figure 13-9, you don't have to do the test — the condensation is obvious. Water droplets that form on the window will run down to where the glass is mounted. If it isn't properly sealed, water will seep behind the paint coating and destroy the surface-to-paint bond. The result? Peeling paint.

Condensation causes paint damage
Figure 13-9

Condensation can also cause water staining on surfaces. Figure 13-10 shows a pocket door in a basement laundry that doesn't have a vent system. Condensation has collected near the top of the door because the cool garage is on the other side of the door. This will require bleaching and then a water-resistant stain.

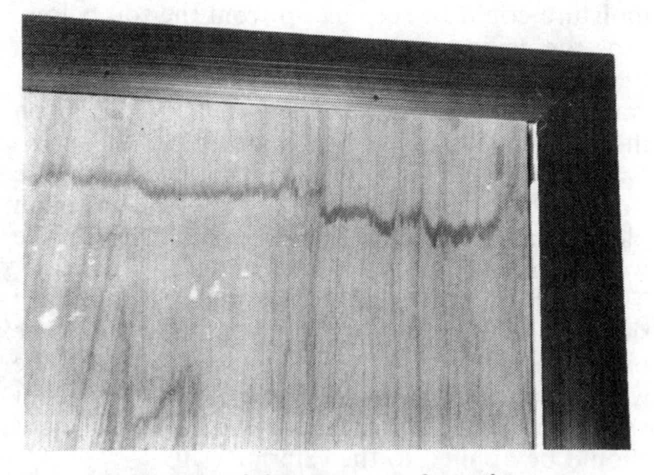

Water staining from condensation
Figure 13-10

Exterior problems— If exterior wood paneling is in contact with the ground, capillary action could be drawing water up into the siding. And it doesn't have to be siding. In Figure 13-11, the wood door sill is in contact with the ground. Dry rot has set in. It'll have to be replaced with metal or cement.

Door sill in direct contact with ground has dry rot
Figure 13-11

Poor caulking may allow wind-blown rain to enter the wall cavities and the living space. Does the slope of the ground around the building drain water toward rather than away from the wall? Are there plants growing where they can touch a painted surface? In Figure 13-12, the plants are fine. But if they're not trimmed back as they grow, they can damage the surface two ways: First, they hold water against the surface. Second, minerals and acids in the plants rub off on the wall, eventually damaging or staining the paint. Some plants actually "eat" or consume the paint coating.

If fences or gates are fastened to the building, are the fasteners sealed? In general, how well is the house maintained? Figure 13-13 is a good example of what can happen to a poorly maintained building. Window caulking deteriorated, allowing water to enter and saturate the wood. Wet wood destroyed the paint coating and is beginning to destroy the window itself. At this stage the wood is nearly unprotected and deterioration accelerates. The owner here won't spend money on a new paint job. The window, or perhaps the entire home, will have to be replaced long before its natural life expectancy is reached.

If you can't find any water sources in the structure itself, question the residents. Do they allow the

Keep plants away from building surface
Figure 13-12

Poor maintenance is costly in the long run
Figure 13-13

sprinklers to water the side of the building? Has there been any flooding? Was there a fire that was put out with water?

Other Water Sources

Some paint damage isn't the result of water leaks or condensation. Consider these possibilities: Was latex paint used in a high humidity area? Oil-base paint would be the right choice. Was paint applied to plaster, spackle, or acoustic coatings before they were dry? Maybe the paint was applied in the rain or when humidity was high, or to a wet surface, or to green lumber.

If these possibilities seem far-fetched to you, rest assured that most painters encounter each of them in the first few years on the job. I did. Many times you'll find several of the problems I've described in the same building. So don't quit looking when you find one or two.

Consider the example in Figure 13-14. This older home has asbestos shingle siding and was last painted with a paint that contained lead. There is evidence of fume staining. There are rust stains near the gutter and downspouts. Moss is growing on the siding next to the steps. The roof is stained because the flashing wasn't painted. There's a combination of mildew and rust stain at the laps. The wood window sills are peeling. Where do you start? The house is worth saving. It's in a neighborhood where property values are increasing.

First, recommend a new shingle job on the roof. But this time the flashing gets painted. The siding can be repainted or covered with aluminum siding. But either way it needs a thorough TSP wash and then a bleach wash. All rusted metal should be replaced. Windows should be stripped to bare wood, primed and repainted. Gutters should be installed over the porch and bay areas. Any rotted lumber, such as in the front steps and porch decking, should be replaced with decking lumber. Obviously, this is going to be an expensive proposition. And most of it could have been prevented if the work had been done right in the first place or if the house had been maintained correctly.

Let's look at one more example, a masonry building. There are several problems in Figure 13-15. First, the brick was painted with an oil-

Many jobs include multiple water-related problems
Figure 13-14

based enamel. Second, the sidewalk drains toward the building. Third, decoration under the wood sill traps and holds water. Finally, the building is in a part of the country where salt is used to melt ice and snow. Moisture and salt have attacked the paint from within and without. You'll need to sandblast the entire brick wall surface and then repaint with a water-based epoxy or latex formulated for use on brick. Another choice would be to leave the natural brick surface uncoated. Also, I would recommend replacing the wood sill with vinyl or aluminum.

The wrong paint, weather and salt all did their work
Figure 13-15

Once in a while you'll hit a real puzzler. On one job I found water stains half way up an interior wall under a window. The window was watertight and in good condition. I couldn't figure where the water could possibly be coming from. I mentioned the problem to the owner's wife. She explained that the prior owner had a planter sitting on a stand under that window. Watering the planter always wet the wall. She pointed out indentations and stains in the carpet that showed where the planter had been. Sometimes it takes a Sherlock Holmes to pinpoint the problem.

Other Water Problems

Water has two other qualities that I haven't mentioned yet. It boils and it freezes. In both these states it expands. When that happens, any moisture that's trapped exerts tremendous force under a paint coating, lifting the coating and forming blisters. Even if the outside air temperature is only 85 degrees, the subsurface temperature can easily hit 212 degrees in direct sunlight. At that point you have steam and no more paint film.

At 32 degrees or less, water freezes and expands, blistering the paint. Since the paint film isn't very flexible at low temperatures, it usually cracks and peels rather than blistering. Check the houses in your neighborhood. If you live outside the sunbelt, the north side of your house is prone to paint problems. The north side gets little sun, most of the rain, and stays the coolest. In Figure 13-16, you're looking at the north side of a stucco-covered porch. Because it never receives sunlight, it's covered with moss and mildew. A spraying with bleach followed by a clear water rinse should solve the problem.

North side of porch never receives sunlight
Figure 13-16

Efflorescence

Every painter has heard of it but very few really know what it is or what causes it. What is it? *Efflorescence* is defined as a change to a powder by loss of water. Figure 13-17 shows what it looks like. What does it do? It creates a surface that's not paintable; paint won't stick to powder. What causes it? Water seepage through a substance that contains minerals, leaching the water-soluble salts from the material to the surface. Salts can be formed from potassium, magnesium, calcium, silicates, or sodium bicarbonate. The salts can also be formed from vanadium, chromium, and chloride. All of these minerals can be found in trace amounts in brick, cement and concrete. Many of these salts are water soluble and will wash off during a rain storm. But some are not so soluble and have to be washed off with an acid or base solution wash. That's covered in Chapter 12.

Good example of efflorescence
Figure 13-17

Even a seemingly dry surface can effloresce. Water from the ground can seep upwards via capillary action to a height of several feet. Dissolved minerals are carried up with the water. It doesn't take much of this contamination to cause efflorescence. Just a one-tenth of one percent concentration of these water-soluble minerals can cause the problem.

How do you eliminate it? First, find the source of water and make whatever repairs are necessary to keep the water out. Then remove the salts from the surface before repainting. Try a heavy brushing or sandblasting, or a muriatic acid wash, a neutralizer rinse, and a clear water rinse.

Just cleaning a damaged surface won't do the job. If the surface was damaged by water or other contamination, then the source of water or contamination must be removed. I've seen so many painters wash down a wall that contained efflorescence, let it dry, and then start to repaint. Six months later the problem is back and so are they, for a costly rework. Either solve the problem or inform the owner that the problem will recur if not corrected.

Figure 13-18 shows efflorescence on the side of a firehouse built with concrete block. The roof drains were clogged and water had puddled on the flat roof. Eventually water seeped through the roof membrane and into the block. A little preventive maintenance could have saved this town a lot of money in repairs.

The best way to prevent efflorescence is to use the correct materials and construction in the first place. Use *dehydrated lime* that is free from calcium sulfate when mixing mortar or stucco. Use clean washed sand and gravel. Keep the mixing

Powdery efflorescence on a masonry wall
Figure 13-18

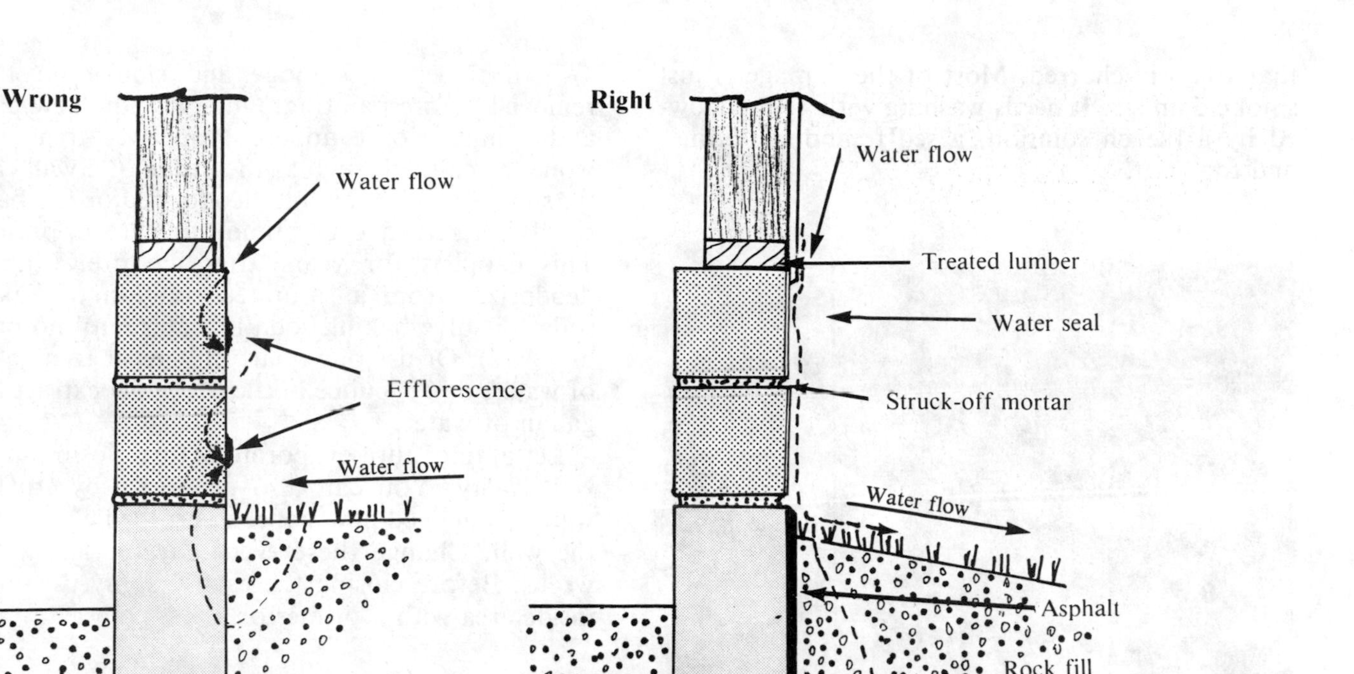

The wrong and right way to build a waterproof foundation
Figure 13-19

tools clean and free from soil contamination. Seal all below-ground and ground-level cement or block with a waterproof or asphalt coating before backfilling with dirt. Install proper drainage and covers to keep the water out. Figure 13-19 shows the wrong and right ways to build a foundation.

Laitance is similar to efflorescence, but it refers to an accumulation of fine particles on the surface of fresh concrete due to an upward movement of water. Before painting, clean it with a muriatic acid solution and a clear water rinse.

Occasionally you'll have to bid on a job that's the result of fire damage. In addition to covering smoke stains, you'll have to repaint surfaces that have been damaged by the water used to put out the fire. The last part of this chapter will cover this important topic.

Repairing Fire Damage

Kitchen fires are common in all parts of the country. In the dry western states, brush and timber fires damage many homes. A brush fire in the housing complex where I live destroyed eight homes and damaged eighteen more. Eventually, most painters get a fire job.

Fire damage is threefold: First there's the charred and burnt roofing, cement, drywall, and lumber. Second, there's damage from smoke in the form of smudge or carbon. Finally, there's damage from the water used to put out the fire.

The water problem is the easiest. Just let everything dry out for a few weeks. There may be no permanent damage. If there is, just follow the procedures for water damage I've described in the first part of the chapter. Smoke and smudge can be washed off with a TSP solution followed with a bleaching. The main problem area is blistered, burnt and charred material. Do you attempt to clean and repair or do you just replace? Cost is the major factor. If the structure was fully insured and the insurance company will pay for it, I highly recommend replacing all burnt or charred material. If the property wasn't insured or if the insurance company refuses to pay in full, here are a few suggestions for salvaging and repairing fire damage:

1) If only the surface is charred, use a power sander to remove all charred material and then apply paste wood filler. Figure 13-20 shows siding

that's lightly charred. Most of the damage is just smoke damage. It needs washing with TSP followed by a bleach solution, a scuff sanding, primer and top coat.

Lightly damaged siding after a fire
Figure 13-20

2) If the charring extends beyond the surface, the only correct solution is to replace whatever is burnt.

3) If wall or roof framing is charred, chisel out the charred area and clean the surface. If the burned area goes over one-third of the way through, it *must* be replaced. Figure 13-21 shows some lumber that was going to be a new deck. The home is now gone but part of the deck lumber can be saved. Power sand the lightly charred areas. Saw off and discard the sections that have over a third of the material charred.

Some of this charred lumber can be saved
Figure 13-21

4) All traces of smudge and smoke must be removed before painting. Otherwise the new paint and primer won't adhere properly. Areas that won't be painted, such as the inside of a wall cavity, should be cleaned and deodorized or the burnt smell will return every time moisture is present. This happens for years after the fire. Make a deodorizer from a saturated solution of baking soda (dissolve baking soda in water until no more dissolves). Or use one quart of vinegar to a gallon of water, or a 5-ounce bottle of vanilla extract to a gallon of water.

Let all moisture evaporate before closing in the wall cavity. You can also deodorize by stuffing water-dampened newspaper or crushed charcoal in the wall. Change these every day for about two weeks. Before closing the cavity, spray the entire inside area with a primer or sealer paint.

5) External smudge and smoke can be washed with a water blaster. Wash the entire building from the roof peaks to the ground. Wet the building first with a hose from the ground up to prevent streaking. Use only a TSP solution on the roof. Use TSP and bleach solutions on the siding. Allow several days drying time and then prime and recoat.

6) Replace all fiberglass insulation in an open wall cavity or attic area that's been exposed to smoke. Fiberglass will retain odors for months.

7) All charred trim lumber should be replaced. If you just sand and fill, it'll never look quite right. Use redwood lumber for exterior trim, and paint all edges before installing.

8) All metal that isn't warped should be sanded to bare metal and then primed and top coated. If it's warped, replace it.

9) Low-cost decorative metal should be replaced. That's cheaper and easier than trying to strip and repaint.

10) Check for out-of-alignment gutters, doors, and windows. Check for putty and weatherstripping damage at doors and windows. Especially important, check for any burnt and now uninsulated electrical wiring. Check for loose or damaged brick work, especially chimneys.

That just about does it. It's mostly common sense. Fire isn't pretty, so be careful that you're not the cause of one. But there is one time when a painter can use fire to advantage. I'll mention it here even though it's slightly off the subject of this chapter. It's a decorating trick used to give a distressed look to furniture, wood plaques, shelves, cabinets or counters. Use a propane torch to lightly singe the wood surface. This will slightly char the softer wood grain and not the harder wood grain. Clean with TSP or mineral spirits and then seal with a thin polyurethane sealer coat. Allow to set hard, scuff sand, and then coat with full strength polyurethane. This works well on soft pine, to give a western look. A note of caution: When burning the wood, keep a bucket of water and some wet rags nearby, just in case.

chapter fourteen

Selecting and Using the Right Tools

I'm convinced that sloppy, careless, amateurish painters give the painting profession a bad name. I'm also convinced that the best way to build a career in the painting business is to do only top quality, professional work — and charge appropriately for the work you do. But I recognize that no book can make a painter take pride in his work. A professional attitude, the desire to become a skilled, quality painter, the character you show in dealing with clients — all this has to come from within. You have to supply it yourself. What I can do is give you the benefit of my experience. The rest is up to you.

It's very hard to do top quality work if you don't have the right tools and equipment or don't use them correctly. This chapter will identify the tools a professional painter should have and describe how they're used to best advantage.

If you're looking for conventional advice in this chapter, I'm sorry. You won't find it here. I believe that much of what's been written about painter's tools and techniques is just plain bad advice. I don't like criticizing what others are teaching, but I just don't agree with a lot of the "facts" being taught. Much of the information in this chapter is my personal opinion. You may or may not agree. That's your privilege. But try following my recommendations before you reject them. Eventually you'll select the best tools and techniques for the type of jobs you handle.

The Tools of the Trade

In this chapter, I'll cover the tools you'll need to make your work professional, and as easy as possible. In the next chapter, I'll go on to the techniques that set your work apart from the run-of-the-mill painters whose only goal is to slap on some paint and deposit the check.

Tarps and Drop Cloths

First of all, I don't like plastic tarps. I prefer canvas or disposable paper. Here are my reasons: Plastic is hard to tape or hold in place, especially if you're trying to cover a wall. Second, paint dries slowly on plastic and delays removal of the tarp and cleanup after the job. Third, if there are wet spots, you take the chance of spreading the wet paint to the very items you wished to protect. Fourth, when paint finally dries on plastic, it has a tendency to flake. These flakes fall off and can get trapped in the finished but still wet paint. Finally, heavy plastic tarps don't bend or form themselves around what you want to protect. Thin tarps (two mil or less) are easy to place, but they rip easily when you remove the tape that held them in place.

If you want to use plastic, I recommend only the four and six mil embossed tarps — they eliminate some of the problems. Use plastic tarps for added protection under a cloth or paper tarp, especially

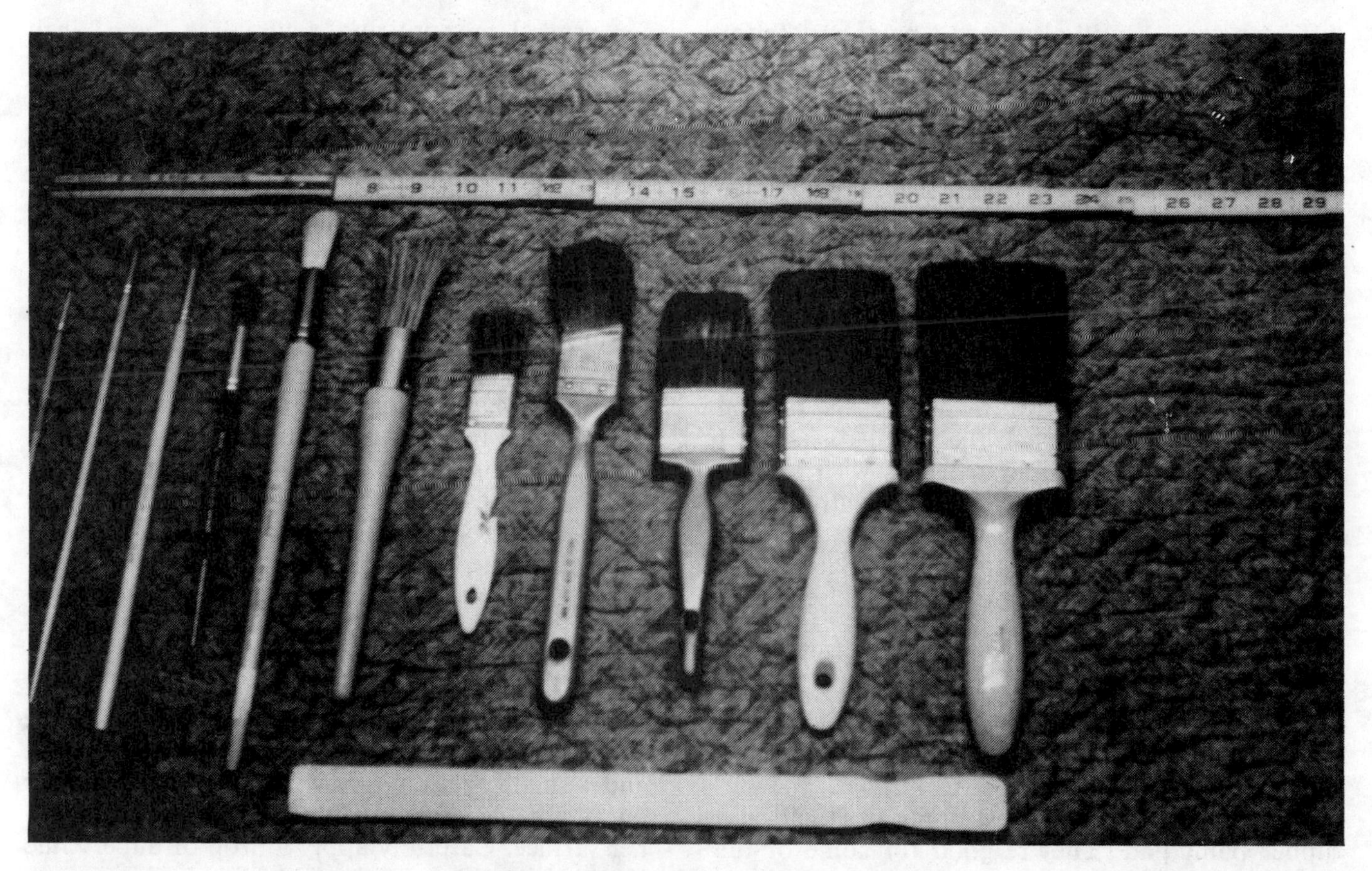

These brushes will outfit you for most jobs
Figure 14-1

when you set up a mixing area. But for most spray work and roller splatters, you don't need plastic. Use cloth or paper instead.

Some painters are tempted to use newspaper as tarps. They're almost free and are readily available. My advice is, don't do it. You don't need the problems. Newspaper ink comes off easily, staining furniture, rugs, carpets, and vinyl. Ink stains can be very difficult to remove. Most paints will pass through a single layer of newsprint to the surface you're trying to protect. If you do choose to use newspaper, then use multiple layers, overlapping each, with a plastic tarp under all. Don't walk or place heavy objects on it.

If you're painting something that's resting on newspaper, move what you're painting at least a few inches before it dries. Otherwise the newspaper will stick to the paint as it dries. If you're painting a flat item that's in full contact with paper, crumple the paper first to help prevent sticking. But for best results, don't use newspaper at all. Use professional painters' tarps, kraft paper, or opened supermarket bags.

Brushes, Pads and Rollers

Remember, if you're a pro, you should be using professional tools. That goes for rollers and brushes as well as drops.

Brushes

In selecting a brush, consider the size and type of material being brushed, and the quality desired. For large surface areas, you'll probably use a 2½, 3, or 4-inch brush. For trim, molding, sash, windows, and other difficult to reach areas, pick a 1/2, 3/4, 1, or 1½-inch brush. Trim brushes are available with straight cut, beveled cut, and angled cut bristles. Try them all to see which you like best. There are also special brushes available for stenciling, cleaning, and touch-up work.

Figure 14-1 shows some of the brushes you'll need to have for most jobs. They include a few artist's brushes for fine touchup work, a couple of cleaning and paint-removing brushes, a stencil brush, trim and sash brushes, and paint brushes of different sizes. You'll also need a paint paddle for loosening solids from the bottom of cans, and a tape or wooden ruler.

Select a brush appropriate for the material being applied. Otherwise you risk ruining the brush *and* the paint job. For latex paints, use a nylon or polyester bristle brush. A natural bristle brush will soak up the water in the latex and become soft and mushy, making it unusable. For oil/alkyd paints, you can use either nylon, polyester, or natural bristle.

Most of the varnishes, polyurethanes, stains, and plastics won't harm manmade or natural bristles. To be sure, check the label. Actually, I recommend applying these materials with a cloth or, even better, a spray. Apply them with a cloth only on smooth surfaces such as loose or new trim lumber, or on furniture. Thin the first coat by 10% to 15% so it seals the wood better. Use it full strength on the next one or two coats.

I'll give the same advice for lacquers. They dry too fast for proper brushing and should always be sprayed on.

Paint Pads

For some materials, the best applicator will be a rubber paint pad. They're good for some of the oils, some stains, and most latex. But don't use these pads with strong solvent-based materials such as shellacs or lacquers — the solvents will dissolve them. There are two types of pads available, all foam and foam with fiber. Use the all-foam type only on smooth surfaces. Fiber pads can be used on brick, stucco, cement block, rough-sawed lumber, siding, and textured walls. Look for pads with nylon fibers at least 3/8-inch long.

For good streak-free results in staining, use a lambswool pad. Buy good quality pads that have a dense field of fiber. Always use a matched paint tray from the same manufacturer to assure the right amount of paint pickup without drips and runs. Your regular 9-inch paint tray will work, but you have to work very carefully. That usually slows you down too much. The idea behind paint pads is to increase painting speed. Using the wrong tray defeats that purpose.

Rollers

I highly recommend the purchase of a new roller cover for each job and for each color change within a job. Charge them to the customer. Covers are available in 3, 4, 7, 9, and 18-inch lengths. Most are standard diameter, but some half-diameter covers are available. Use the 3 and 4-inch size for trim work, the 7-inch for lap siding, and the 9-inch for general purpose and walls. Use the 18-inch roller cover for very fast full-wall coverage. You'll have to go to a store selling primarily to contractors to find an 18-inch roller. Most other rollers are available in paint and hardware stores.

Be careful in selecting the right roller cover for the job. The glue that holds the nap to the core on some rollers will dissolve in certain solvents, releasing particles of nap onto your newly-painted surface so it looks like tiny threads running through the coating. If the plastic sleeve on the roller cover states that it's for a certain type of paint only, believe it! Many lower-cost roller covers use cardboard tubes for cores. If you use these for latex paint, the cardboard will soften and fall apart. Others use plastic cores that will soften or dissolve in oil/alkyd or other solvent paints. *Read the label,* and follow the instructions on the sleeve.

Roller frames and handles also have their problems. Look for frames that rotate freely and are threaded for an extension handle. As soon as you unpack a new roller frame, remove the little plastic end cap on the non-handle, or free end of the frame. What you'll probably see is a center shaft and a small "keeper ring." It looks like a washer but it's beveled and may have teeth cut into the inner surface. Carefully apply a drop of Super Glue to this ring. Rotate the frame several times to assure free action. Here's why. When the keeper ring gets wet from paint or solvents, it can twist off very easily. This lets the frame come off the handle shaft, usually in the middle of a job and with a full load of paint on the roller. Then you've got a real mess to clean up. Glue adds just enough friction to keep the ring in place.

Pressure-Fed Roller Systems

Pressure-fed paint roller systems are a recent development, but they work well and can save a lot of labor and time. They do have one major drawback: the cost of the special roller covers is high compared to most of the standard covers. Pressure-fed covers are made to feed the paint from the inside of the core through to the outside nap. To clean, you just pump the cleaning solvent through the system as if it were paint. That takes a lot of cleaner, so I suggest using some sort of reclaim system.

The simplest reclaim system is just two buckets. Hold the roller in the first bucket to get rid of the very dirty solvent, which you'll discard. Then place the roller in the second bucket to "reclaim" the not-so-dirty solvent. Use this as a first rinse next time. As with any roller cover you plan to reuse, always use it on the lightest colors first and advance progressively to the darker colors. Are these covers to be considered disposable? Yes, if you can use one cover to do the job of three standard covers or more. And since you aren't continuously dipping into a paint tray, you should be able to cut painting time by 1/3 to 1/2. It's a balance between roller cover cost and painting labor. Only you can decide if the time saved is worth the extra expense.

Roller Trays and Liners

Another tip from the booklets is to line the paint roller tray with foil. It's supposed to make for easier cleanup. It does work, providing you use heavy-duty embossed foil that's wide enough to line the entire tray up and over all sides with one single piece of foil. If you have to use two pieces, the paint will seep under them and defeat the whole purpose. Use heavy duty foil. Lighter foils tear easily or lift out of the tray with the paint-wetted roller.

If you want to use paint roller trays, I suggest you use *plastic tray liners* instead of foil. They cost less than a dollar, work well, and are disposable. Be careful, though. Some aren't deep enough to touch the bottom of the tray and they'll collapse when filled with paint, spilling paint all over. A piece of wood cut to size and placed in the bottom of the tray under the liner will prevent this problem.

The Five-Gallon Bucket

But for the best results, I recommend not using a roller tray at all. Instead, use a five-gallon plastic bucket with a removable roller rack installed. This helps prevent the splashes and accidental spillage that are almost inevitable with a tray.

Here are a few paint-can tricks that make cleanup easier and help prevent spills when using one-gallon cans or five-gallon buckets:

1) Puncture the can's rim lip with a series of holes, using a hammer and nail. This will let any paint that settles there drain back into the can.

2) Glue a paper plate to the bottom of the can. This catches drips or runs on the sides of the can.

3) Keep a roll of plastic cling-wrap with you. Cover the lid opening whenever you take a break. This keeps the paint from drying. Also put a layer of plastic across the opening before re-applying the lid. Same reason, plus it makes re-opening of the can much easier.

4) When the job's done, you may have some paint left over. There are a couple of ways you can save it. Transfer it to a small baby food jar, pickle jar, or other container that has a tight lid seal. Label the jar as to paint type, color, and job number. If there's quite a lot of material left over, just leave it in the original can. Seal the can by brushing a layer of paint around the lip. This does two things: It removes any puddles of paint that will splatter when you try to hammer the lid down tight, and it forms an airtight seal after the paint dries. If you want, you can also lay a membrane of plastic wrap over the can before applying the lid.

While we're on cleanup tips, let me slip in a few more. If I'm taking a long break or an overnighter, I'll use the plastic wrap to cover my brushes and rollers. No, I don't clean them, I leave them wet with paint. Starting up again is fast and easy — there's no need to wring out cleaning solvents. A sandwich bag that zips shut also works well.

The plastic cover that comes on the roller cover is also reusable. I open them carefully at one end and save the cover until the end of the job. Then I merely slip the plastic cover back onto the roller and pull the roller off the handle. That leaves clean hands and easy disposal for disposable roller covers. Plastic bread wrappers will also work well. Just be sure the plastic wrap and the paint are compatible. If they're not, the plastic can melt.

Cleaning Brushes, Pads and Rollers

Cleaning techniques are the same for brushes and paint pads. For stains, polyurethanes, varnishes, plastics, or oil/alkyd paints, your best cleaning solvent is mineral spirits. For water-based stains and latex, use mild soap in water. Clean and rinse well, then wrap in wax paper or aluminum foil, taking care not to bend or distort the bristles. Store in the refrigerator or another cool place. Never leave brushes standing on end in a container. The bristles will bend out of shape, making the brush useless. Also, as the solvent cleaner evaporates, the dissolved residue will harden in the brush bristles.

I've found that brushes that have been improperly cleaned sometimes can be restored by soaking in commercial liquid sander containing toluene. I use a product called Sandeze.

To clean excess paint from a brush during painting, most experts recommend that you dab, not wipe. Tap the bristles lightly against the inside lip of the paint can rather than wiping along the rim. In theory, this prevents air bubbles from forming at the tips of the bristles. In theory this is probably true, but I question the practice on one count. It splatters paint both in and outside the paint can. You'll have to be your own judge on this.

Brushes wet with paint are always a problem. Where do you put them when you need to take a short break? Some suggest gluing a small magnet to the brush handle so you can stick the brush on the side of the paint can. It's a nice idea, but there are some catches. First, some manufacturers are starting to use plastic cans. Second, the paint drips from the brush down the sides of the can.

Here's an idea that works better. Buy stiff, large-mesh screening that will hold its shape when formed. Cut a semicircle about three inches larger than the lid of the can. Form this over the lip of the can in the shape of a half basket, leaving half of the opening uncovered. There's room to reach into the can with the brush. But you also have a resting place for the brush on top of the can where paint will drain from the brush back into the can.

Here's another idea: Cut hardware cloth so it fits over a rectangular baking pan and bend it down into the pan. Fill the pan with solvent. Then leave the brushes to soak on their sides, not on the bristles. The screen lets residue fall through to the bottom of the pan, away from the brush. A Teflon-coated baking pan makes cleanup easier.

There was a time when roller covers were expensive and were worth cleaning. My experience is that you can't truly clean a roller cover economically. Yes, there are methods. Yes, there are people who claim it can be done. And yes, there are spinners manufactured for just that purpose.

But why bother with the added cost, time, and effort when low cost disposables are available? A poorly-cleaned roller cover that has hard, caked-on paint is useless for quality painting. It'll leave streaks and voids, and give poor coverage. The old paint in it may dissolve and discolor the new paint. Do you want to get maximum use from an old roller cover? Use it for applying paint stripper, driveway sealers, etc.

Did you ever forget to clean the roller cover and later find it stuck to the roller frame? To remove it, slit the cover with a sharp utility knife down the full length. If a knife won't work, try a hacksaw. Once it's split, pry the cover off the frame with a screwdriver blade. If this doesn't work, soak the cover in solvent for 24 hours. It should soften enough for removal.

Spray Equipment

Airless Systems

The marketplace has been flooded with low cost airless spray paint systems designed for the do-it-yourselfer who does one paint job a year. Don't waste your money. Rent a professional unit until you can afford to buy one. They cost $500 to $600 used, $1,200 to $1,700 new, and rent for about $60 a day. To save money, plan your work to minimize the time you'll need the airless unit. Get all the prep work done the day before you rent it so it's in use all day, not sitting idle while you scrape a wall.

There are electric and gas-driven models, in hand-carry and portable wheel-around units. For interior work, I like the hand-carry electric type. For exterior work, I find the wheel-around gas units work best.

A word of caution: Pressure at the tip is very high and can be dangerous. *Don't point the spray at yourself or anyone else.* The pressure is high enough to inject paint under the skin. Paint has chemicals that can cause painful swelling and inflamation. Don't risk injury or possible death. Keep the tip pointed away from people.

Conventional Spray Equipment

If you're going to do texture work on a small scale, consider buying a minimum one-horsepower compressor. It should use 120 volt power and draw no more than 15 amps. Many older homes don't have 230 volt power required for running some of the larger compressors. Even a 20 amp load may be too much for the circuits. A good tank-type spray outfit with a one or two cylinder compressor will cost about $400. If you're planning on doing a lot of stain, lacquer or metal painting, I suggest you make the investment.

If you plan to do mostly heavy enamel or latex work, save your money and go directly to an airless. The heavy-bodied paints need a lot of thinning to work in a conventional spray system, sometimes to the point where you have to respray several times to get coverage. A good airless spray rig will accept paint as thick as mud and still give you fast one-coat coverage. A good professional quality airless rig delivers one-third of a gallon per minute or better to the surface.

Spray Shields

If you're going to spray paint, you'll need a spray shield, not a paint guide. A paint guide is something else. It's a small blade 6 to 24 inches long with a handle and is used when working with a brush. A paint guide is well worth the money if you're doing a lot of trim work with a brush. But for spraying, use a shield. It's 36 inches long, 12 inches wide, made of metal or plastic, mounted on a 36-inch handle. It allows you to spray walls along ceiling lines and trim without the overspray getting on the ceiling, or any other protected area.

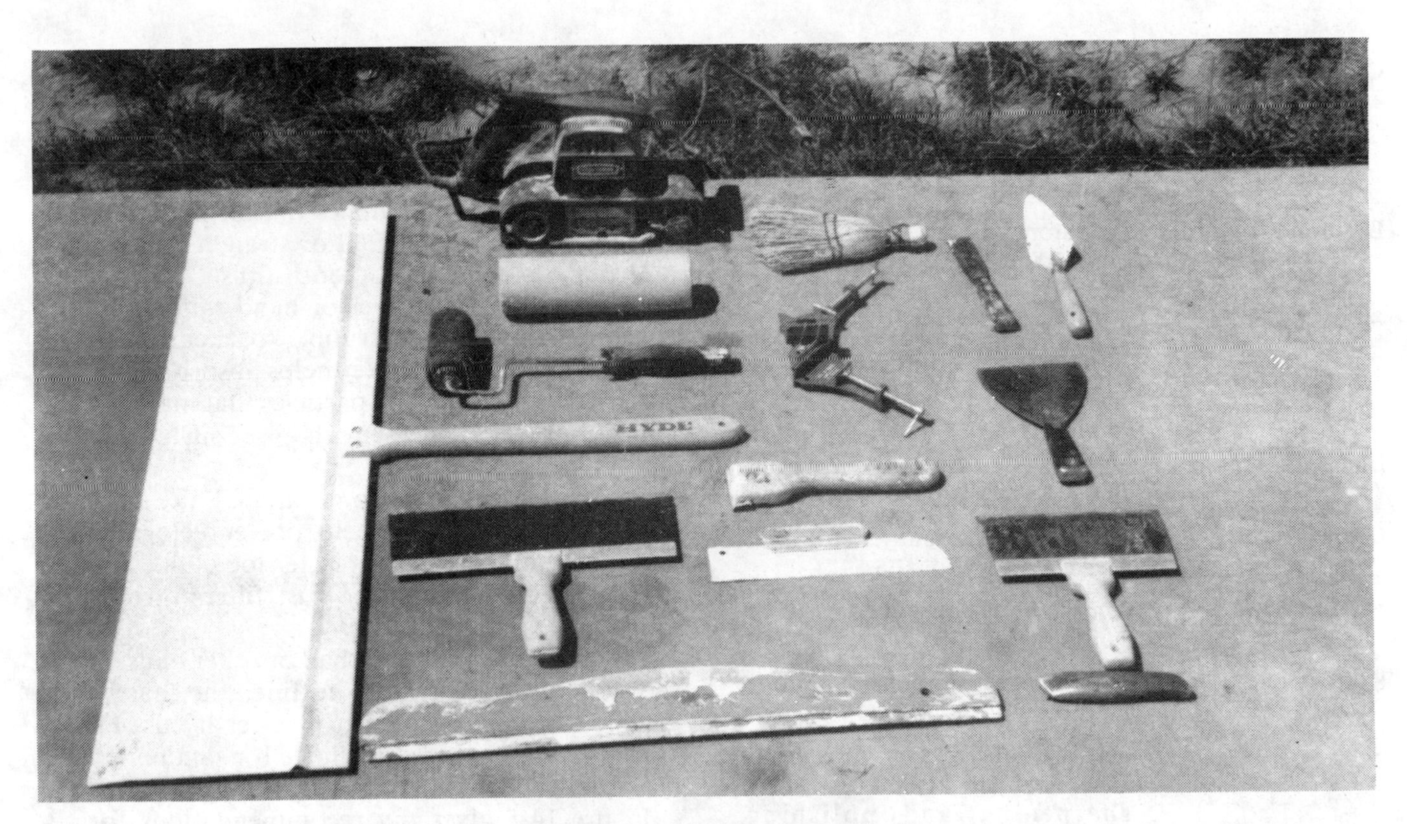

Spray shield, paint guide, drywall knife, scrapers and other small tools
Figure 14-2

I recommend a metal spray shield rather than plastic. The plastic shield is thicker and doesn't get into tight places as well as a thinner metal shield. The plastic is easily cleaned, but I don't think that outweighs the disadvantage of thickness. I use a plastic shield only for spraying acoustic ceilings. The thicker plastic shield automatically leaves a 1/8-inch paintable border line.

No matter what type of guide or shield you use, paint builds up on both sides. You'll have to clean them often during use. Both a paint guide and a spray shield are shown in Figure 14-2. The white foam roller shown in the picture, with serrations every 3/8-inch, is made for acoustical ceiling work. The serrations trap loose particles, preventing them from falling everywhere.

Tools to Remove the Old Paint

Sandpaper, one of the painter's most common tools, usually isn't considered a tool at all. But it is, of course. It's an often misused, abused and poorly selected tool. Let me take a minute of your time explaining what every professional painter should know about the abrasives we use every day.

Abrasives

There are many ways to smooth a surface. Most are based on the principle of rubbing a hard surface against a soft surface to wear away and smooth the soft surface. Even little kids know how to rub the end of a stick against a rock to create a point on the stick. In primitive societies the people learned to smooth and shape the rock itself by rubbing it with sand. Around 700 B.C., man discovered the diamond, which is hard enough to cut and polish most other materials. The final step was inventing the process of bonding abrasives to a flexible backing. Sandpaper was born. Today we use synthetic abrasives and adhesives, but the concept is the same.

The professionalism of your paint job depends in part on the correct selection and use of abrasives. The best paint job in the world can't hide a rough, poorly prepared surface. And you can't select the right abrasive until you know their properties. Look at Figure 14-3, a chart of common abrasives.

Corundum (Al_2O_3)	A natural form of aluminum oxide from alumina (Al_2) and oxygen (O_3). Includes the gemstones of sapphire and ruby. Rockwell hardness #9.
Diamond, natural	Gemstone used as an abrasive since 700 B.C. Rockwell hardness #10.
Diamond, synthetic	First used in 1955 as an inexpensive replacement for natural diamond. Hardness #9.5 to #10.
Emery	A dark granular form of corundum.
Garnet	Red silicate, usually dark red. Hardness #6.5 to #7.5.
Pumice	Very light volcanic glass used for polishing.
Rottenstone	Decomposed siliceous limestone used for polishing.
Sand	Mixture of siliceous materials first used in 21,000 B.C. for sharpening and polishing weapons.
Silica (SiO_2)	Silicon dioxide found naturally in agate, flint, and quartz.
Silicon carbide (SiC_2)	First Synthetic, trade name Carborundum. Formulated in 1891 by Edward Acheson.
Tripoli	Skeletal remains of minute organisms. Known as jeweler polish.

Guide to common abrasives
Figure 14-3

If you're just painting walls and exteriors, you'll have little use for powdered abrasives. But if you want to do fine cabinetry, furniture, metal, or ornamental work, they're a must. They're the only way to give a smooth-as-glass hand-rubbed finish to fine wood or metal.

For painting buildings and equipment, however, the various sandpapers will do most any job required. Always start with the highest grit number that will accomplish the job in the shortest period of time. There's a very human tendency to want to skip a step or two in any job. Don't. If the surface requires sanding, then sand it. To skimp here is to spoil an otherwise good paint job. Don't skip grit sizes either — it only makes the job more difficult. Finish each job using the highest number grit that meets your requirements for surface smoothness.

For hand sanding, start with the higher number grits. For machine sanding, use lower number grits. The speed of the machine sander causes an abrasive action that matches that of hand sanding using higher numbers. Belt or straight line sanders are your best bet for a smooth finish.

When using any sander or hand sanding block, be sure there's a soft rubber backer under the abrasive sheet. This backer helps absorb shock and acts as a cushion for any particles that may clog the abrasive sheet and scratch the surface being smoothed.

For plaster or rough wood sanding, use an open-type abrasive sheet to help prevent clogging. A special sanding sheet for plaster looks like a white plastic screen. The plaster dust filters through so the grit doesn't clog.

Abrasives are ground in ball or roller mills, much the same as paints are. The finer the material is ground, the higher the grit number it has. Figure 14-4 gives the grit sizes available for sandpaper and other abrasives.

Figure 14-5 gives my recommendations for selecting abrasives and grit sizes for various sanding and finishing jobs. This is only a guide, as each material and surface differs in the degree of smoothness, and each user has a different feel or touch.

When you're using any abrasive, let the abrasive do the work. Applying too much pressure or force can gouge or mark the surface. Use a coolant if possible, usually water. Heat generated by sanding can destroy the abrasive, and possibly the surface being sanded. For fine polishing work, always use a lubricant such as water or oil to help reduce scratches.

Liquid Sanders

For paint removal and feathering, I prefer liquid sanding fluids over abrasives. They do the job quickly and easily, allow very good feathering, and will *amalgamate* many paints. By that I mean they'll bring many paints back to life by cutting through dirt, grease, grime, and surface oxidation. They tend to heal small blemishes and cracks, and they prepare the old paint for the application of new paint.

There are a couple of disadvantages, however. Many liquid sanders are toxic and flammable. Follow the precautions on the label.

Machine Sanders

Machine sanding can remove most paints. But I don't recommend removal of lead-base paint by

Type	Grade	Grit
Aluminum oxide paper	0	80
	2/0	100
Color: Light backer sheet with	3/0	120
gray-black grit	4/0	150
	5/0	180
	6/0	220
	7/0	240
Aluminum oxide paper, no-clog	0	80
	2/0	100
Color: Light backer sheet with	3/0	120
gray-black grit	4/0	150
	5/0	180
	6/0	220
	7/0	240
	8/0	280
	9/0	320
		400
Garnet cloth, no-clog	2	36
	1½	40
Color: reddish	1	50
	½	60
	0	80
	2/0	100
	3/0	120
Garnet paper, cabinet paper	1/2	60D
	1/0	80D
Color: reddish	2/0	100C
	3/0	120C
	4/0	150C
Garnet paper, finishing paper	5/0	180
	6/0	220
Color: reddish	7/0	240
	8/0	280
Garnet cloth, molding cloth	1/0	80
	2/0	100
Color: reddish	3/0	120
	4/0	150
	5/0	180

Type	Grade	Grit
Wet-dry silicon carbide	F	320A
	VF	400A
Color: Black	XF	500A
Color: Non wet-dry SC paper is	XVF	600A
light with silver grit	XXVF	1200A
Sanding disk	Fine	
	Medium	
	Coarse	
	Extra-coarse	
Belts, cloth	2	36
	1	50
Normally garnet	½	60
	0	80
	2/0	100
	3/0	120
Pumice	Fine	
	Medium	
	0 F	1F
	2/0 VF	2F
	3/0 XF	3F
	4/0 XVF	4F
Steel wool	Fine	0000
	Extra fine	000
	Very fine	00
	3	
	2	
	1	
	1/0	0
	2/0	00
	3/0	000
	4/0	0000
Nylon web pad	Fine	
	Medium	
Rottenstone	Super fine	
Tripoli	Super-super fine	

Abrasive grades and grit numbers
Figure 14-4

sanding or sandblasting. It's too easy to inhale the dust and lead particles. Instead, use chemical removers. Just remember to wear eye and body protection. The chemicals can cause severe burns and even blindness.

For sanding, many sources recommend a *rotary* or *disk* sander. In my opinion, disk sanders are appropriate only for smoothing welded joints on metal prior to finish sanding. They're hard to control, leave cut marks, and on wood the rotary action cuts too many of the wood fibers. An *orbital* or *belt sander* is a better choice. They're almost always easier to use and leave a smoother finish. Use the orbital sander on "orbit" for rough sanding and on "straight-line" for finish sanding. Figure 14-2 shows a belt sander, along with some of the other tools you'll need.

Surface	Abrasive recommended	Grit to start with
Bare wood		
rough	Garnet cabinet paper	150
smooth	Garnet finishing paper	280
very rough	Garnet no-clog cloth	120
turnings	Garnet molding cloth	180
machine use	Garnet no-clog cloth	120
machine use	Cloth belt	120
machine use	Sanding disk	Fine
Softwood (fir, pine)	Garnet finishing paper	220
Hardwood (oak, walnut)	Garnet finishing paper	280
Bare metal		
weld bead	Aluminum oxide no-clog	100
weld bead	Aluminum oxide disk	Coarse
surface	Aluminum oxide paper	180
aluminum	Aluminum oxide no-clog	220
Composition board	Aluminum oxide paper	180
Plastic		
rough cutting	Aluminum oxide no-clog	240
scratches (optical use)	Tripoli	SSF
scratches	Plumice	4F
to be painted	Aluminum oxide no-clog	400
Sealer		
wood or metal	Silicon carbide	400
Primer		
wood or metal	Silicon carbide	400
Enamel		
wood or metal	Silicon carbide	400
Lacquer		
wood or metal	Silicon carbide	400
Finishing of top coat on fine furniture or autos	Silicon carbide wet-dry	600
Finishing of top coat on cabinet projects	Aluminum oxide no-clog	400
Final finish of top coat on furniture or autos	Rottenstone or tripoli	SF
Piano finish	Tripoli	SSF
Cleaning or scuffing	Scotch Brite	Fine
Paint		
removal, chemical	Steel wool	4/0
removal, machine	Belt or disk	120
removal, hand	Aluminum oxide paper	120
Wallpaper, scoring	Garnet no-clog cloth	80
Wet sanding	Silicon carbide wet-dry	400

Recommendations for abrasive use
Figure 14-5

Wire Brushes

Wire brushing, either by hand or by machine, is often recommended for removal of loose paint. If you do wire brush, use only stainless steel brushes. Steel, aluminum, and brass brushes will themselves abrade, leaving a residue on the surface. The residue from a steel brush can rust. The residue from aluminum and brass won't stain, but it does change the surface texture and reflection, leaving what looks like a discoloration of the paint applied over the brushed area. This also happens if you smooth a surface with steel wool. Many professional paint stores carry stainless steel brushes.

Water Blasters

Water can also be used as an abrasive to remove paint. You can buy a water blaster for about $2,500 or rent one for around $60 a day. If you have a building with hundreds of square feet of chipping, peeling paint, a water blaster is the best choice. Just hook the machine to a standard garden water spigot. It converts 40 to 60 P.S.I. pressure in the water line to a pulsating 2,000 to 4,000 P.S.I. water jet. When guided at an angle across the building surface, the water will pop paint off fast. It's also ideal for general cleaning of stucco and aluminum sidings.

There are two disadvantages. A highly pressurized stream of water is dangerous. Don't aim at yourself or anyone else. You can easily injure someone by accident. Second, don't get too close to the surface being cleaned. The pressure can easily cut a hole through the side of a wood or stucco building, and do it as fast as a saw could.

On wood surfaces, allow a few days after cleaning for the wood cells to dry. You've injected water under the surface and into the cell stucture. This can create another problem: Any paint not removed in the initial blasting may come loose as the wood dries, requiring a second water blasting or at least some scraping or sanding.

If you never used the water blaster before, you're in for some fun. It's easy to make the paint fly. Doing the job by hand would take days instead of hours. But be ready for the initial kick-back when the nozzle is first opened. There's a lot of pressure and the unit does have some kick. Unless you intend to replace the glass, don't even think about aiming it at a window. At high pressure, begin spraying 30 to 40 inches from the surface and slowly bring the water stream closer to the surface at a 15 to 20 degree angle. When the paint starts to fly, you're close enough. At low pressure settings, stay about 3 to 5 inches from the surface. For cleaning dirt, chalk, and grime, 12 inches is about right.

The units are fairly large and heavy. So plan to have a helper available when loading or unloading from your truck or van. Most units have wheels and can be moved easily by one person once on the ground. You'll need between 50 and 100 feet of 3/4'' or 5/8'' garden hose and at least 50 feet of high pressure hose. Also rent a 3 to 4 foot spray extension.

If you own or plan to buy a water blaster, consider putting it to use between paint jobs. Rent it out for cleaning house trailers, siding and roofs, commercial buildings, heavy equipment, and trucks. Rental income from your blaster can be a nice sideline business. Many water blasting units are self-siphoning or can be attached to an aspirator so they spray chemical cleaners along with the water.

Removing Paint with Heat

Many painting authorities recommend removing exterior paint with a propane torch and a scraper. Yes, it works. But here are my objections:

1) You're going to spend a lot of money on propane.

2) You risk setting yourself and your customer's property on fire.

3) Most paint contains some chemicals that, when burned, give off toxic and harmful fumes.

4) It's doesn't work on rough surfaces.

5) There are better and safer ways to remove paint.

If you're removing paint from stucco or cement, I recommend either sandblasting or water blasting. If you're removing paint from reasonably smooth wood or siding, use an electric *strip heater* paint removing machine. They cost from $40 to $90 and produce heat in the 800 to 1,600 watt range. Notice that's *heat,* not flame. The cost of that much electricity is only pennies per day, not dollars as with propane. The units are a little slower but do a very effective job of softening the paint for removal by scraping.

Masking with Painters Tape

Most painting authorities advise painters to use masking tape. What they should be telling you is to use *painters tape.* Masking tape causes too many problems. First, it's slightly embossed and allows paint to seep under it. Second, the paper will absorb paint by capillary action, sucking paint into and under it. Third, if allowed to remain on the surface for more than a few hours, it dries out. Dry tape tears and cracks when removed. Fourth, it has too much adhesive. The longer it stays on a surface, the stronger the bond becomes. It can pull the paint off with it, or leave a residue of glue on the surface. Then you have to wash it off with solvent and risk washing off some of the paint. Figure 14-6 shows the kind of damage tape can do to drywall. It took off not only the paint, but the drywall paper as well. This will have to be replastered before it can be repainted.

Tape damage to drywall
Figure 14-6

Easy-mask is a nationally distributed painters tape. It doesn't have the disadvantages of masking tape. I recommend it highly. The cost is high compared to masking tape. But it leaves a clean edge-line, is easy to remove, and saves manhours. You don't believe me? Look at Figure 14-7. This decorative strip will have to be repainted because the masking tape left a very rough edge-line. This painter didn't save any money by buying the cheaper masking tape.

When painting next to a hard glazed surface such as ceramic tile, a bathtub, a sink, or windows, it's usually easier not to mask or take the time to cut a proper edge. Either wipe the wet paint off with a damp lint-free cloth or allow the paint to dry. Once it's dry, you can use a flat blade screwdriver or a razor scraper to remove the excess. The only precaution is to avoid painting the grout — it's very difficult to clean.

Rough edge left by masking tape
Figure 14-7

When painting windows, most instructions will tell you to paint onto the glass by at least 1/16 inch. This is to seal the glass and window putty joint against weather. There are two or three ways to do this. Some painters have a good eye and a steady hand and can brush the paint onto the glass in a 1/16 inch, even, straight line. You're not that good? Me either. The second way is to use painters tape to mask the glass and leave the 1/16-inch border. Or third, just paint onto the glass and wait until the paint has dried somewhat. Then hold a 1/16-inch wide straightedge against the putty and scribe a line with a sharp razor. From this line scrape the hardened paint inwards towards the center of the glass. Use a single edge blade in a blade holder designed for scraping, available in most stores for under a dollar. I find this last method is usually faster and cheaper than masking, and the results are very good.

There's also a new method on the market. I've seen the product but haven't tried it yet. But the manufacturer claims it works, so give it a try. It's a liquid masking designed especially for masking windows. You wipe it on the glass, paint as usual, and then wipe off the overpaint. Dried paint and masking are supposed to come off, leaving a clean glass surface. The manufacturer is Wagner and the product should be available in larger paint stores.

Caulking

Most authorities suggest that you caulk everything before painting. The reason is that caulk is basically a sealer. It fills cracks so air and water can't leak through. Figure 14-8 shows what happens to a paint job that's not properly caulked. And the caulk itself gets additional protection from the paint. On exterior work, I agree. Generally you'll caulk first, then paint. But there are exceptions. Silicone-base caulk is more durable than oil and latex base caulk. But silicone-base caulk usually can't be painted, so you must paint first, then caulk.

It's hard to apply caulk in a true straight line. This means you'll seldom get a good paint edgeline when painting on caulk. Inside a building, caulk doesn't require the protection of paint since weathering isn't important. You use caulk on the interior primarily for its decorative value, filling voids between a wall and a door, window, or counter, for example. If you paint first, then caulk, you'll usually get a better looking, more acceptable job. The exception is around a tub or shower. Caulk first, then paint so that water and steam don't "weather" the caulk.

It's a judgment call, in my opinion: caulk first and then paint, or paint first and then caulk. Which is going to give the most protection and look the best? If caulking will be covered by trim, the choice is easy. Caulk, paint, then add the trim.

If adjacent surfaces are the same color as what's being caulked, caulk first, paint second. In my house, the tub/shower is almond color. I painted the walls in an antique white to blend the wall and tub together as one unit. That's why I caulked first and then painted.

If caulking is worn or cracked, it should be removed and replaced. Scrape off the old caulk, brush the surface to remove any small particles, and then prime. Now recaulk, being sure to fill all voids, cracks, and seams. Allow the caulk to set up for at least four hours, preferably 24 hours. Then complete your painting.

Some manufacturers recommend letting caulk set up for four weeks before adding paint. That's usually not practical. I've never had trouble painting after 24 hours if painting is done carefully. Flow the paint on liberally but not to the point where it drips or runs. Use a light touch on the brush when coating the caulk so the caulk doesn't pull or sag away from the surface. Always paint at least 1/16 inch past the caulking so you get a good seal. Most new caulking will require a minimum of two coats of paint or stain for adequate coverage.

Colored caulks are available from some manufacturers. These require set-up time. Don't get them wet for 24 hours after application. Figure 14-9 shows one brand of colored caulk and a caulking gun.

Improper caulking causes checking and flaking
Figure 14-8

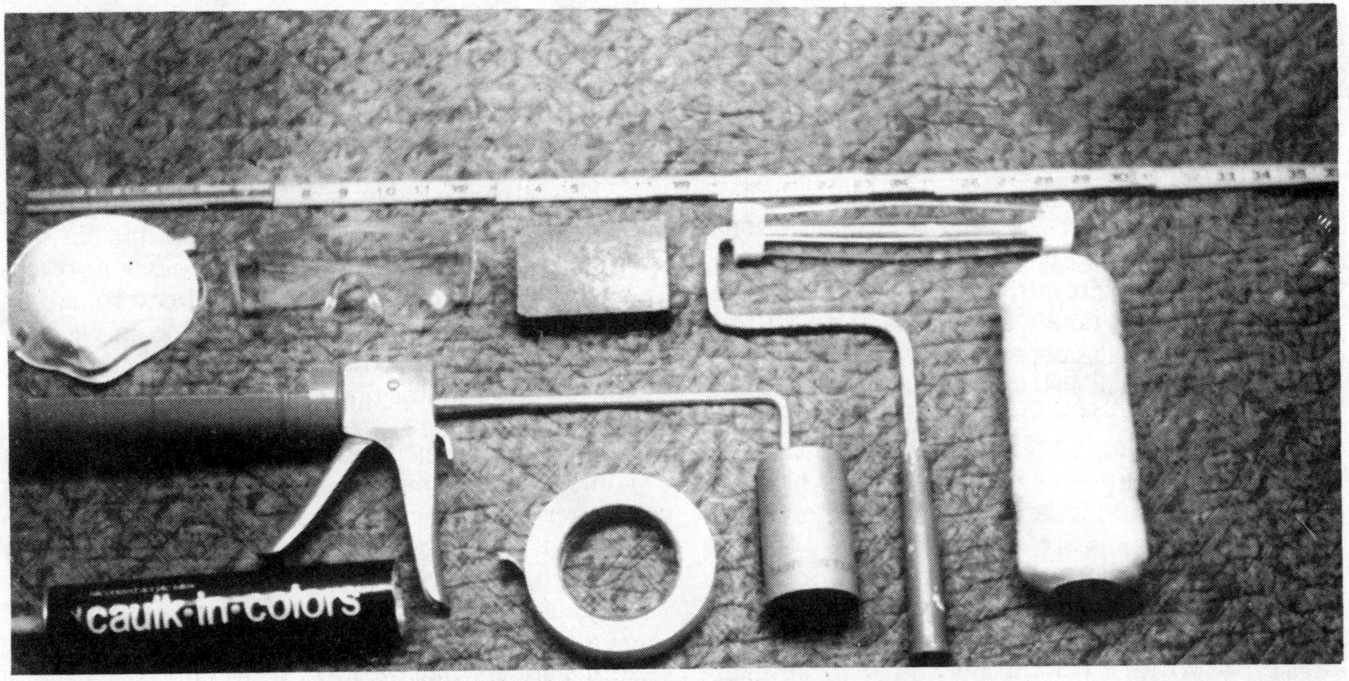

Basic tools for caulking and roller painting
Figure 14-9

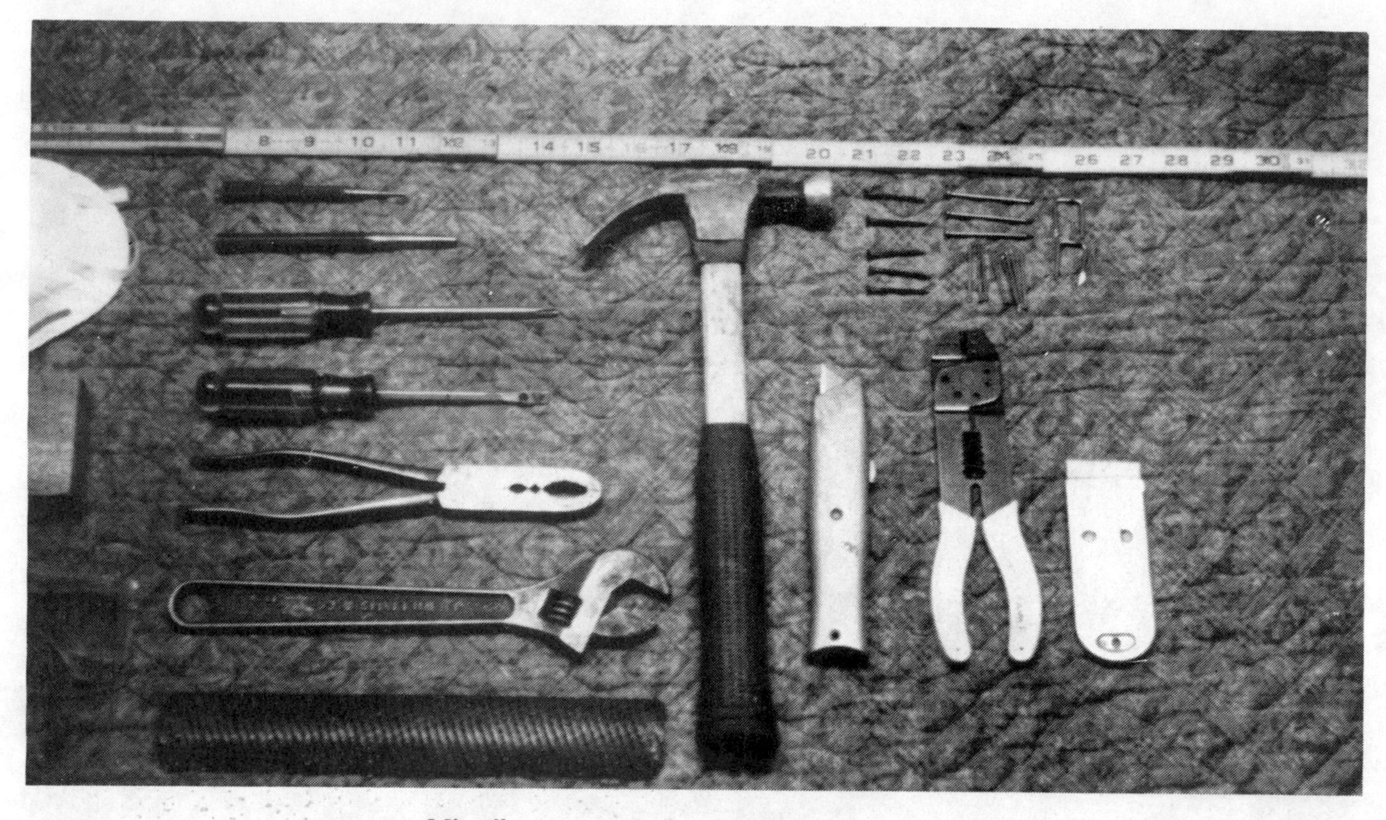

Miscellaneous tools for the nonpainting jobs
Figure 14-10

Miscellaneous Tools

You're going to need a few general-purpose tools in your tool kit. Include a few phillips and straight blade screwdrivers, a hammer, a punch for resetting nails, a caulking gun, pencils, a utility knife, extra blades, a small coping saw or keyhole saw, and some rasps and files. Don't forget a sanding block, a power sander, a few clamps, some assorted nails, neoprene gloves, a putty knife, scrapers, pliers, wire strippers, a drill and bits, sandpaper, and a tape rule. Buy a 25 to 50 foot 3/4'' wide rule that's coated with Mylar. It will last longer and be much easier to use than a short, narrow rule.

Plan on spending at least a few hundred dollars on this assortment of hand tools. Buy the best you can afford. You'll save money in the long run. Some of these tools are shown in Figure 14-10.

One tool you can make at little cost. Solder an alligator clip on the end of a wire coat hanger and use this tool to relight the pilot lights you turned off. (You *do* turn off the pilot light of any gas-operated appliances in the area being painted, don't you?)

Did I mention a box of matches and a flashlight? I also keep a tripod and a set of floodlights with me. That may be the only way to find paint ''holidays'' in a dimly-lit room.

Toss a box of cotton swabs in your kit. They make good disposable brushes for touch-up painting or for cleaning spray gun tips. I also carry a box of wood toothpicks in my tool kit. For small indentations, such as finishing-nail heads in wood trim, you can make a tiny brush from a toothpick. Chew the end of a wood toothpick until the wood softens and feathers and use that feathered end to dab on paint. Toothpicks also work wonders in cleaning out softened paint from corners and other tight spots without damaging the wood surface.

Speaking about cleaning, what about cleaning yourself? Most painters leave the job with paint splatter over their hair and skin. Cleanup is much easier if you apply cold cream and hair spray before painting begins each day. Apply cold cream on your exposed skin and hair spray on your hair. You'll be surprised how easy cleanup becomes. I've seen painters put on long sleeve shirts, gloves, a hat, even a ski mask before beginning work. But they get uncomfortable and hot before too long.

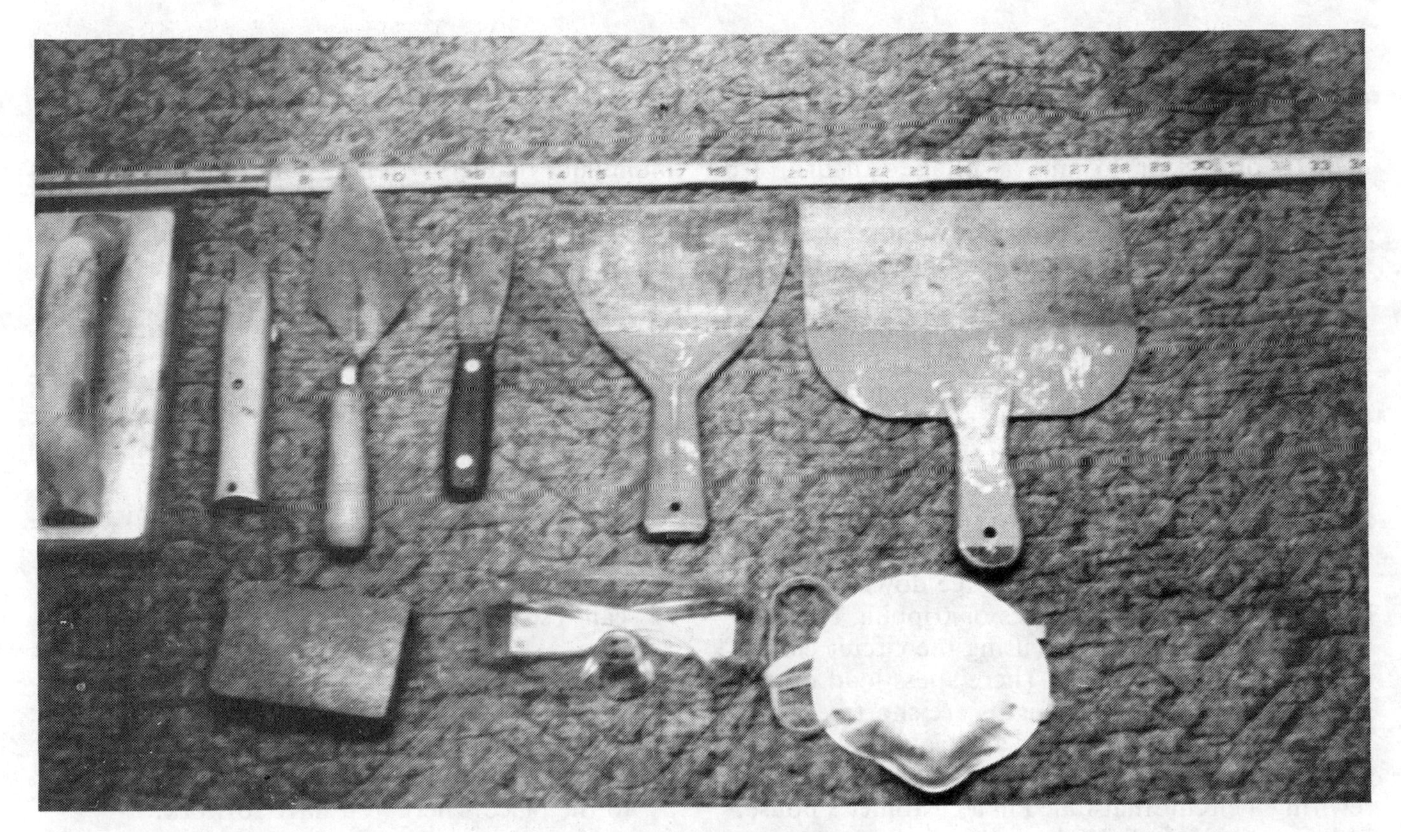

Plaster and drywall repair tools
Figure 14-11

Bulky clothing can cause accidents and will snag on wood, furniture, trees and bushes.

Tools for Window and Wall Repair

A window putty "V" tool is great for reputtying windows before you paint them. Don't forget some patching cement and spackle for repairing walls. You'll need a small mixing tub. The plastic container soft margarine comes in works perfectly.

Every painter makes plaster and drywall repairs. Have a flat sanding block for large areas, a utility knife for undercutting cracks, a pointed trowel for applying new plaster, a putty knife for smoothing, a cold chisel or two for chipping out old plaster and cement, a broad-blade smoothing knife, a grit-covered sanding sponge, eye protection, and finally, a dust mask. Figure 14-11 shows the basic wall repair tools.

Landscaping Tools

If you're going to be painting exteriors, you'll need a few extra tools in your kit. Most homes have tree branches or bushes growing against some wall surfaces. A trimmer or saw comes in handy. If you have to do any trimming, be sure to include the cost of this labor in your bid. And let the owners know you'll be cutting some of their plants. You may be surprised at their reaction if you hack off a limb of some ratty-looking shrub that was in your way, without telling them first.

If you have to cut large tree limbs, seal the cut to prevent insect damage. Sealer paint made for this purpose is available at better garden centers. If you don't want to take the time, or you think the work is dangerous, tell the owners what needs to be done and let them contract it out to a qualified company. Either way, the plants should be trimmed back from the paint surface or you'll have a call-back in a few months.

Some painters carry a pick, shovel and hoe oı rake with them. If you paint a foundation or wall, the coating should go down below grade level. You'll have to dig away several inches of soil to paint below ground level. Make sure this work is in your bid. If you're painting a fence, a grass cutter comes in handy.

Electrical Tools

On interior work, you'll have to remove switch and outlet box covers. Keep a small circuit tester with you to be sure the electricity is turned off when you think it is. You'll also need some pliers.

Extension cords are going to be required if you're using floodlights or electrical spray equipment. Buy good ones — the orange, waterproof, 25 amp, exterior type. Why 25 amp when the equipment is only 8 or 15 amp? Voltage can drop as much as 20% from one end to the other in an underrated extension cord. When your compressor starts up it may draw 60 to 120 amps for a fraction of a second. With cheap cords, the amperage draw increases substantially, and that may be too much demand for the circuits in an older home. You could easily burn out the wiring. And the amperage loss from a cheap, underated cord can cause your compressor motor to burn out. That's very expensive compared to buying a good extension cord in the first place.

Always start your compressor with the air valve open to cut the starting amperage down to a safe level. If you're blowing fuses or tripping breakers with your compressor, try using the circuit closest to the distribution panel. There's less load on the system when the circuit lengths are shorter.

If you think poor wiring is a trivial problem, you should have been with me the day I spent 14 hours rewiring a circuit that failed in a customer's house. I blew out half the lights in the house a second after my compressor went on. Unfortunately, the problem wasn't a tripped breaker. A wire had burned out somewhere between the service panel and the receptacle I was using. I had to rewire the entire circuit. It ran above the ceiling on the first floor and below the carpet on the second. All that rewiring was done at my expense.

Scaffolding and Ladders

Unless you're well over six feet tall, you can't reach the ceiling line of a wall and place paint accurately. I'm six feet tall and can reach the ceiling with handle extentions. But they don't allow trimming in at corner joints. If you need some type of platform to reach the ceiling line, consider adopting my system. I've attached a sheet of 1/2-inch plywood to fill in one whole side of an eight-foot aluminum ladder. The plywood should fill in the entire height and width of the rail side of the ladder opposite the steps. Use screws to attach the plywood every 6 inches along each rail. Countersink the screws so you don't trip on them. I lay this lightweight, portable platform across two low sawhorses so I'm about 18 inches off the floor when standing on the platform. The platform is wide enough to move around on easily.

While we're on the subject of ladders, here's my advice: Buy the best you can afford. Wait if you have to. Rent a good one when you need it until you can buy one of your own. The lightweight aluminum ladders aren't built for continuous heavy use. Buy heavy aluminum or, better yet, wood. Extension ladders for use outside can be aluminum, but buy the best quality. In the long haul, it'll save you money. For work more than 20 feet above ground level, rent a 30-foot ladder from an equipment rental yard.

Here's more advice, on ladder safety:

- Keep metal ladders away from power lines.
- Always set ladders on dry, level ground.
- On hillsides, dig out a place so the ladder feet can rest flat.
- When using a ladder, always climb holding the rails, not the rungs.
- Don't overextend your reach. Move the ladder to the place where you have to work.
- Buy only heavy-duty ladders.
- Look for wood ladders that have a wire rung under each wood rung.
- Keep your ladders in good mechanical shape at all times.
- Never stand on the top two or three rungs, and *never* stand on the fold-out shelf.
- Hook your paint can to the ladder to prevent it from falling.

You can also rent scaffolds, like the one shown in Figure 14-12. But don't set it up like the picture shows. This painter is taking unnecessary chances. He's set up very close to the power lines and didn't install the top safety rails. He should have invested in a few extra platforms to have a wider area to work from. There's no sense risking injury to save a few dollars. The cost is very reasonable, considering the ease of working and the safety you're buying.

One person can assemble a 4 x 10, 15-foot high unit in about fifteen minutes. The trick to assembly is not installing the wheels until you get the first section together. That keeps the pieces from rolling away from you. I always draw a crowd when I put

Rent scaffolds for high work
Figure 14-12

one together. People just don't believe that one person can put together such a big unit. Start with an end section, add the two side rails, then the other end section. Install the cross bar and one platform. Now put the four wheels on and lock to the end sections with locking pins. Lock the wheels tight so the unit won't move. Now from the platform, lift the second layer end rails into place and then the side walls. Put another platform on this section and do the same for the third section. Move the platforms up one section at a time until they're on the top. Finish off with the top safety post and rails. Removal is just the reverse. It's not as hard as it looks.

Being able to stand on a 4-foot wide, 10-foot long platform makes for easy work, especially if you're scraping and painting gutters. Just make sure the area around the building will allow you to move the unit to where it's needed. If not, there are narrower end sections that you can rent. But I suggest you rent the widest units you can. They have a lot more stability. Most 4-foot wide units will fit into a full-size truck bed or into a cargo van.

Your Truck or Van

Speaking of trucks, I recommend a full size pickup. When your business grows, you'll need capacity for ladders, compressor, airless spray system, paints, mixing unit, shaker, tarps, and all the miscellaneous tools. A small pickup may be too small. I prefer vans because they're protected from weather and vandals. You can't lock the bed of a pickup truck. For extra protection in a van, drill holes through the inside door and wall frames and insert metal pins through the holes. Even if the locks are sprung, the pins will keep the door from opening. Since you can't pin the driver's door, use a security lock on it. An alarm system is also a help. During the time I was writing this manual I had a $750 power generator stolen. Believe me, I'm more security conscious now.

If you do any paneling or hang wallboard, you'll need a truck bed that holds a 4' x 8' sheet of paneling or gypsum board. Small imported pickups are cheap and give good gas mileage. But those are the only advantages I can think of. You'll get better use out of a full sized pickup or van.

Your company vehicles are tax deductible capital investments. But you carry the burden of keeping good records and all receipts. We talked about depreciation back in Chapter 3. Also, your truck or van can be a traveling billboard, proudly displaying your company name, insignia and phone number.

chapter fifteen

Surface Prep and Painting

I'll start this chapter by emphasizing again that every good paint job starts with a clean, dry surface. Don't open a can of paint until you've done a good job of preparing the surface. I'll admit that prep work is no fun. But there are some easier, better ways to get any surface ready for coating. That's what I'm going to emphasize in this chapter. I'll share some tips and techniques you can use to make preparation and painting faster, easier, better-looking and more durable.

Mixing the Paint

Let's start with the mixing. The hardest and *least* effective way to mix paint is with a wood paddle that paint stores give away. Paddles are great for scraping solids or pigments from the bottom of a can, but not for mixing the paint. Instead, buy an inexpensive mixing attachment that fits on an electric drill. It's fast, easy, and does the job correctly. But here's a word of caution: *Don't* use an electric drill for mixing oil or flammable solvent-based paints. The sparks generated by the motor brushes can ignite the solvent fumes, causing a fire and possible serious injury.

The best way to mix paint is with a shaking machine like the ones used by paint stores. But these machines are expensive. Few painters need or want to invest that much money to get paint mixed. So what do you do? You *box* or *chase* the paint. Pour the paint into another container, scrape the solids from the bottom of the can, then pour the paint back into the original container. Do this eight to ten times. It's fast and easy and the paint will be well mixed.

For mixing several cans of the same type and color of paint, I suggest you use five-gallon plastic buckets. They cost only a few dollars each and are well worth the price. Mix all the paint you need for the job at one time to get more uniform texture, sheen, and the best color uniformity. Pour the paint slowly to avoid splashing. Always have newspaper or drop cloths under the mixing area.

Cleanup Tips

No matter how careful you are, there will be spills. Always keep a supply of paper towels or rags and the proper cleaner or solvent handy when you're painting. Cleanup is easier if you do it right away. In some areas, like under paint cans, you know there will be spills or splatters. Spread out brown or white paper, whichever will contrast with the color of the paint you're using. You can see the paint splatters, and you're less likely to walk through them and track paint all over the area.

For a major spill, use a dust pan or scoop to control the paint and pick up as much as you can. Discard this paint. *Don't* put it back in the can. It's probably contaminated with dirt or dust and can ruin your paint job. Wash the area with the proper cleaner and blot until dry. Don't rub. Rubbing forces paint into the surface being cleaned.

For a spill on carpet, like the one shown in Figure 15-1, quick work and the right cleaning agent can prevent a disaster. For latex paint spilled on a carpet, grab the paint with a paper towel and pull straight up. When you've picked up as much as you can, begin washing with soap and water, using a clean towel. Repeat the washing as many times as necessary. Blot up the excess water and fluff the carpet fibers when the paint is all out.

Oil-based paint spilled on carpet
Figure 15-1

For oil-based paint on carpet, use the same technique with mineral spirits instead of soap and water. For small drops, let them dry undisturbed. Then cut the drops off the carpet fiber with scissors. I advise against saturating a carpet with solvent. Some carpet binders will dissolve in mineral spirits or other paint solvents. A few will dissolve when soaked with water. Use just enough solvent or water to take up the paint, and blot up the excess immediately. Before the carpet is completely dry, fluff the fibers to prevent them from sticking together.

On a hard, smooth surface like a waxed floor or a laminated plastic counter or window sill, remove latex spills with a towel and water. An oil-based paint spill is sometimes better left to dry. When fully dry it will flake off easily. If you use a solvent like mineral spirits on a waxed surface, you'll probably remove the wax along with the paint. That means rewaxing, of course.

If the spill was on cement, I suggest an immediate drowning in the proper solvent. I always keep a water hose close at hand when painting outdoors with latex. Cement is porous and stains quickly if the paint is allowed to sit for more than a few seconds. If the paint does dry, try chipping it off or use a paint remover such as methenol chloride, followed with a water rinse.

Protecting Surrounding Areas from Paint Splatter

The easiest and safest way to keep paint off of what you don't want to paint is to cover it. I use inexpensive plastic garbage bags to cover items like bushes and shrubs, flowers, lamps, TV's, ceiling light fixtures and so on. Be sure you unplug electric appliances if you cover them. The heat generated by electric fixtures can melt the plastic or damage the appliance.

Some items, such as recessed light fixtures, can't be completely removed. The lens and frame of most square recessed ceiling lights you find in homes do drop down for bulb changing, however. The frame and lens are on wire spring hooks. Pull the frame straight down. It should clear the ceiling by about six inches. Clean the area, and if it's very dirty, spot prime. After the ceiling is painted and the paint has dried, give the light a push. The frame and lens will pop back into place. Larger fixtures should be covered with a trash bag before painting begins.

As a general rule, white electrical wires are "cold" or neutral, green wires are the ground and normally carry no current, and black wires are "hot." In some three-way, four-way and split circuit situations, however, you may find that the white wire is hot. You may also see blue, orange, or red wires. These are usually switched lines. Here's my point: To be safe, treat all wiring as "hot" wiring and potentially dangerous.

One final point. Mask the switch or socket any time you remove a cover plate. A professional painter never leaves behind a receptacle like the one in Figure 15-2. Getting paint on these looks bad, produces a difficult clean-up job, and can even render the fixture useless. Paint, especially spray paint, can ruin a duplex receptacle. That may mean calling in an electrician at your expense.

Remove What You Don't Paint

I'm a lazy type of painter. Maybe you are too. I don't like to do a lot of fine cut-in work. Sometimes that means I have to do a little extra work. For example, I think it's worth my time to remove most of the items I'm not going to paint. I'm not just talking about the pictures and furniture, but also the light switch and outlet cover plates, door knobs and hardware, light globes, mirrors, and so on. If it's not to be painted, remove it.

Wash all the washable items in soapy water, rinse well, pat dry, and set them in a safe place for reinstallation after the paint has dried. Keep screws and other small hardware in bowls or a small plastic bag. Label each as to where they belong.

Why go to all that trouble? First, all those clean, shiny fixtures will make your paint job look that much better. Second, you'll save time in cleanup after the painting is finished. Third, the painting will be done faster because there's less cutting in around small items. Fourth, if you leave them in place and get paint on an edge, you're just asking for a callback. If they aren't removed for painting, eventually the owner will remove them for some reason. That will probably crack or chip the paint, ruining your good paint job. Finally, it's just good customer relations. You do want referrals from satisfied customers, don't you? Spend a few minutes working with a regular and a phillips screwdriver. You'll be pleased with the return.

This duplex receptacle and cover need replacement
Figure 15-2

Preparing and Painting Interior Surfaces

Now that the room's been cleared, with all the unpaintable fixtures removed or covered, it's time to start the actual preparation and painting. I'll cover the interior work first, beginning at the top: ceiling repair and painting.

Repairing and Painting Ceilings

The first step in interior prep may be repairing the acoustic ceiling. If the "popcorn" is loose, damaged, or puffy, scrape it off. Apply acoustic patch and let it dry. Acoustic patch is available at most paint stores for about $2.00 a quart. It comes in two colors, "new white" for fairly new ceilings and "old white" for discolored ceilings. It's ideal for patching holes left after lamp and plant hangers have been removed.

When painting a ceiling, look for possible damage. Kitchen and bath ceilings usually have small yellowish grease and soap spots. Wash them off with TSP and water. Acoustic ceilings may be brown from smoke, or stained from water. Vacuum them gently, then spray with a fine mist of bleach. Be sure to cover floors and walls first, or the bleach will make white spots on them.

For painting smooth ceilings, the roller is usually best. Paint acoustic ceilings with an airless. If you must roll an acoustic ceiling, use a "serrated" foam roller designed for acoustic ceilings.

The ceiling will usually be white to improve light distribution. If the walls aren't white, try this trick. Add one cup of the wall paint to the white ceiling paint. The paint color added will soften the contrast between wall and ceiling color without reducing the light level. One cup to a gallon is plenty. Experiment with this to see if you like the effect. But remember, the paints must be of the same type. Don't try mixing oil/alkyd and latex.

Preparing and Painting Interior Walls

The first step in repairing and prepping interior walls is patching the plaster or wallboard, if necessary. You won't often find many interior walls in perfect condition and ready to paint. For example, you're likely to find dents in the wall behind a door where the doorknobs hit the wall because the door stops are gone. I usually carry a supply of door stops and put one in to keep the doorknob from damaging *my* paint job. Sure, I charge for them — cost plus my mark-up. It's just a little extra service I provide for my customers. For goodwill you might call it. It also happens to put a little extra profit in my pocket.

Repairing plaster walls— If you're faced with holes in an old plaster wall, first wet the surface to be plastered. This helps keep the new plaster from drying out too quickly. Water in the patch will be absorbed by the old wall. Skip this step and you won't get a good bond. But you probably *will* get a call back from an unhappy customer. Next, cut away any loose plaster so the patch has a firm base. Finally, apply the patching material.

Wet-sand the new plaster within a few hours of application to smooth and feather it. The sanding block should be rinsed in clear water frequently to remove plaster buildup that will scratch the surface. I recommend using a grit-coated rubber sanding block. They're available at larger paint stores.

Fully dried plaster has to be dry-sanded with a power sander. Special open-grid sanding sheets are available for sanding plaster. They look like plastic wire screening.

Corner bead warp is a relatively rare problem. But you'll see it occasionally. The paint and plaster will show a hairline crack approximately 1¼ inches from a wall corner. Either the metal corner bead wasn't secured right or it has warped. This causes the bead to push outward and split the paint and plaster. The cure is to sand the surface down to bare metal. Then either "V-cut" the warped section out or renail it. You'll need a pair of tin snips to make the V-cut in the corner bead metal. Finally, apply a new coating of plaster, sand, prime and top coat.

Hairline cracks in paint on cement or plaster are probably the result of changes in the subsurface. The plaster or cement probably wasn't mixed right or had too few solids as a binder. As it dried, stress in the material caused the cracking, just like mud will if it's left to dry in the hot sun. If you just paint over this surface, the paint will be drawn into the cracks. That doesn't solve the problem. Since the material is structurally weak, it should be chipped out and replaced. Since that may not always be practical, an alternate solution is to fill the cracks with plaster, caulking, or texture paint, then prime and recoat.

Before painting over new plaster, let the plaster cure for 30 to 60 days. Then prime with plaster primer or PVA (polyvinyl acetate). Then you're ready to top coat. When I say new plaster, I mean a new lath and plaster wall. Repairs and small patches can be primed and top coated in a few hours or days.

Repairing wallboard walls— I find that premixed drywall compound works better than patching plaster for filling cracks and holes in interior walls. It comes in paste form, goes on smoothly, and dries in 4 to 8 hours. It can be wet or dry sanded to a fine feathered edge. Patching plaster, on the other hand, dries rock hard and is difficult to sand to a feather edge. Whichever you use, a light coat of white primer is needed to seal the surface before painting. The white primer coat will also make it easier to see if you've done a good feather sanding job. If not, resand, reprime, and then paint.

For painting over new drywall repairs, use drywall primer or PVA after the spackle has cured for two or three days.

Preparing splotchy or cracked walls— Splotchy, dull or glossy walls are caused by using the wrong primer and sealer before painting. Painting over the problem won't solve it. You need to sand or chemically degloss the surface, apply the recommended primer/sealer, and then top coat.

If the old paint has cracked in long thin straight lines, it's not the fault of the paint. The drywall wasn't applied correctly, or the wall framing may be shifting. The most common problem area is where walls join or where the wall meets the ceiling. Cracking along these joints means the drywall tape has pulled loose. The only solution is to sand off the old tape and do the job right before trying to repaint. Figure 15-3 shows the proper procedure.

Occasionally you'll see a really bad wallboard job. For example, the ends of sheets may not be attached to anything, or too few nails may have been used. The edge of every sheet of wallboard must be glued, nailed, or screwed to wall framing every 6 inches. If not, you have what's called a *floating drywall joint.* This type of joint cracks when even a little pressure is applied. Someone leaning on the wall or even normal expansion and contraction due to temperature and humidity changes will crack the joint. The only real solution is to remove the drywall, install proper backing, and replace the drywall. Then, after applying joint compound and sanding, prime and top coat.

There's another common problem I've seen in "do-it-yourself" drywall projects and old homes that were drywalled when drywall was still new. The problem is a total lack of joint compound over the drywall tape. The only solution here is to scuff-sand the wall and apply the joint compound, as shown in Figure 15-3.

Clean before painting— Many homes have wall registers mounted high on the walls. Remove the register grilles before painting. Clean the grilles and the wall area around where they mount. The reason? There's usually accumulated dust, oil, and soot in these areas. It comes from the air circulated by the heater. Your brush or roller will pick up this contamination and transfer it to everything you paint, leaving you with a messy cleanup job.

Remember to turn off all heating and cooling equipment before starting to paint. HVAC ducting usually has dirt and dust that can contaminate your wet paint if the fan kicks on.

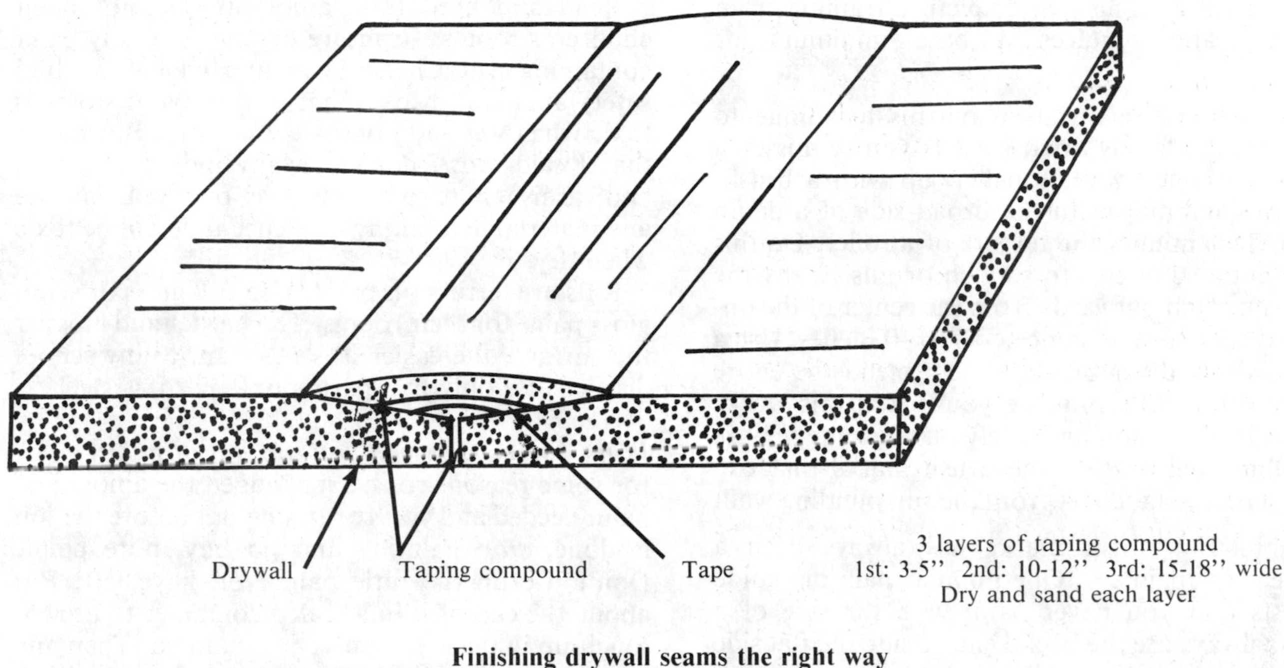

Finishing drywall seams the right way
Figure 15-3

Most homeowners do open their windows from time to time. Airborne dust that enters the building collects on the ceilings, woodwork and wall surfaces. This dust must be cleaned off before you repaint. I've also found interiors that were painted with exterior paint. Exterior paint "chalks" and new paint doesn't stick to chalk. These walls should be dusted and then given a light wash with a liquid sandpaper before repainting.

There's another important source of dust in most rooms: carpets. They should be thoroughly vacuumed before painting begins, especially in the areas usually covered by furniture. Then cover them with drop cloths before you paint. Dust also accumulates on window sills, on door and floor moldings, and in cracks. Use an old toothbrush to move dust out of cracks before you vacuum. A toothbrush won't scratch the surface the way a wire brush would. Another good dust brush is a soft paint brush with the bristles cut off at about half the original length.

Avoiding "picture frame" walls— I've learned to avoid giving rooms the picture frame look that's inevitable if you paint the corners and edges with a brush first and then fill in the rest with a roller. The brushed areas won't match areas where paint was rolled on. Rolling over the brushed-on strips won't help much. By the time you go back to roll, this brush strip is partially or fully dry. New paint from the roller now adds a second coat over the brush coat. This results in a double layer of pigment and a darkened area that emphasizes the picture frame.

One solution is to have two painters working. One rolls while the other follows close behind, brushing. If the time between brushing and rolling is short enough and the film depth is approximately the same for both operations, the coatings will tend to blend.

Unfortunately, most of the time you won't have two men working from the same can in the same room. So here's a technique that works for me. I roll paint up to the corner as close as I can and in three foot sections. Then, with a wet trim roller, I roll as close to the corner as possible. This gets me to within a half inch of the corner. Most rollers will leave an "edge trail" of wet paint that is somewhat thicker than the rest of the coverage. Using a wet brush on edge and at an angle, I start at this edge trail and sweep the paint to the corner in small semi-circles. Then I dab the area with the brush straight on. The bristles give the wet paint a texture that's nearly identical to the rolled area.

If two adjacent walls are going to be the same color, I run a wet roller right into the corner. This squeezes the paint into the corner, filling the joint line. The side of the corner painted should be the wall you first see when entering the room. Then do the adjacent wall. When both sides of the corner are painted, go back to the first wall. This time keep the roller 1/2'' away from the joint. If there are any voids in the joint line, use the edge of the brush to fill in. This avoids picture framing from brushing and requires a bare minimum of brushwork.

You can use a variation of this technique to touch up an already dried wall. If you've missed a spot on a rolled wall, touch it up with a brush. Finish by dabbing with the broad side of a damp brush. This mimics the texture of a roller. Do this very lightly so as not to scratch or push away the wet paint. Dab outwards from the center of the unpainted spot for a distance of 6 to 10 inches. Use a progressively dryer brush to help blend the entire area. With a little practice you'll find this works very well. The dabbing closely matches the texture of rolling, eliminating the brush marks that can make a patch stand out from the surrounding wall.

Eliminating brush marks has always been a problem for painters. One popular painting guide suggests that you never paint with the side of a brush, always use the broad end. Using the flat side puts less wear on brush bristles and reduces brush marks or *fingering.* But in practice, there are times when using the side of a brush is best. The extra wear will be undetectable unless you have a very expensive brush or a favorite brush that you expect to use for years. You're more likely to wear a brush out with improper cleaning or physical damage than use it up by painting with a side or edge.

Painting to reduce missed spots— Do you look at a wall and see a smooth surface? Most surfaces you'll paint aren't smooth. Plaster and wallboard warp and bow, especially between wall studs. Shine a flashlight beam horizontally across a length of wall sometime. The light pattern will reveal all the high and low spots. Your roller has to cope with all these defects if you're going to get complete coverage. A medium nap roller will cover the low spots better than a short nap roller would. Try rolling on paint in a "M" or "W" pattern with horizontal and vertical overlaps to fill the low spots. Work from the dry surface into the wet surface to eliminate runs, sags, and roller marks.

After you've finished each wall, take a few seconds to examine it closely. Reroll any paint misses or voids with a damp, not saturated, roller. This is especially true when using latex paints. Even if the wall is thoroughly wetted with paint, latex tends to separate in spots as it dries, allowing the old paint to show through. These spots are probably caused by handprints or excess paint or primer in the earlier coat. Examine the wall in good light. These voids are sometimes hard to see in poor light and at some viewing angles.

Read and heed this safety tip: Do not paint children's rooms, furniture or toys with any paint containing lead. Check the paint can label for child safety approval. Most paint sold in paint stores is toxic when wet and nontoxic when dry. But notice that I said *most*, not all. Many industrial paints and varnishes are toxic when wet or dry. Don't use any material in a family residence that will be toxic when dry.

Kids are terrors on paint. Use full gloss or semi-gloss paint for their rooms. The harder and glossier the surface, the easier it is to clean. Many school districts use only two-part epoxy, hard as steel.

What do you do when you run out of paint?— If, for some reason, you've misjudged the amount of paint needed and you're running out before the job is done, *stop* painting and go buy more paint. Don't use up the little paint you have left. Put about 1/2 cup of it in a sealed container to use for touch-up in the area you've just painted. Then mix what's left with the newly purchased paint. With luck, you'll get a good blend of texture, sheen, and color.

As you finish each wall, make a quick mental calculation of whether you have enough paint to finish the next wall. Above all, don't run out of paint in the middle of a wall. A small color, texture, or sheen difference from one wall to the next won't be noticeable. But it will be real obvious if you change paint in the middle of a wall.

Preparing and Painting the Kitchen and Bath

You'll want to remove all the drawers before painting cabinets. So that you don't get paint on the contents, ask the owner to empty all drawers first. This protects the contents from accidental paint splatters and protects you from accusations if something turns up missing. It also makes painting the drawer fronts and cabinet surfaces much easier.

When the drawer is empty, set it up on end so the front of the drawer is a horizontal surface. It's easier to paint this way. Paint is less likely to drip, sag or run. To do a good job, you'll have to remove the drawer pulls before you paint. If the drawer is sitting on the floor, it won't accidentally slide shut, effectively gluing itself closed as the paint dries. Prying a drawer open after it gets stuck shut will ruin the paint job.

As a little extra service, I usually carry a small can of oil or grease with me. A little drop or two on the slides before you replace the drawers makes for a happy customer.

Most painted cabinets, baths, and kitchens are painted in semi-gloss enamel. Most are also very dirty or greasy. Grease from hands, soap, food spills, and cooking has coated them with a dull film. Liquid sander works wonders. It both cleans and deglosses so your new paint bonds to the old. Liquid sander is clear in color and difficult to see if it's spilled or splattered. Cover everything that you don't want to paint or replace. Another good cleaner is clear nonsudsing ammonia. Both chemicals smell bad, so warn your customer ahead of time that there will be an odor.

Removing and preventing mildew— Anything that looks like dirt but doesn't come off with soap and water may not be dirt at all. It's probably mildew. Chapter 11 tells you how to remove mildew. When repainting, add one ounce of mildewcide to each gallon of paint to deter regrowth of the mildew. It won't affect the quality or color of the paint. Buy the chemical in quart or larger sizes for the lowest price and add the cost to the selling price of the paint. But let your customer know that you're adding the chemical. They may not want you to if they're sensitive to some chemicals, have children, or have pets in the home.

Mildew is most common in unvented rooms or rooms where the occupants aren't using the vent fans available. Point this out to your customers, and suggest they use the vent fans regularly. You don't want to get a rework call because of fungus growth.

There's another chemical you may want to add to avoid an irritating problem. If insects use your freshly painted surface as a landing field, mix a small amount of insect repellent into your paint. Any commercial insect repellent will do. Mix in one-half teaspoon for each gallon of paint. It shouldn't change the color or quality of the paint. There are commercial repellents manufactured especially for paint. But they cost about $16.00 a quart. If you're only going to use one ounce per gallon of paint, that works out to about 50 cents for each gallon you treat — a small price to avoid insect problems. As an extra benefit, these commercial additives will control insects for several years. That's a good selling point. But again, check with the occupants before using it.

Removing and replacing the toilet— The toilet will be a nuisance in every bathroom you paint. There are three ways to deal with this obstruction. The first is to just not paint behind the tank. But that's a poor option for a professional painter. The second is to use a special roller that's about one-half the diameter of the standard roller — if you can find one. The third option, and the one you should choose, is to remove the toilet.

You'll need a bucket, sponge, plenty of newspaper, a replacement wax seal, a rubber cone washer kit and an adjustable wrench. First, turn off the water supply to the tank. Flush the unit two or three times, then lift off the tank cover. Sponge out any water that remains. Don't worry about water in the bowl, it will stay there. There's a bolt located on each side of the base near the floor, probably covered with a round white cap. Pry the caps straight up and off. Remove the nuts from the exposed bolts. Unhook the water line where it connects to the tank. Now lift the tank straight up and off and move it to the pile of newspapers. Cover the hole in the floor with several sheets of newspaper also. Then get on with your painting.

To replace the toilet, first replace the wax seal with the new one. Make sure it's the correct type, as there are several. Now there should be a mark on the floor where the bottom flange of the toilet sat. Apply a bead of caulking to this flange mark. Move the toilet back into place and lower it over the bolts. You did remove the newspaper from the hole? Replace the nuts and tighten. Don't overtighten as you can crack the base. Connect the water inlet line, using the new rubber cone washers. Sit on the unit and gently rock back and forth to set and level it. Retighten the nuts, apply a ball of caulking to each, and press the white caps on. Turn on the water and check for leaks. You're done.

Why bother with the caulk around the bottom flange? Two good reasons: it prevents air leakage into the house and keeps water out from under the toilet. Water puddled under the toilet will eventually rot the floor.

Preparing and Painting the Living Room

The main problem in a living room will probably be the carpet. We've already talked about the dust, but the second challenge with a carpet is how to move it out of the way so paint doesn't get on the fibers along the wall. You can use a paint guide to push back carpet fibers as you paint the baseboard molding. But as soon as you move the guide, the fibers spring back into the wet paint.

Instead of a guide, I recommend using masking tape — the 2-inch wide variety. With the sticky side away from the wall, slip it between the carpet and the molding. Then smooth it down onto the carpet. It takes a little practice. Now paint your baseboard and let it dry. Remove the tape carefully so it doesn't lift the carpet off the tack strip or pull out too many carpet fibers.

An alternative to masking tape is an old product that's back on the market. These are 3-foot strips designed and formed to be used as carpet protectors. They used to be made of cardboard, but the new ones are plastic and are reusable. They slip in place between the baseboard and carpet and are left there until the paint is dry. They're easy to use, they work, and they're cheap! You'll find them in most paint stores, at about $4.00 for a pack of ten.

Another way to handle this problem is to pull the carpet from the tack strip and fold it back out of your way. But unless you have a kick pad or carpet stretcher, I don't advise it. The tack strip keeps the carpet under tension to reduce the chance of bumps and wrinkles. Take it off the tack strips and you may have a hard time getting it back on.

Removing baseboard— In some cases, it may be necessary to remove the baseboard and later reinstall it. It's hard to remove baseboard without damage. If pried off incorrectly, most baseboard will crack or break without warning. If this happens, you have two choices. Either glue and fit the pieces carefully during reinstallation and then touch up with paint. Or you can replace and repaint the baseboard.

What's the best way to remove baseboard to minimize breakage? First, free the molding from the painted wall by cutting along the top seam with a utility knife. This cuts the paint and prevents it from chipping when the molding is removed.

Next, you have a few choices. You can loosen one end and slip a wooden wedge between the wall and the molding. Carefully hammer the wedge along the joint until the molding is free. The nail heads will pull through most baseboard materials. You can remove the now exposed nails and go about your painting.

If you plan to cover the floor molding with carpet, try prying the molding off. Pry from the bottom upwards and outwards. This will leave marks on the bottom edge of the molding. But the top or visible edge won't be damaged. The new carpet should cover the bottom area and hide the marks.

The best way to remove molding is with a nail punch. Drive the nails through the molding, if you can find them under the paint.

When replacing the molding, always renail at points different from where it was nailed before. Fill old nail holes with filler and touch up the spots with paint. Finally, check the baseboard for scuff marks. Remove any you find with a solvent such as toluene or xylene. Paint won't adhere to scuff marks like the ones in Figure 15-4.

Remove scuff marks from baseboard before painting
Figure 15-4

If you really want to have some fun with moldings and other trim items, try this. You can create what looks like beautiful hardwood from old softwood. I've made imitation walnut from fir 2 by 4's. Sand the surface smooth and then paint with a solid coating of gloss black enamel. Allow to dry a few days. Using a medium or light brown gloss enamel, brush on the wood grain. Use very little paint and very light strokes until the desired effect is obtained. For imitation cherry or red mahogany, mix a small amount of red into both the brown and the black paint. It takes skill and practice, of course, and will work only if you use oil-based paints oı the same type and from the same manufacturer. If done right, the result will be wood that looks like highly polished hardwood. Experiment on scrap lumber until you've mastered the technique.

The railing in Figure 15-5 is pine 2 x 2 and 2 x 4. The beautiful walnut grain comes from a coat of full gloss dark black paint. After it dried for a few days, it was lightly brushed with a full gloss brown.

Installing new trim— Here's a tip to follow when installing new trim. Any time trim lumber is cut, and this includes the newer vinyl plastic trim "lumber," you will have a visible cut edge. Butting the pieces of trim together at corners and around doorways will leave a joint that's a little rough. Try painting the cut ends before you install the trim. Then do your final painting of the joints. That should make the rough edge almost disappear. For the same reason, paint the ends of prefinished trim that won't be painted after application.

Before you install any trim or paneling, always paint the surface under the trim the same color as the trim. This way any seams that aren't quite tight will be masked.

"Walnut" railing is really painted pine
Figure 15-5

Here's how to paint a piece of trim or other lumber on all sides. Screw an eye hook into one end and hang the piece from a rafter or tree branch. Now you can paint all sides at once. A paint mitt or lint-free rag dipped in paint works well. Just dip and rub. When it's dry, unscrew the hook and touch up the end.

Painting a Wall Adjacent to Wallpaper

That last tip was easy; this one's a little trickier. When you paint a wall where the adjoining wall is covered with wallpaper, the wallpaper may act like a sponge, sucking up the wet paint and spreading it into the fibers. Masking can help, but even this isn't effective with some wallpaper. The only sure solution is to leave an unpainted strip 1/4-inch wide next to the wallpaper and cover this gap with moulding.

First do a test on a hidden portion of the wallpaper to test its absorbency. You may be lucky. But don't count on it until you've done a test.

If you're painting one wall and plan to paper an adjoining wall, do your painting first. Feather the paint onto the wall that's to be papered for a distance of about 6 to 10 inches. You'll have a clean paint-to-paper joint that doesn't show any old wall surface or require molding to hide a gap.

Before wallpapering, paint all your door, floor, ceiling, and other trim moldings, feathering the paint onto the wall to be papered. Allow the paint to cure for a few days to insure a good solid base for the wallpaper. The paint should be fully dry so you can wash wallpaper paste off during the wallpapering.

Painting over Wallpaper

You may be asked to paint over wallpaper. I don't recommend it, but if you must do it, at least test the paper first. Is it a foil, flock, vinyl, or a metallic? If so, forget trying to paint over it. The job will be a disaster. *If* the wallpaper is paper and *if* it is properly sealed and *if* it has the right ink, then maybe it can be painted.

The first test is for color fastness. Using a clean white cloth soaked in water, rub the surface. Now do the same with another clean white cloth soaked in mineral spirits. If the color came off with the water but not the mineral spirits, paint with an oil/alkyd paint. If the color came off with the mineral spirits but not the water, paint with a latex paint. If no color came off, then use either paint. If the color came off with both the mineral spirits and water, then you must do one of three things. Choice one is to seal the surface with shellac and then paint. The second option is to panel the wall and forget about painting it. Finally, you may have to face the worst: strip off the old paper.

Removing Old Wallpaper

The trick here is to know your wallpapers and wallpaper adhesives. There's a right way to remove each type of adhesive.

Nonwashable wallcoverings include corks, burlaps, tules, hemps, and grasscloths. They are natural fiber materials that are bonded to a paper or rice paper backing. These coverings are applied to the wall surface with cellulose or wheat paste adhesives and can be removed with water or steam.

Vinyl wallcover is plastic and can be washed. It probably has a backing sheet of either paper or cloth. Most vinyls are strong enough to be removed by peeling. They'll peel in full continuous strips. The adhesive will have to be removed with an organic solvent such as xylene or with mineral spirits. Vinyls are put up with non-water soluble adhesives. Since the vinyl doesn't breath, water soluble paste would mildew under the vinyl.

The cloth-backed vinyls will almost always peel off in full sheets. Loosen the top edge and drip solvent between the wall and the cloth. Once you have a loose piece large enough to grab on to, just pull downwards in a firm, even motion.

Now that I've given you some guidance on the type of adhesive that's used with each type of wallpaper, I'll advise you not to believe it. I've been describing proper application, not what you're likely to run into. Test the adhesive with both water and solvent to find out which cuts the adhesive best.

Foils fall into the same category as vinyls, except they are usually more difficult to remove. The aluminum foil top sheet has a tendency to peel from the backing sheet. It usually peels in small rather than large or full strips. Aluminum is different in that it has a grain direction. If peeled in the right direction, it comes off in long strips.

After the foil is removed, score or rough sand the paper before soaking it in solvent. *Be careful!* These solvents are highly flammable.

Sometimes you'll run into an imitation foil. This is paper that's been printed with metallic ink or paint to resemble real metal. Test what you think is foil by wiping the surface with xylene. If the foil dissolves, it's really ink or paint. If not, it's aluminum.

If it's a printed foil, first wash the surface with solvent, then rough sand and soak. Use a broad blade spatula for the actual paper removal. A drywall knife (trowel) also works very well. Gold tones are almost always inks. Gold foils are too expensive to use as wallpaper. If you find a gold foil, let me know and I'll help you strip it. I'll even cart it away for you.

Flocks are wallpapers or vinyls or foils that have rayon or nylon fuzz bonded to the surface to create the design. It would seem that this flocking would be the weakest point and the place to soak with water or solvent. In fact, it's just the opposite. The background colors are printed over the entire surface. Then water-resistant glue is applied where the design is to be, followed by the flocking. This makes the flocking the least desirable place to start. Determine what the background is, paper, foil, or vinyl, and proceed to remove accordingly.

Hand prints are silk-screened patterns on blank stock. You can recognize them by their imperfections or pattern mismatches. Many hand prints are made with oil-based paints or inks. You must get through this coating before you can remove the backing paper. Paint remover, liquid sander, or just plain hand sanding is generally required.

Machine prints usually are somewhat flat in tone and appearance. They may contain up to twelve different colors and are usually free from mismatch. Most can be removed by scoring or rough sanding the surface and then soaking or steaming with water.

With the exception of the cloth-backed vinyls, the trick to paper removal is breaking through the surface inks, paints, or foils so that the solvent can go to work on the adhesive. Use rough grit sandpaper, number 100 or less. Most instruction manuals will tell you to knife-cut the wallpaper at 4-inch intervals to break through the surface coatings. I can't recommend this practice. It's just too easy to cut through the backer and damage the wall surface.

Wallpaper removal is messy work. If you're working over a carpeted area, be careful to avoid soaking the carpet. Most carpets use latex as a binder and jute as a backer. Both latex and jute are easily damaged by soaking in water or solvent. Keep a lot of plastic or paper garbage bags and a bucket of clean water on hand.

Start peeling the paper at a top corner. A squeeze bottle with a long bent snorkel is best for applying solvent to the area between the paper and the wall. Be sure the bottle isn't made of plastic that dissolves in the solvent you plan to use. Let the solvent soak into and dissolve the adhesive, then peel some more of the paper. Continue this until the sheet is removed.

Commercial wallpaper solvents are available that are formulated to dissolve the paste. Most wallpaper stores can supply these solvents. You can rent wall steamers from some wallpaper stores and most rental yards. The hot steam makes the underlying paste soft and pliable to make removal easier.

Be careful in older homes. Excessive solvent can damage wall surfaces, especially water on plaster.

Let me describe another problem I've run into. Unfortunately there's no way of anticipating this problem. Here's how it happened to me. The previous contractor had applied up to one-quarter of an inch of sizing and paste to flatten a rough wall surface before applying the wallpaper. Naturally, the stripping I did removed this phony wall surface — leaving me with a wall that wasn't suitable for painting. It took a couple of days of spackling and sanding to correct the problem. Needless to say, I didn't make much money on that job.

I mention this so you can be prepared. Include in your contract a clause stating that the bid is based on a subsurface that's flat and paintable. If there are defects that have to be repaired before painting can begin, charge an hourly rate for extra labor and for extra materials used.

Some builders used plywood as a subsurface for wallpaper. If you encounter this, it's best to suggest a re-wallpapering. Painted plywood looks bad on interior walls.

Here's a final caution. Many walls were sealed with shellac before the paper was pasted on. This shellac should be removed before painting and *must* be removed if you plan to use urethane or lacquer. Wash with alcohol if you suspect that shellac is present. Shellac will turn whitish in color when washed with alcohol.

Remove shellac with a solution of 50% water and 50% ammonia. Remember to allow a day or two of drying time after stripping and before painting. This is especially true if you plan to use enamel. You don't want to trap water in the wall surface. That's sure to cause future problems. If the wall is unpainted, proceed as for new work, priming the surface before painting.

And one final wallpaper tip. If there are loose seams in wallpaper that your customer wants left on the wall, you can sometimes reset them with a steam iron. Place a damp, lint-free towel over the seam and heat it with the iron set on "steam." This will usually soften the paper and glue, rebonding it to the wall.

Stripping Wood for Recoating

You're bound to find situations where you must strip wood of all its old finish before you can recoat. This is common with cabinets, bookshelves, window sills, railings and furniture. Here are a few quick ideas.

When stripping wood-framed windows, a good commercial paint stripper that contains methenol chloride works best. Cover and mask off everything that's within five feet of the work area. Then brush the stripper on in a heavy coat. Allow 10 to 15 minutes for it to work. Use paper towels and #000 steel wool to lift the residue from the surface. After the old top coat and stain are removed, clean with a rag dampened with mineral spirits. This type of stripper requires no water rinse. The stripping action will stop after about fifteen minutes. But there will be a waxy residue. The mineral spirits take care of that.

Stripping paint from sash around glass coated with solar film can be a problem. Most strippers will dissolve the solar film. But a commercial stripper made from methenol chloride shouldn't damage solar film on glass — if the solar film was installed properly. If the chemical gets between the film and the glass, the film will shrink, distort, and lift. The stripper doesn't affect the solar film. It eats at the bonding glue. Do *not* use masking tape on solar film-covered glass. The tape has too much adhesive and will lift the film when you try to unmask. Painters masking tape is a better choice. There's one sold under the trade name *Easy-Mask*. It has a very light adhesive on one edge. Place that adhesive edge toward the wood you're stripping.

Here's a safety note. Methenol chloride is as clear as water. It's hard to see splatters. So take the proper safety precautions. Wear rubber gloves, eye protection, and keep an ample supply of clean water handy.

I don't recommend wearing a long-sleeve shirt, though. If splatters get on the shirt, it's too easy to transfer them to other items that can be damaged. If you're not wearing long sleeves, you'll know right away if you're getting messy. As soon as a drop of the chemical hits your skin, it starts to burn. Just wet the spot with water to stop the irritation. The chemical is somewhat toxic, so keep the area ventilated. Keep a few paper bags handy for disposal of the steel wool and paper towels.

If you do get into stripping furniture and other finer woods, use the same procedure as when stripping windows. Don't use a stripper that requires water to stop the stripping action. Water will raise or lift the grain of the wood and make sanding and finishing more difficult. Be careful of the glue joints on both the windows and furniture. Some glues will dissolve in some paint strippers. If you have to reglue a joint, be sure to wipe off all excess glue. Stain and many paints don't stick to dried glue. When it finally comes time to top coat the item, you may have light spots at the glue joints if you're not careful.

Painting Multipane Windows

Multipane windows have several smaller glass panes framed together to make up one large window. These are a pain to paint. Here's a trick that can save some time. Have your local print shop cut you a stack of #20 bond paper that's 1/2-inch smaller in both length and width than the glass is. Tape this paper to the glass when you're masking. It prevents a lot of splatters that have to be cleaned up. You can even spray paint the window sash, saving a lot of hand brushing.

I've also used glue-sticks to hold the paper to the glass. Just placing a dab in each corner works well. If you're using glue instead of tape, cut the paper only 1/4-inch smaller than the glass.

Tips for Clear-Coating Wood

Here's a trick to use when you're coating wood with clear plastic or polyurethane. Aerosol cans of these materials are expensive and only provide a thin layer of protection. Most wood that's been stripped and restained will need several heavy coats of plastic to get the required smoothness and depth of shine you want. Use liquid plastic from a can, brushing it on with a good quality brush. Drying takes a long time. Be sure you're in a dust-free, bug-free area. Cure each layer for several days and sand between layers. I recommend a minimum of three coats on most items.

It's hard to get a mark-free final finish when brushing. While the last brush coat is still wet, spray the surface with aerosol of the same type plastic. The spray will level the brush marks and leave a quality finish that requires little if any final sanding or rub out. Flow the plastic on liberally. But be careful of drips and runs. Practice on some scrap to get the feel of this.

Here's another tip. Don't use exterior or spar type varnish for interior wood or furniture projects. The longer drying time makes it more likely that your project will pick up dirt, dust, and insects. But there's one exception: I recommend exterior varnish for window sills on the sunlit sides of the house. Exterior varnish contains U/V inhibitors that keep sunlight from destroying the plastic-to-wood bond. That makes it worth the long drying time.

Preparing and Painting Exterior Surfaces

Now let's leave the freshly-painted interior and cover a few exterior problems and solutions. I'll start with the most common and most severe problem: deteriorated wood windows.

Repairing Wood Windows

Cracking and peeling on exterior sills and window bottoms are a frequent problem. This area gets more sunlight, rain, dirt and soot than any other part of the building. It's also the area that usually gets the least surface prep and attention. It's likely to have many coats of paint, no primer, and cracked or missing caulking — all of which lead to premature paint failure. Moisture from dew, condensation, and rain enter behind the caulking. The wood soaks the water up like a sponge, destroying the bond between paint and wood from within.

The sun's bleaching action causes paint discoloration. Ultraviolet radiation destroys the paint binders. Air pollution, dust, and airborne chemicals attack the paint's surface layer, reducing its resistance to weather. The sun bakes the sills both directly and from reflected heat from the glass. This causes expansion and contraction of the paint layers and subsurface at different rates, pulling them apart.

The solution? Easy. Start over. Strip all paint and caulking to bare wood and glass. Use a chemical stripper and lots of it. A pull scraper and plenty of steel wood are musts. Recaulk using window putty. Apply a light coat of oil to the wood where you're going to caulk. This helps the caulk stick because oil in the caulking isn't pulled out into the wood too quickly. Wait a few days for the caulking to cure properly before painting.

If the wood is cracked or split, use wood filler, filler paint, or some caulking to fill the cracks. Large splits can be repaired with exterior grade glue and finishing nails. Then prime, let dry and

The least expensive solution may be aluminum siding
Figure 15-6

top coat. I recommend a good grade of oil stain or enamel.

If you use paint rather than stain, I highly recommend adding some penetrating liquid to the paint. First, coat the bare wood with several coats of the liquid and allow it to seep in. Mix some into both the primer and the paint. The penetrating liquid will help the paint disperse deeply into the wood, giving a better bond. Two types of penetrating liquid are available, one for oil-based paints, the other for latex. The most popular are Penetrol for oil, and Floetrol for latex. These are both trademarks of the Flood Company. Other manufacturers offer similar products.

Make sure you paint over the window glass by at least 1/16 of an inch. This weather seal will help prevent future problems. If you can, repaint with light colors or white. This helps reflect the sunlight and heat so your new paint job will last longer.

Why not just replace the entire window? Usually because that would be too expensive. Most windows are installed directly on the wall studs before the exterior wall siding is installed. It's a major task to replace wood windows.

If the window has to be replaced, consider aluminum replacement windows. An aluminum siding dealer will have a stock of prefinished window trim. You cut and fit the window to size and nail it in place using aluminum nails. In some cases this will be the best and least expensive solution, even if the cost is high. You pay more for materials but far less for labor. On windows that need heavy paint stripping or that were painted with lead base paint, consider doing the entire job with aluminum replacement windows.

The house in Figure 15-6 offers another possible solution. The owner solved his painting problems forever by deciding not to paint. He covered the old siding with building paper, insulation and aluminum siding. The result is a new-looking building, low maintenance for dozens of years and added weatherproofing.

Preparing and Painting Under Eaves and Porches

The second biggest problem area of buildings is probably under the porches and eaves. This is a protected area and you'd think it wasn't subject to

Protected areas under eaves need special help
Figure 15-7

Consider aluminum covering for problem areas
Figure 15-8

paint damage. No sun, no rain, no weather. But that's exactly where the problem develops. The area collects airborne salts and contaminants that rain can't wash away. The contamination remains and eats at the coating. The area is also subject to isolated water damage from roof leaks and water spilling out the wrong side of gutters. See Figure 15-7.

Trying to scrape or strip paint from under the eaves is nearly impossible, especially if the rafters are exposed. Instead, use a water blaster and an airless spray rig. The water blaster will remove most of the loose paint and the salts or other contamination quickly. Allow a few days for drying and then scrape off any remaining loose paint. Now use some paint stripper to remove and feather any remaining paint.

Before priming, mix some penetrating liquid into your primer and spray. When the primer dries, apply a top coat that has some penetrating liquid added. An airless spray gun will cover the surface quickly and apply the paint deep in corners and cracks. If there are any large cracks, such as where the wood meets the siding, these should be caulked before the painting is begun.

After the job is done, advise the homeowners to hose this area off occasionally to reduce the damage from pollutants.

To avoid all this trouble, aluminum may be the answer. The porch ceiling in Figure 15-8 was refinished in aluminum. It looks good and will last a long time. The installation includes "breather panels" that keep moisture from collecting behind the aluminum and doing damage inside.

Areas Subject to Moisture from the Inside

In older homes that weren't properly insulated or don't have a vapor barrier, moisture from kitchens and bathrooms seeps through the walls and gets behind the exterior paint. This will loosen the paint-to-surface bond and cause the paint to peel. The remedy is to strip and refinish as you would any other area. Then install small plastic or aluminum vent plugs available in most hardware and lumber supply stores. Drill a hole in the wood siding and tap the plug in. This gives the warm moist interior air a place to exit to the cooler, drier exterior air. If there's no better way, the moisture will try to escape through the new paint, resulting in paint failure and a callback from unhappy customers. To stop paint peeling from the area above clothes dryer vents, add an extension cap to the vent. This will help expel the moisture out and away from the painted surface.

Ground Level Damage

The next most common trouble area is the foundation at ground level. Concrete or block that isn't protected against moisture encourages water damage, rot, and insect damage that can eventually lead to structural failure. Dig down several inches and clean the surface. Let it dry and then apply a waterproofing paint or a tar-based product. Coat to at least two inches above the ground level. Let this coating cure for a few days and then paint. Grade the backfilled soil so it slopes away from the building. That will reduce water puddling at the foundation. Figure 15-9 shows siding damaged at ground level by exposure to water.

Paint at ground level is subject to water damage
Figure 15-9

Balconies, Patios and Gutters

On two-story buildings that have balconies or patios over the garage, check for cracking where the balcony floor joins the building wall. These joints should be waterproofed with a tar-based product.

Rain gutters should be cleaned before being painted, especially if you intend to spray paint. The air blast from the spray will lift dirt out of the gutters and into your new paint. Also, be sure the gutters are dry before trying to paint them. Use a fish oil or galvanized primer before applying the top coat. Use acrylic paints on aluminum and either acrylic or oil-based paints on steel. Oil will be best as it seals the surface against moisture penetration.

Preparing and Painting Metal

Galvanized metal (iron with a thin zinc coating) carries with it a few old wives' tales. "Etch with acid or vinegar and water before painting." Or "It can't be properly painted." Or "Wait six months after installing, it needs to weather." These are but a few. The truth is that all it requires is a good washdown with kitchen detergent and water. What you have to do is remove the manufacturing oils that are on and in the surface. This is true of most unpainted metal. If you want to be doubly safe, wipe with degreaser, mineral spirits, or an organic solvent. After the surface is cleaned, rinsed, and dried, apply a primer and then top coat.

When priming metal, use tung, linseed, or fish-oil based primers. As long as the oils remain on the metal, they won't rust. These oils are referred to as *long oil.* Long oils have a penetrating quality that cuts through rust and into the metal's surface pores. Since they are also *drying oils,* they absorb the surrounding oxygen, retarding future rust formation. They tend to dry to a hard finish that excludes further air and water penetration. If you can't find these oils, use a primer that has zinc chromate. Apply at least two or three coats, then top coat.

Clean as much of the rust off as you can before priming. Use a wire brush or sandpaper. Naval Jelly (phosphoric acid in paste form) will also work most of the time. Be careful, though, as it can attack some metals.

There's a better choice than a paint brush when painting gutters, ornamental furniture, railings, pipes, or wire mesh fencing. Use a paint glove. This glove has a rubber liner and a mohair outside cover. You dip your hand into the can of paint and then just rub the paint on the surface to be covered. It's fast and easy. Spray painting will also work. But you'll waste more paint than you use when spray painting rail, pipe, fence and the like. If the item being painted is open but reasonably flat, such as a wire fence, you can use a long nap roller. A pass down both sides will usually do the job.

Preparing and Painting or Staining Siding

The first thing you need to know about siding is the material it's made from. Aluminum siding is normally factory-coated for life. But after several years the surface takes on a dull chalky look from paint oxidation. Before you decide to repaint, try giving the surface a good washdown. Use a sponge mop with a TSP water mix, starting at the bottom and working your way up. The TSP water mix is described in Chapter 11. If the washdown isn't successful, at least you've completed the first step of the painting process, cleaning the surface.

If you decide to paint aluminum siding, give the surface a very light scuff sanding, prime and paint with enamel paint. Use nonchalking paint if the siding is installed above anything you don't want to be stained, such as dormers, roofs, and the concrete foundation.

Vinyl siding over 15 years old should be replaced. You may be able to restore or paint it tem-

Wood siding can discolor under a clear coat
Figure 15-10

porarily, but sunlight, moisture, heat and cold have already started to destroy the chemical structure. If you want to try to restore or paint it, read the section on vinyl siding in Chapter 12.

Wood lap siding and trim sometimes have knots that can ooze sap for years after the siding is installed. There are two ways to cure the problem. The first is to remove the knot, clean the hole that remains with mineral spirits, and then fill the hole with plastic wood filler. Don't do this if you plan to stain, of course. It only works under opaque paint. The second method is to clean off the sap with mineral spirits, coat with cut shellac, prime, and paint. Once again, do this only if you're painting. If you plan to use stain, the siding and trim should be nearly knot free. If it is, all you can do with oozing knots is clean them with solvent and stain, or use opaque stain.

When painting wood lap siding, it's important to paint the underside of each lap. That's easy if you're painting with a brush or spraying. But it can be a problem when you're using a roller. I recommend using a small foam roller to make lap painting easier. The roller I use looks like a doughnut on a handle, but the edge is cut in the shape of a "V" to fit in the lap joint. If you can't find one of these, try using a long-nap 3 or 4 inch trim roller, painting with the side of the roller. Keep the roller wet with paint and rub, not roll, along the lap joint. The remaining siding can be rolled with a 7 or 9 inch standard medium-nap roller.

If you're rolling rough cut or rough finished siding, be sure to have a few extra covers on hand. The roughness of the wood will wear out the nap very quickly.

If the siding is wood shake or wood shingles, I suggest using a colored penetrating stain instead of paint. There's a product on the market called CWF, clear wood finish, or Aquatrol, a trademark of the Flood Company of Ohio. The manufacturer claims the product protects the wood and can in some cases reverse the aging process that turns wood black. Some painting companies specialize in applying this product to wood shake roofs. It might be a good sideline for your painting business.

Avoid clear finishes on exterior wood— I recommend against using clear finishes on exterior wood or siding. Ultraviolet radiation in sunlight will pass through a clear finish to both destroy the paint-to-surface bond and turn the wood gray or black. This is also true for varnished surfaces. The varnish will crack and shatter when exposed to sunlight. In Figure 15-10, the protected area under the

Ivy can actually eat away at the paint
Figure 15-11

overhang has retained its natural color, while the unprotected area over the door has turned dark from exposure to sunlight.

On siding and trim, use only paints and stains that are especially formulated for exterior wood. For clear coating, there are a few penetrating oil clear finishes that work well, but they generally have to be redone every few years. If you're going to use stain, test it on some inconspicuous area first. Stain affects different wood different ways. The color you get depends on the type of wood, its age, and how the wood was finished previously.

Avoid using semi-gloss or full gloss enamel on exterior wood. When it comes time to repaint, the entire building will have to be deglossed with sanding liquid.

When painting or staining any siding, cut or tie all plants away from the building so they don't brush against the wet paint or the siding after the job is completed. Ivy towers are fine — but not on the side of a painted building. Ivy and most plants will stain any paint and can actually take root on the painted surface. Ivy will eventually ruin not only the paint but also the siding, making expensive repairs necessary. What should have been a simple wash-down prep before painting has turned into a difficult paint stripping job in Figure 15-11.

Stain techniques— Here are some tips for successful staining of exterior siding. First, use semi-transparent oil stain only on bare wood that's free from pencil marks, manufacturer's code marks, and any other discoloration. Apply semi-transparent stain in two or more coats on smooth wood.

Use solid color stain over old or previously coated wood to cover marks, blemishes, and discoloration. You may be able to lighten some woods with a bleaching solution. There's a recipe for bleaching in Chapter 11.

Why use an exterior stain and not a paint? A stain will give a smooth, mellow, look to the sur-

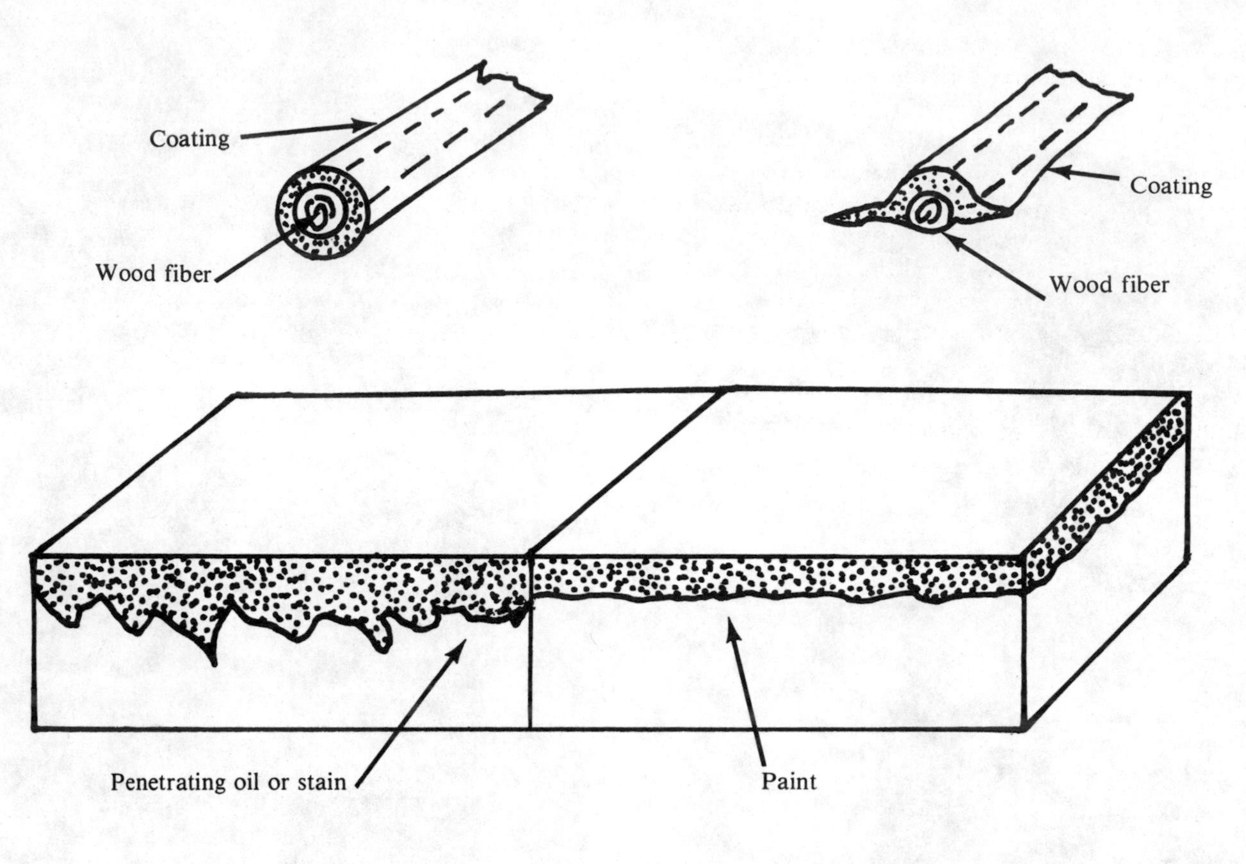

The difference between penetrating stain and paint
Figure 15-12

face. It has excellent hiding power, good color retention, and very good durability. Oil stain surrounds each fiber in the wood with a coating. Oil stains don't seal as a paint does, nor do they just lie on the surface.

Figure 15-12 shows how stain or penetrating oil seeps deep into the surface to coat each wood fiber. This allows moisture to move in, around, and out of the subsurface without affecting the coating. A paint coating would seal the surface and block movement of moisture. This makes oil stain best if you want to prevent blistering and peeling. Stain will retard the bleeding that occurs with most wood. It also lets the natural texture of the wood show, if you're using semi-transparent stain. An opaque stain hides more of the natural wood texture.

Stain is available in many different colors and shades, and most people think it has a more natural look than paint. There are some weathering stains that help wood weather quickly to a full natural look.

What is stain? Most formulations use dye (rather than pigment) mixed in oil or water. The most common oil is linseed, but others may be used. Note this caution when using stain: Since stain is a dye, it will dye almost anything it touches. This includes cement, concrete, block, and brick. Cover or mask off anything you don't want stained.

Some stains change color as they dry. They go on as one color, dry to a second, and slowly change to a third or fourth in the next few days. That's why it's important to allow several days of drying time when testing stain colors. Be sure your customer likes the results. Once stained, the wood will remain stained. To darken the color isn't too much of a problem. Just use a darker stain or else paint. But to lighten the stain you'll have to bleach the entire surface before restaining. Get it right the first time. You may want to have your customer approve the test color in writing.

On rough lumber siding, the best way to apply stain is with an airless spray rig. Proper spraying will work stain into all cracks and pores. A single heavy spray coat or several lighter coats will be required. Use a good professional spray unit. If you're using a roller, rough siding will tear the nap and leave threads on the surface. You can use a good quality brush, but that's doing things the hard way.

Unlike rough wood that absorbs stain well, smooth, well-sanded wood tends to repel stain. New smooth wood should be scuff sanded or weathered before applying stain or top coat. I recommend a weathering time of four to six weeks for plywood, up to six months for board lumber. Allow at least 24 hours of drying time for oil stains before applying the second coat. The stain shouldn't get wet during that period. Latex stain can dry in as little as 30 minutes, but it also should have a day or two to cure properly.

We've covered a lot of ground in this chapter — more than you can master in a single sitting. Don't be concerned if you didn't absorb it all or expect to forget most of it by tomorrow. The important thing is that you've been exposed to the information here and know where to find the answer the next time a key painting question comes up. That's when the suggestions in this chapter will pay big dividends.

This chapter offered painting tips for both new construction and recoating of previously painted surfaces. In the next chapter I get more specific — and go into more detail — on one very important topic, painting new construction.

Painting New Construction

Painting new construction can be either easy or hard, depending on when you, the painter, get into the act. Offering suggestions early enough can prevent some expensive mistakes, both for you and the builder. This chapter will suggest ways to speed and simplify painting new construction — and help you anticipate problems so they don't become mistakes.

Recently I built a 3,000 square foot, two-family duplex. During construction, I noticed again and again that the building materials industry still hasn't standardized what it produces and sells. For example, each load of lumber we received was a slightly different size than the last. Quality grading was rough, at best. Worse still, most of the lumber was green, even though we specified S-Dry.

As a house painter, I used to criticize builders for some of their workmanship — workmanship that made for very difficult painting. Now I have a little more sympathy for builders. Sure, there is some shoddy work being done. But many of the problems are caused by non-standard materials. Wet lumber warps and repels the paint coating. Lumber installed improperly warps, cups, and splits. Bad trim lumber cracks, checks, bleeds, and has knots.

But errors in design and construction still cause many problems. For example, poorly designed corners and wall intersections cause split drywall seams. Rooms that aren't standard sizes require too much cutting of drywall and paneling. Using the wrong nails can lead to structural difficulties and paint problems. Poor planning and workmanship means painters have to work with corners that aren't square and make awkward transitions between colors and textures on adjoining walls.

Construction Techniques That Make Painting Easier

When we built, we did the smart thing. We studied first, then built, preventing most (but not all) of the problems. Many of the problems were solved by learning the hard way, by doing it wrong first and then doing it right the second time.

Do's and Don'ts

Here are some of the do's and don'ts I picked up from the house I built.

Do use "hot-dipped" galvanized nails for siding.

Do use aluminum nails for cedar and redwood, to prevent rust staining like that in Figure 16-1.

Do use redwood for exterior trim lumber.

Do dimension areas so they can be tiled or painted easily.

Do include natural break points for color or material changes.

Using the wrong nails causes wood stains
Figure 16-1

Splattered cement peels off, taking paint with it
Figure 16-2

Do use the nailing technique recommended by material manufacturers.

Do the weatherstripping as you build.

Do careful cement work and yard grading.

Do figure on a primer and a minimum of two top coats for everything.

Do try to paint all trim lumber before it's installed.

Do allow time for proper weathering and drying of materials.

Do check your rule, square, and level for accuracy before you start the construction job. Even new tools from a leading manufacturer can be defective or inaccurate.

Do check with your building department for accepted construction techniques in your area. Get it in writing. We received several verbal OK's that were later rescinded. That's an expensive way to do business.

Don't try to paint over form lumber.

Don't try to use pine or fir for trim lumber.

Don't splash cement on areas to be painted. Figure 16-2 shows why. The workers pouring the patio slab splashed cement on the block wall. It was painted over, and six months later the cement splash peeled off, taking the paint with it.

Don't depend on primer to do your job. Clean all manufacturing marks off of wood and all oils off of metal.

So much for the do's and don'ts. Now look at Figure 16-3. It shows some bird's-eye and side views of wall construction framing. Section A is the standard butt of an intersecting wall to a continuous wall. The asterisk shows the location of an optional 2 x 4 stud or small 2 x 4 spacer blocks. Sections B and C show acceptable variations for this construction.

In section D, I suggest using the 2 x 6. If 1 x 6 or plywood is used, glue and nail it or the board may break loose when you apply the drywall.

The arrangement in section G gives a good surface for attaching the drywall. The insulation can be installed after outside siding is installed. This may not be permitted under all building codes.

In section H, an optional outside corner design, I show a small filler to make up the size difference of newer 2 x 4's. The newer ones seem to be only 1½ by 3½ inches or less after sanding when lumber is finished on four sides.

In section K, the standard procedure is to use a 2 x 4. But a 2 x 6 or 2 x 8 works better because they provide a surface for attaching the ceiling drywall.

Section M shows the drywall installed and the baseboard molding attached. If the room is to be carpeted, the molding can be installed 1/2 inch off the floor. In rooms to be tiled or vinyled, don't install the molding until after the tile or vinyl is down.

The double sill in section N will result in less warping of the wall. It also provides a backer for

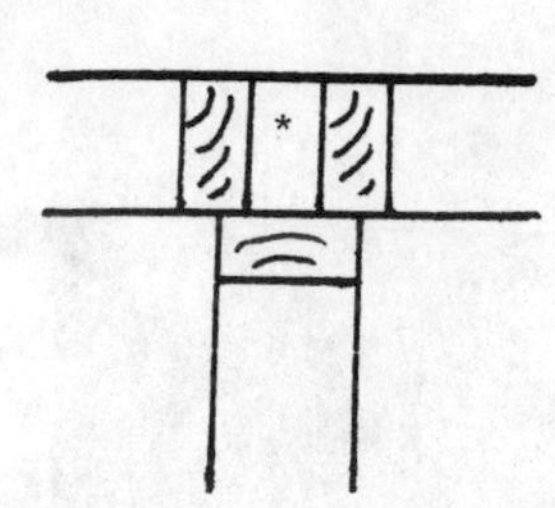

A Standard wall intersection

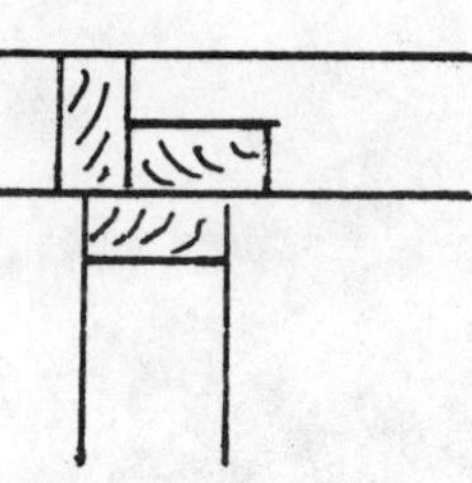

B Variation on A

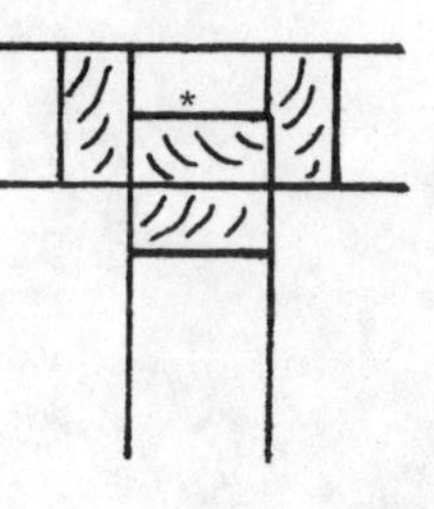

C Variation on A

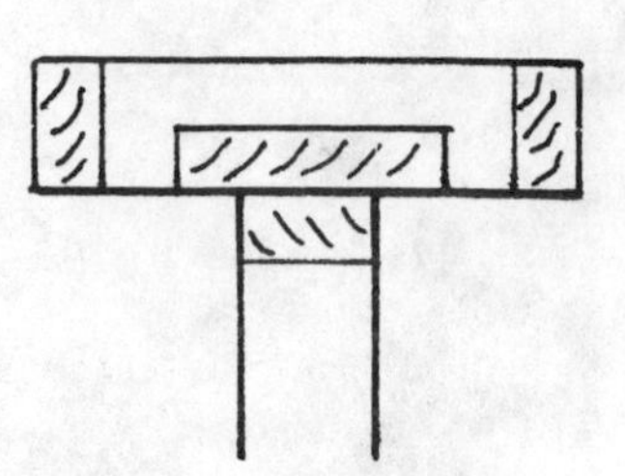

D Variation using plywood, 1×6 or 2×6 as backer for drywall

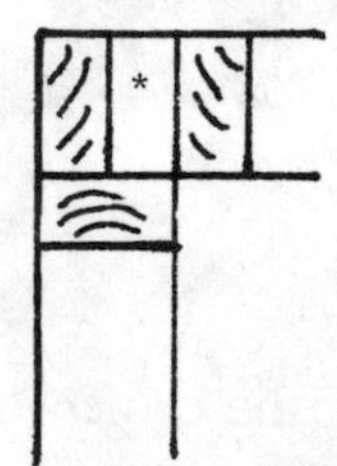

E Standard outside corner design

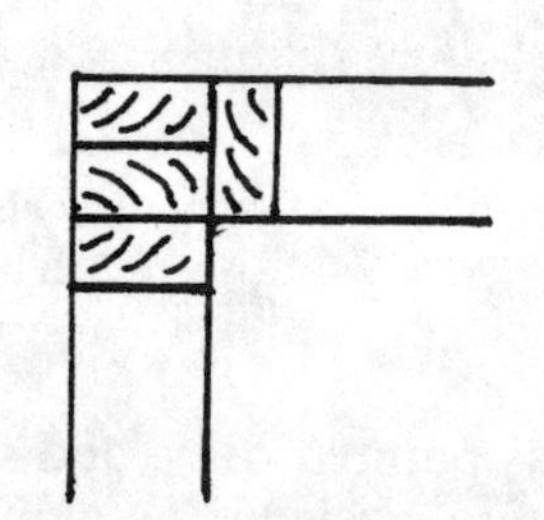

F Strongest outside corner design

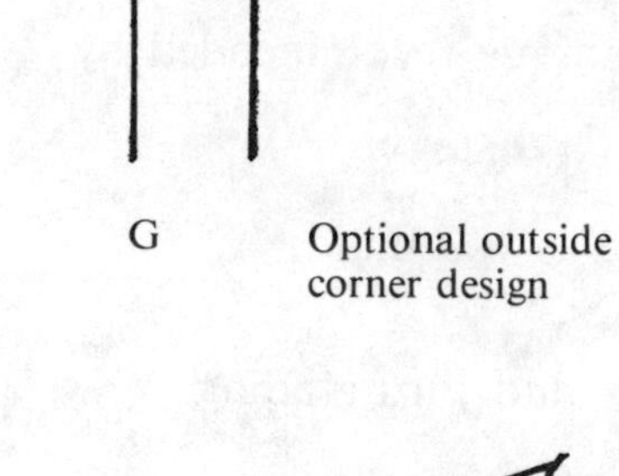

G Optional outside corner design

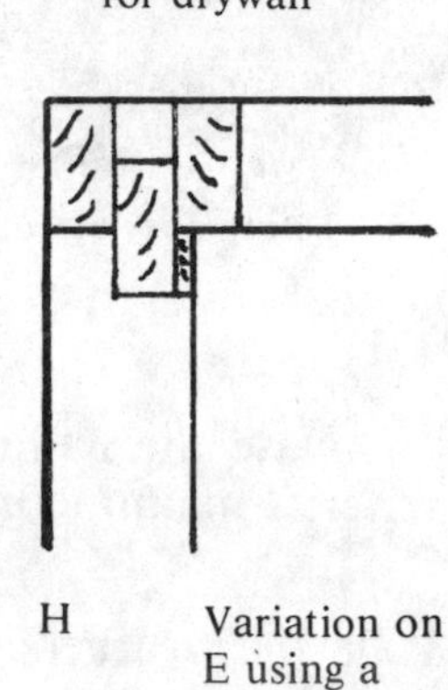

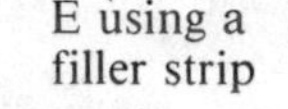

H Variation on E using a filler strip

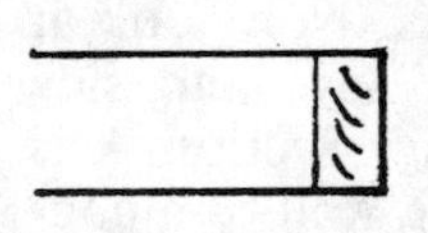

I Wall terminating at a doorway

K Top plate to outside wall

L Top plate to outside wall

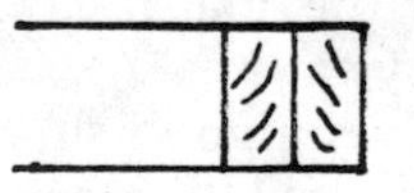

J Stronger variation of I

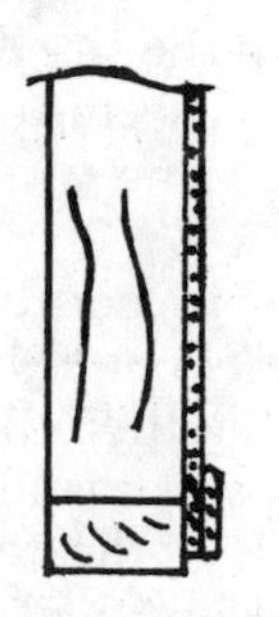

M Standard bottom sill

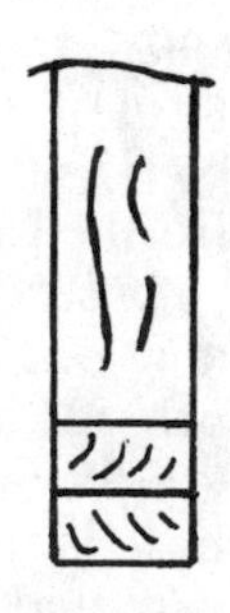

N Better bottom sill

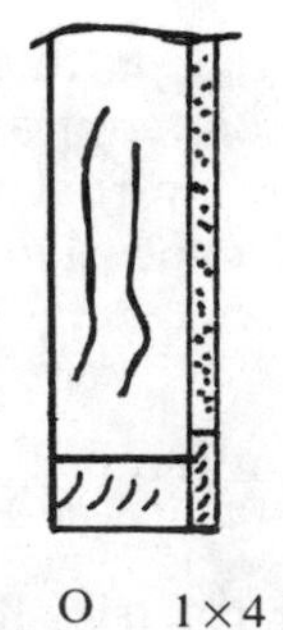

O 1×4 backer at bottom sill

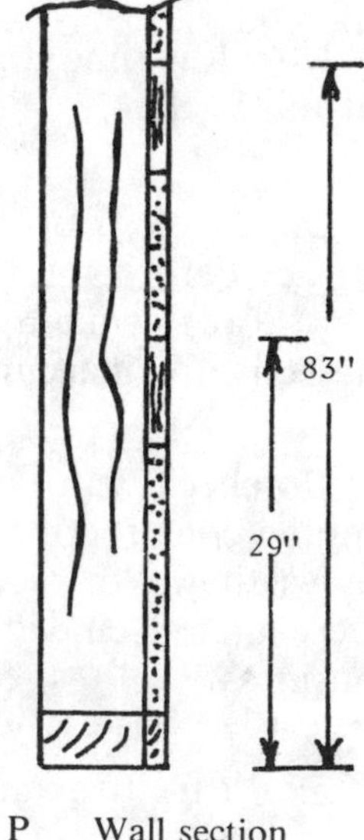

P Wall section in kitchen showing plywood strips for cabinet hanging

Wall framing techniques
Figure 16-3

drywall and molding. Unfortunately, you won't find this second 2 x 4 sill in most homes.

Section O shows a 1 x 4 backer nailed in place with the drywall butting to it. This isn't common in homes that are drywalled. But it's essential in walls that are to be plastered. This strip of lumber is called a "ground" and acts as a reference point when the plasterer levels the wall surface.

Section P shows a kitchen wall. Plywood strips have been installed on the wall where cabinets will be placed. Screw the cabinets to these 8" wide strips. Otherwise you would have to find a stud for each screw placed. These plywood strips are essential if you're installing prefab modular cabinets. The drywall should be green board (water-resistant gypsum wallboard) that meets ANSI-A108.4. It's sometimes referred to as tile-backer board.

In closets, consider placing a single 8-inch strip of plywood mounted at 65 inches from the floor. This provides a backer for shelves and rods. All seams should be filled with spackle or caulk to prevent outside air and insects from entering the home.

Here are a few more construction techniques that can make your job easier: First, there's a growing trend toward using inexpensive drywall backer clips in new construction to replace 2 x 4's. I don't like them. They create floating drywall seams that damage easily. Second, builders like to install 12-foot sheets of drywall horizontally along a wall. They usually don't use blocks or a backer at the seam joints. The result will be split horizontal drywall seams.

You say you don't care, you're a painter, not a framer? You should care! If the frame isn't right, the plaster or drywall isn't right, and your paint job gets a big crack from floor to ceiling. If you don't care about that, you're in the wrong business. Even if it isn't your fault, anything that destroys the value of your work should concern you — and every craftsman that takes pride in his or her professionalism.

Floating drywall seams open and crack the painted surface. Rusted or improperly set nails cause nail pops and poor plaster or paint adhesion. Lumber that cups will collect water that leads to wood rot. Improper venting or installation of insulation and vapor barriers will lead to paint peeling. I could go on and on.

Even if you're never going to build a house yourself, it's to your advantage as a painting contractor to counsel with the general contractor. Point out the problems. Suggest ways to prevent these mistakes. Learn your job and help the general do his better. Save him time and money and you'll get a call every time he needs a painting bid.

Building with Painting in Mind

If you're building or adding on, or have any influence with someone who is, keep this point in mind: It takes room for painting. A 9-inch roller needs 11 inches of clearance space. Leave that much clearance over doorways, under electrical outlets, under cabinets in kitchen and bath, over tubs, and over appliances. A brush needs from 1 to 2 inches of clearance. Examples: Door molding to wall corners, toilets to cabinets, cabinets to wall corners, window frames to wall corners. Plan accordingly. All windows and doors should have moldings that become the natural break line between paint and no paint. A lip of 1/16 to 1/8-inch is required.

Acoustical ceilings should be undercut where they meet the walls. Room corners should be square. Kitchen cabinets should be at least 12½ inches higher than the counter tops. That's at least three 4-inch tiles plus a cap high.

Toilet tanks should be 2 inches away from wall. Air grilles and light fixtures should be easily removable. Time and time again I've seen builders design areas that are impossible to paint. Don't contribute to problems like that.

A Problem to Watch For

I've found that many sheets of drywall, especially green board or water resistant drywall, have material voids under the paper surface. When you paint it, especially if you use a latex, paper above a void will tend to bubble. Examine the drywall carefully. Don't use sheets that have these voids. If it's already installed, use a razor knife to cut out the paper and fill the void with spackle, spot prime, and then paint. If you don't take the time to do it before you paint, you'll have to do it later on a callback. The paper will eventually blister or show obvious defects that ruin the paint coating.

Begin Early in the Construction Process

If you're in new construction painting, there are several problems you'll face over and over again. You're at the end of the list of subcontractors. The painting phase of construction always seems to be one of the last items thought of and scheduled — and paid for, I might add. Try to contract for the job early in the construction process. An early start will benefit you *and* the building contractor.

Your job is much easier if you can start *before* trim, roofing material, gutters, downspouts, windows, doors, appliances, fixtures, hardware, electrical covers, carpets, vinyl, and cabinets are installed. A skilled painter with a good quality airless gun can cover 7,000 to 10,000 square feet of surface in an eight-hour work day.

If everything that won't be painted has either been removed or isn't installed yet, there's less danger of slip-ups and overspray. In fact, overspray can be an advantage. For example, at the roof line, exposed ends of roof sheathing will get a seal coat by accident. That will help prevent water damage. Window and door openings will also get extra protection from your overspray.

It's a real advantage for the siding to have a seal coat under the trim and molding. This area is a prime candidate for water collection and dry rot. Before the trim is installed, the siding can be painted without tedious shielding or masking. Both sides of the trim lumber can be coated for extra protection.

If you've worked around construction for several years, you know that painting early increases the risk that the surface will be damaged before building is complete, and repainting will be needed. I agree that's true in some cases. It depends on the circumstances and type of building construction.

First, remember that most unfinished building materials require two or more coats for proper coverage and protection. Why struggle with masking or shielding during both coats? If you spray on a color-matched primer coat before the trim is installed, you only have overspray and masking problems during the final coat, after the trim is up. Trim that's been prepainted needs only one coat on the exposed surface. The edges are already completed. The final coat can be done in minutes with a trim roller and extension pole.

Siding that's been prepainted is already nailed. This nailing is now under the newly installed trim. The next stud nailing line should be 16 inches away from this point. This gives you a noncritical surface to paint for the top coat. You don't even have to paint near the trim. Just blend or feather toward it.

If you apply both coats of paint after the trim is installed, you'll have to mask or shield. There will almost always be uncoated spots where trim meets siding due to the unevenness of both surfaces. You'll have to touch up those spots with a brush, a time-wasting task. Painting the siding and trim before installation eliminates this touch-up work. Most overspray problems and the tedious cleanup they cause will also disappear, especially those at the roof line and door and window openings.

Scheduling early painting requires cooperation from the builder. But it can result in a better quality job at a lower labor cost. You can keep the savings as additional profit or use them to lower prices, generating more contracts.

Here's a point in your favor when asking the general contractor to schedule early painting. Many contractors use prefinished panels and sidings and understand the need to work carefully around these materials during construction. If they used prefinished siding, why can't they let you do a little early finishing yourself?

If you do a lot of exterior siding painting, you might want to look into some of the new automatic sheet painters. They're not cheap, but they do allow you to feed a whole 4 x 8 foot or 4 x 10 foot sheet of unpainted siding in one side and get a fully painted sheet out the other. Many lumber companies now have these units and offer this service.

Scheduling Painting in New Construction

Try to have the building contractor schedule your outside work while other subs are doing inside work, such as plumbing, electrical wiring, or drywalling. Do the opposite for the interior painting: Schedule it when cleanup and landscaping are being done. This minimizes conflicts between trades.

On inside work, plan to arrive two to three days after the wall texturing and spackling have been completed, but before the trim and acoustical ceiling work has started. The two to three day wait gives the joint compound time to dry completely. Wallboard that's still wet will ruin your paint from the inside out.

You'll probably have to start inside before the acoustical ceiling is applied. Most ceiling material manufacturers recommend applying a white primer coat before their product is sprayed on. Any acoustical ceiling overspray that gets on your newly painted walls should wipe off easily without damaging the paint below — especially if the walls are white. And most new construction for rental or sale is usually finished white or off-white. If the walls are to be a color other than white, either paint the final color after the acoustic spray has been applied or hope the acoustic contractor respects your work and properly masks the wall areas.

Many cabinets made by cabinet shops and some made by prefab shops don't have rear panels. The wall surface serves as the cabinet back panel. This is a good reason for painting that surface before the cabinets are installed.

The doors, door trim, and baseboards are generally the last items installed. If your contract includes coating these items, begin working on them two to three weeks before the installation date. Doors and door trim may have to be clear coated. Most clear coatings other than lacquer require two to three coats and two or three days drying time between coats. Give yourself time. It's also a good idea to do this work in a clean, dust-free work area.

Carpet, vinyl, and electrical cover plates should be installed only after all interior painting is finished.

Most contractors use prefinished cabinets. But if you're doing the finishing, schedule cabinet finishing at the same time as the trim and doors. They're much easier to coat before installation.

Tract builders usually have many units in progress and will usually be cooperative and flexible in scheduling the work. So will most owner-builders. It's the small builder who's most likely to give you problems. He's under more pressure to get the work completed. Explain how you can save him time and money while giving him a better quality product. He'll probably give you a chance to prove it.

Some builders seem to be plagued by poor quality workmanship, needless delays, and people stumbling over each other due to poor scheduling. Others, doing the same type of work in the same area and with many of the same subcontractors, have few or none of these problems. There's no excuse for incompetence. It doesn't take much extra effort to do it right. Good supervision and proper scheduling will make your work and the work of everyone else on the job much easier. The result is a better finished product.

Painting New Stucco

A building finished in stucco breaks some of the rules I've just given you. The trim should be installed first — either during the framing or after the scratch coat and before the dash coat. (The *dash* is the final color coat.) Then you can spray the trim without any need to mask or shield. Any overspray will be covered by the dash coat. Your job is made easy because you don't have to deal with the irregular edges where the dash meets trim. The trim makes it easier for the stucco crew to apply the dash coat. It becomes a border or dam for them to work against.

Painting New Plywood Siding

Paint manufacturers usually recommend that new plywood be sanded before painting. This conflicts with the plywood manufacturers' advice. They claim that sanding cuts the softer fibers and produces poor results. Instead, they recommend that the plywood be allowed to weather a few weeks in the sun and rain, or an occasional water spray. This opens the pores of the plywood and allows the paint to sink in, giving better adhesion. I suggest that you follow the plywood manufacturers' advice. After all, it's their product.

Air and Moisture Leaks

Air infiltration through walls and around windows and doors causes paint failure, high heat bills and higher cooling bills. Put a stop to these leaks before you paint. You have the opportunity to save a bundle in wasted fuel bills and reduced maintenance costs. Make the most of it.

Leakage around baseboards and window trim probably accounts for half of the heat loss in most homes. Remove the molding and use caulk or urethane foam (available in one-part formulation in aerosol cans) to fill the cracks. If moldings are to be painted the same color as the walls, reinstall and paint with the wall. If moldings are to be a different color than walls, paint the walls before reinstalling. Run a light bead of caulk along the back side of the molding before installing.

Weather Seal

When you're painting the outside of a house, always take time to weather seal the surface, windows, and doors. All you need are a caulking gun and a few tubes of good quality caulk recommended for outdoor use. Inspect all areas where trim, electric, or plumbing lines are protruding or attached to the walls. Check around door bells, doors, windows, attached fences and gates, lights, sill plates (where the house is attached to the foundation), corners, and vents. Air can leak through a crack as small as 1/100th of an inch. Most paint failure is from moisture being trapped in walls behind the paint. So take the time to caulk everything. Then wait at least a few days so the caulk can set before painting.

While painting, remove outdoor light fixtures and paint behind them. Don't disconnect the fixtures, but do be sure the electricity is off. Replace the fixtures while the paint is still wet. The paint will dry to form a weather seal around the fixture.

Painting Doors, Jambs and Trim

Instruction for painting doors, door jambs, and door trim is covered in every painting manual I've ever seen. Painters have been following these directions for years. But do they tell all? Not really. For instance what color do you paint the jambs if the door is the same color as the wall or siding? What color do you apply if the door is coated with clear spar varnish? Which do you paint first, the door trim or the wall? Should the door be painted the color of the wall or of the trim — or should it be another color? What can be done to a door if the owner doesn't want it painted?

Are there any black letter rules? No. But here are some practical suggestions:

Exterior doors that are white or same color as siding: If an exterior door is white or the same color as the siding and the trim is another color, paint the jamb the same color as the trim. Look at Figure 16-4.

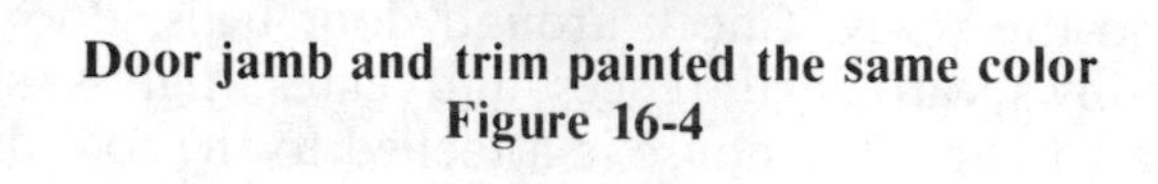

Door jamb and trim painted the same color
Figure 16-4

Exterior doors that aren't white or the same color as the siding: If an exterior door is a different color, paint the jamb the same color as the siding.

Jamb painted the same color as siding sets off the door
Figure 16-5

If you've got a natural clear-coated door with white siding and brown trim, for example, paint the jamb white. This sets off the door and shows its true beauty. Figure 16-5 shows a good example.

Utility doors: For utility doors that shouldn't stand out, paint the door, jamb, and door trim the color of the siding or wall. Figure 16-6 shows a door painted to blend in rather than contrast. The arrow leads us to another tip: Always paint the top surface of the door trim. Dust and dirt collect there, and it's much easier to clean if it's painted.

Blending a door into a wall surface
Figure 16-6

Interior door, clear coated: For an interior door that's clear coated, I like the jamb and trim to be coated to match. They may also be painted the same color as the wall if you wish.

Interior door, different color than wall: If the interior door is painted a different color than the wall, paint the jamb the color of the wall. The trim color is optional.

Interior door, same color as wall: If an interior door is painted the same color as the wall, paint the trim and jamb a different color to make them stand out, or the same color to blend in. It depends on what you want to accent or hide.

Trim color same as wall: Paint the door, jamb, and trim before painting the wall if the trim color is the same color as the wall. The reason is that you'll be painting with a semi-gloss or full gloss paint. It's easier to clean up paint splatters and hand prints from a gloss paint than from a flat wall paint. And when painting the trim, you don't have to worry about being too careful at the trim to wall joint. Just paint right onto the wall. It's easier to brush or cut into this break when painting the wall.

Trim clear coated: If the trim is to be clear coated, do it before installation, or remove it if it's already installed. If you can't, then coat it first and let it dry before painting the wall. Any splatters will wipe off with a rag and the proper solvent when the wall is painted.

Doors you don't want to paint: If you don't want to paint the door, you can do one of several things. If the wood is hardwood, stain and wax it or just wax it. If the door is softwood, you could wallpaper it. Utility and standard grade softwood doors don't look very good unless they're covered with something. Paint, wallpaper, veneer, metal, or plastic can be used. More expensive premium grade softwood doors look good when clear coated.

Metal doors: Metal doors are usually shipped with a primer coat. This coating is intended to last only 60 to 90 days. Then it should be cleaned and reprimed. Top coat these doors soon after they're received. Thirty, sixty, or ninety days may have already passed while the doors were in storage or transit.

Painting the edges of the door: If the door is a different color on each side, what color will you paint the edges? Paint the back edge the color of the door *away* from the hinge pins. Paint the front edge the color of the door *on the side where* the hinge pins are attached. This way, the edge of the door visible when the door is open will be the same color as the side of the door the viewer can see.

Raised-panel doors: If the raised panels will be a different color than the flat portions, paint the flat sections first. Let them dry completely. Then paint the raised portions. Any drips or splashes can be cleaned off the flat part with a rag and the proper solvent.

Cabinet doors: When painting cabinet doors, paint the cabinet frame first. Be sure to paint the front facing interior edge. Then paint the front side of the door and finally the back side. Keep the door open until the paint is dry. Be sure to remove the pulls before painting. If the hinges are decorative and not to be painted, it's best to remove the door and the hinges. Place the hardware on a shelf in the cabinet it came from or in plastic bags marked to identify which cabinet they belong to. You may not get them properly reinstalled if they're mixed up. Screw threads and hinge screw hole alignment will vary from hinge to hinge.

Bi-fold doors: When painting bi-fold doors, don't forget to paint the edges, especially the ones at the fold point.

Louvered doors: It's best to remove and spray paint louvered doors. Brushing takes too much time. A roller generally won't work well on the louvers. Many louvered doors have a bottom adjusting pin that turns with a screwdriver or wrench. Loosen the pin and lift the door from the hinge pivot to remove for painting.

Sliding doors: Sliding doors can be painted in place or removed. Be sure to paint both sides. Be careful when moving the door. Many sliding doors are on shallow closets. The paint on the back of the door may rub on the closet's contents. Remove the door by unscrewing the door guide on the floor. Then tilt the door toward you and lift up while standing in the closet. This should release the top rollers from their channel. Some doors will require the removal of the bottom glides or glide pins.

Doors with glass inserts: Be careful when painting doors with glass inserts. Even though the glass may clean easily, the glass-to-molding seal will not. If the insert is leaded glass, you can easily have a mess. To remove an insert, look for staples, nails, or screws, usually on one side only. The other side is generally glued in place. Carefully remove the trim. The insert should pop out. Most inserts are sealed with clear silicon sealer. Make sure you reseal it when you put the insert back in. If you're just painting the door, consider masking the insert rather than removing it.

General Door Painting Tips

Be sure to degrease and degloss the door surface before trying to repaint. Figure 16-7 shows what happens if you don't. The peeling is caused by the lack of adhesion between layers. This is a common problem with doors, since they're usually painted with a semi-gloss or full gloss enamel. Use a chemical deglosser or liquid sander, followed by a scuff sanding. If the door's already peeling, you'll have to remove the entire previous coating before repainting.

Should you brush, roll, or spray paint doors? Personally, I prefer spraying because it leaves a smooth, mark-free surface. My second choice is rolling. I'll use a 3-inch trim roller before I'll use a brush. Rolling is still faster and neater than brushing and produces a better-looking door.

Brushing doors is too slow. The paint tends to set before you can complete the job. The new wet paint gets pulled across tacky paint, leaving a rough surface. Save the brushing for cut-in work or touching up spots a roller or sprayer missed.

Paint peeling because the old enamel wasn't deglossed
Figure 16-7

Staining Redwood and Cedar

Tannic acid in redwood and cedar rises to the surface when the wood is moist. This acid reacts with metal in nails to cause staining. Here's how to correct this problem. First remove the source of the internal moisture (leaks, poor ventilation, or whatever they may be.) Clean the surface with a solution of 50% water and 50% denatured alcohol. Let it dry for two or three days. Then prime and top coat. If the stain wasn't removed or if it comes back, sand to bare wood, apply a wood sealer, prime and top coat again.

Always use aluminum nails on redwood and cedar to prevent this problem. If the project calls for the wood to be screwed together, use brass wood screws. They're more expensive, of course, but your project will last longer and require refinishing less frequently.

For finishing redwood or cedar, use only products recommended for these woods. Many paints and varnishes will react with the tannic acid and discolor quickly.

Painting Wood Trim and Decks

Backprime finish carpentry items such as trim lumber or shutters before installation. Paint all sides, front, back, and edges, with a sealer or enamel. Here's why backpriming is needed. Occasionally, water from rain or sprinkling or condensed moisture in the house interior will get behind these items. If that water soaks or leaches into the wood, peeling paint and rotting wood will be the result. Your backprime keeps moisture out of the wood. Be sure all surfaces of all wood trim are sealed.

When repainting a building, check behind the shutters to see if they were backprimed. If not, explain to the owner that you want to remove the shutters and give them this protection. Figure the cost into your estimate.

Trim lumber should have rounded edges. Use a sanding block or file to do this before painting. This will improve paint adhesion and eliminate sharp edges that can splinter or catch on clothing. Sanding off a splintered edge after painting will ruin your paint job.

Construction Tips for Wood Decks

If the job includes a wood deck that will be painted or stained, paint or stain all joist and subdeck lumber before applying the decking. Once the decking is in place, it's impossible to paint the deck supports. Any wood not painted becomes an entry point for moisture, and the first wood to rot.

Also paint or stain all decking lumber on all sides before installing it. Otherwise the sun will dry the lumber and it will absorb moisture from below the surface. Moisture traveling up through the deck will cause the top coat to peel. Use several thinned coatings and let them dry between each application. Use an oil-base stain or a polyurethane floor and deck enamel.

After the deck has been assembled, wait a few months. Then clean, degloss, and top coat with full strength material. Recoat every two years or when water no longer beads on the surface. The wood will turn gray if clear coated or not coated at all. An unprotected deck absorbs water. Wet wood rots, destroying both the surface deck and the deck supports. The only solution at that point is to replace the rotted lumber.

Decks can be made from fir or pine. But for longevity I recommend pressure treated lumber. You can buy special decking lumber that has been treated to retard decay and has beveled edges to

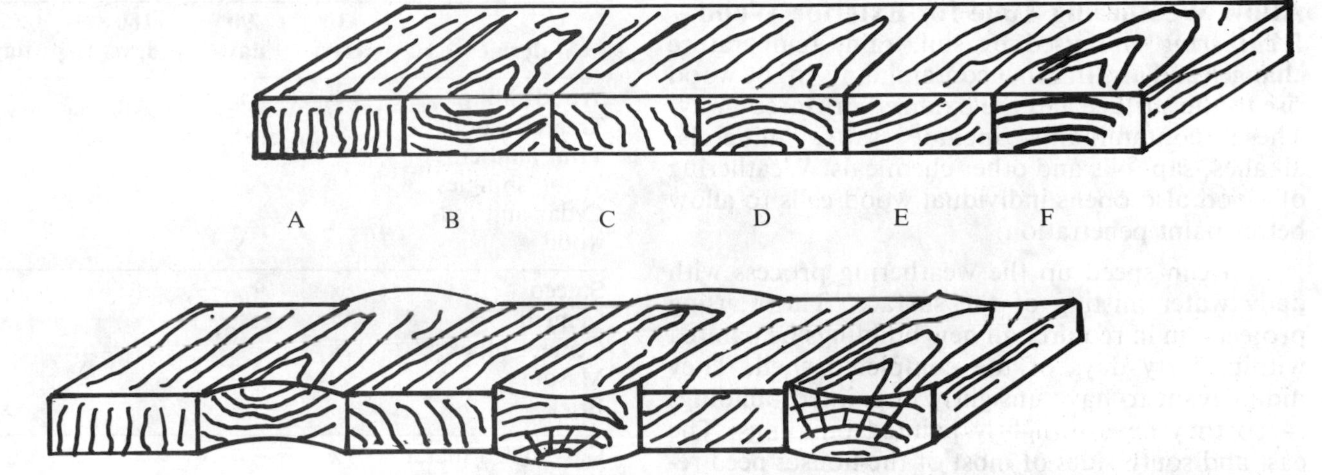

Incorrect installation of deck lumber
Figure 16-8

promote water drainage and eliminate edge splinters.

Install the lumber with the longest annual rings down, as shown in Figure 16-8. The top drawing shows a newly installed floor or deck. The bottom drawing shows the same floor or deck just a few seasons later. The radial sawed lumber (A) shows little shrinkage. The plain sawed lumber (B, D and F) has cupped. The tangential lumber (C and E) has started to shake or split. Board D has warped and is beginning to check. It's also collected some water and dry rot has begun. Board F has checked. Most of this could have been prevented by proper selection, sealing, and installation of the lumber.

Install decking lumber with the U-shaped end grain pointed toward the ground. This prevents cupping and promotes water drainage.

Use flooring or decking lumber that is rated S4S, surfaced on four sides, and marked MC-15 or KD. KD or kiln-dried lumber has less than a 15% moisture content. S-Dry is 19% and S-Grn is over 19%.

Use hot-dipped galvanized nails. Some lumber yards sell vinyl-coated and electro-galvanized (EG) nails and will recommend them for exterior use. I find that the vinyl-coated nails pull out easily. The vinyl coating remains in the wood, the nail doesn't. The EG nails have a very thin galvanized coating that doesn't wear well in wet locations. The hot-dipped nails are rough in texture and don't come out easily. If you're going to use screws in the decking, I suggest you use only stainless steel or brass screws.

Figure 16-9 shows most of what can go wrong with a wood deck. The deck lumber is cracked and discolored. The cut ends of the planks weren't sealed, so water entered and eventually caused the cracking. Water also puddled on the cupped boards and mold and mildew set in. The newel post was fastened to the wood with steel bolts that rusted and caused some staining. The bottom rails collected and dripped water, causing the light areas just under them. The solution at this point is to replace split or cupped lumber, sealing it completely before reinstalling. Replace the bolts with stainless steel or brass. Wax the finished decking, especially under the rails.

This wood deck has problems
Figure 16-9

Allow Weathering Time for Exterior Wood

Weathering is caused by sun, rain, temperature changes and air. Internal contaminants in the wood rise to the surface and are oxidized or rinsed away. These contaminants include salts, minerals, alkalies, sap, oils and other chemicals. Weathering of wood also opens individual wood cells to allow better paint penetration.

You can speed up the weathering process with daily water misting of the surface. The housing project I'm in requires all new buildings be painted within thirty days of the completed shell. They didn't want to have unsightly unpainted buildings — so they have unsightly painted buildings. The east and south sides of most of the houses need repainting.

In my home town in northern California, it takes six to nine months of weathering before wood siding is ready for paint. Wood painted before it's weathered properly cracks as it dries, allowing new wood to show through the paint coatings. The result: Most houses need a new paint job in less than a year.

Figure 16-10 is a table of suggested weathering times for various building materials. It ranges from 30 days for stucco and metals to a year for cement and brick. Of course, the local weather, amount of sun and rain, the original dryness of the material, the amount of contaminants and the type of finish will all influence the weathering times.

Weathering time also gives wood time to dry. Painting applied over wet lumber will peel off because there's no adhesion. When painting old buildings, this generally isn't a problem. The lumber has had ample time to dry or season.

New construction is a different story. The lumber mills don't like to store lumber any longer than necessary. Kiln-dried lumber is both hard to find and expensive, about double the cost of wood that's not kiln-dried. Many builders take what they get, whether wet or dry.

Material	30 days	90 days	180 days	360 days
Wood siding		x		
Treated wood		x		
Trim lumber		x		
Wood shingles			x	
Cedar and redwood		x		
Stucco	x			
Cement				x
Cement block	x			
Masonry	x			
Brick				x
Galvanized metal	x			
White rust metals			x	

Suggested weathering times for building materials
Figure 16-10

As a painter, your reputation is on the line on every job. Paint over wet lumber and both your reputation for quality, and your pocketbook will suffer.

How do you tell if lumber is too wet to paint? One way is to drive a 16-penny nail into it. If a little droplet of water forms beside the nail, it's too wet. A more reliable test is with a tool called a *moisture meter*. It can measure moisture content from 6% to 20%. It's a simple multimeter or ohmmeter that's been specially calibrated to read the moisture content of wood. I've seen them for sale at about $125 at 1987 prices. That isn't cheap. But it can settle an argument and may save you an expensive rework. I suggest you not paint any wood that has a moisture content over 12%.

If you want to buy this device, contact Constantine at 2050 Eastchester Road, Bronx, New York 10461. It's offered as a "Mini-Ligno" by Ligomat. Some paint stores may sell similar devices at competitive prices.

Spray Painting

It's hard to beat spray painting for speed and quality — if you do it right. But to do it right, you have to understand the equipment and the theory behind spray painting. If you don't, you can cause more problems than a fox in a hen house. In this chapter, I'll describe the equipment available and how to use it — and suggest ways to avoid the most common spray painting problems.

Types of Spray Equipment

Spray painting can be a mystery to the beginner. There are so many different types of spray equipment on the market. What's best for you? Here's my answer: It depends on what you're going to spray. Everything? Metal only? Latex only? Walls, trim, furniture, or maybe cars? There are nearly as many types of spray paint jobs as there are spray paint equipment combinations. Each item requires a different type of spray gun or spray gun tip. Each takes different pressures, a different feed system.

Conventional Spray Systems

Just what is a spray gun? How does it work? A spray gun is a device that atomizes a liquid and then ejects the atomized liquid onto a surface in the form of droplets. Conventional spray guns use air pressure or air suction to eject the droplets.

Droplets are formed when paint is forced through the nozzle tip or *orifice.* Liquids have a certain amount of "stiction." That's my word for their tendency to adhere to a surface. As the liquid flows through the orifice, it tends to stick to the edges of the orifice. If we aspirate or pressurize that liquid, it will *atomize* into small droplets.

Aspiration occurs when a flow of air in front of a liquid pulls the liquid with it. The airflow creates a vacuum to its rear and sides. The liquid is then sucked or aspirated along by the vacuum.

Since we're working with rather low pressures and vacuums, the liquids must be thin, or of a low *viscosity.* Viscosity describes a liquid's degree of resistance to flow.

The Air Compressor

The pressure system for a conventional spray unit is an external air compressor. I don't recommend buying anything under a 1/2 horsepower unit. A 1 or 1½ HP unit is best for field work. Use a 5 or more HP unit for factory production work. You need a lot of air at steady pressure for high production painting. A small 1/4 or 1/3 HP unit can supply high pressures but not high volumes for extended periods.

Choose a unit that includes an air storage tank or buy a compressor that has this tank built in. The tank helps maintain a steady flow of air at the right pressure. For consistently good spray painting you need the right combination of pressure in PSI (pounds per square inch) and volume in CFM

Conventional paint spraying system
Figure 17-1

(cubic feet per minute). A unit that produces 100 PSI but only 4 CFM is as useless as a unit that produces 100 CFM at 4 PSI.

I recommend a minimum of 7 CFM at 40 PSI. A larger unit is better for field work, say 15 CFM at 50 PSI, *if* it meets these two conditions: First, air pressure and volume can be adjusted to suit the material being used. Second, it doesn't have over a 1½ HP motor. That's the largest motor that most home and office wiring can handle. In a factory situation, any voltage and current can be brought right to the motor control box with conduit. That makes it safe to use motors with 5 HP or more.

Tank-type units should be equipped with an automatic shut-off and an overpressure safety release valve. Look for both a tank pressure gauge and an outlet pressure gauge. The compressor needs an air pressure regulator, of course, and should have a separate outlet cutoff valve.

Install quick disconnects at the tank-to-hose and hose-to-gun connections to save time and effort. You can buy these at most automotive supply houses.

I usually add a 10-foot length of very flexible high-pressure hose between the end of the manufacturer's hose and the gun inlet. That makes for easier gun handling.

Consider buying two filters: An air filter for the inlet air duct and a condensation filter if you're doing production line or lacquer work. Portable units should be equipped with large rubber wheels so it's easy to wheel them around the job site. To help you load the unit on and off your truck or van, I suggest you build a small portable ramp. Compressors are heavy.

Figure 17-1 shows a tank-type air compressor for general painting. You can also attach a hopper gun to it for applying texture and acoustical coatings. I modified this one to include an extra output valve. The original outlet valve was part of the regulator.

I had to change and adjust the pressure each time I turned off the output air. The second valve makes this unnecessary.

I also added a quick disconnect at the output and hose junctions. The small gun in the extreme lower left corner is a "blow-off" gun for dusting. The extra hose coiled on the ground adds length and flexibility. Also shown are a conventional spray gun and safety glasses.

Types of Conventional Systems

The conventional spray system comes in four basic variations: a bleeder or a nonbleeder, each available with internal or external mix. What's the difference?

A *bleeder gun* has a continuous flow of air. No paint flows until the trigger activates the paint port, feeding paint through the port into the air stream. The bleeder requires steady air pressure at a constant volume. The main advantage of a bleeder system is that there's an almost instantaneous start or stop of paint flow in response to trigger action. The airless and the nonbleeder units have a slight delay, which can upset your hand-to-eye coordination until you're accustomed to it.

A *nonbleeder gun* has no air flow until you pull the trigger. When you do, both air and paint ports open and spraying begins. The advantage to nonbleeder spray guns is that air flow stops when you're not painting. Many times, when you're not painting, it's annoying to have air flowing. Nonbleeder guns turn off the air at the gun, or in a fully automated operation, at the compressor-to-gun connection.

An *internal-mix nozzle* is identifiable by its cone shape. The paint and air are mixed in the nozzle or mixing cup and ejected through the orifice. The internal-mix nozzle is primarily used for heavy-bodied or slow-drying paints, like heavy enamels, urethanes, and latex. The paints, being heavy, must be pressure fed. The internal mix nozzle uses a pressure feed system. More about that later.

An *external-mix* nozzle usually has two "ears" extending from the nozzle. The air is released and the resulting vacuum pulls the paint through the orifice where it mixes in the open air. This works well for thin, lightweight, fast-drying paints like lacquers and stains.

The feed systems are also different. You can choose from a pressure feed and a siphon feed.

A *pressure feed* gun uses part of the air flow to pressurize the paint in the paint cup or canister. This pressure forces the paint through a tube and into the air stream. All internal-mix nozzles are pressure fed.

A *siphon feed* gun is open or vented to the outside atmosphere from the canister. The internal air stream creates a vacuum over a pick-up tube. This vacuum sucks paint out of the canister and into the air stream. External-mix nozzles may be either pressure or siphon fed. Siphon feed systems require thinner liquids because the vacuum formed by the air stream is weak. A weak vacuum can't pull heavier paints from the canister into the air stream.

Most of the newer conventional spray guns have all the necessary valves, ports, and adjustments so you can set them up as bleeder or nonbleeder guns with either internal or external paint mixing. You buy only one gun and use it for any type of spray painting.

Thinning the Paint for Conventional Systems

One last comment about conventional spray guns: I don't suggest using them for covering large surfaces with latex paint. Use an airless gun instead. Here's why. Latex must be thinned considerably to work in a conventional system. This thinning reduces paint opacity or *hide* so much that you're not likely to get satisfactory results. But if you have to spray latex with a conventional gun, thin the paint as little as possible.

To thin just the right amount, use a *viscosimeter*. That's a plastic or metal cup with a calibrated hole in it. Dip the cup into the paint and note the time it takes the thinned paint flowing out the hole to turn from a stream to drips. Then check this time against Figure 17-2. Thin or thicken the material as required until it matches the time in the chart. When it matches, you've got the right viscosity for a conventional spray rig.

Material at 72°F	Time (seconds)
Lacquer	20
Rubber base enamel	20
Auto enamel	25
Spar varnish	25
House or trim paint	30
Semi- or high gloss enamels	30
Floor and deck enamels	30
Latex	45

Viscosimeter time chart
Figure 17-2

Airless Spray Systems

Eliminating the air from the spray system (hence the name "airless") lets it spray paint of a higher viscosity. There are two ways to spray paint without using air pressure. The first way is to spin a wheel while the paint drops onto it. Centrifugal force moves the paint to the outer edge of the wheel, where it breaks free in the form of small droplets. When the wheel is surrounded with a housing and a gate or doorway, paint will exit in the form of a spray pattern. This is the method used by some of the low cost airless units sold for do-it-yourselfers. It works, but it's not very controllable on some of the paints you'll be required to use. Leave wheel-type airless systems to weekend painters.

The second method pushes the liquid through the orifice with high liquid pressure. This is the method used in a professional airless system. The pressure comes from a piston in a cylinder, a metal wheel revolving in a drum-type cylinder, or from a mechanically flexed diaphragm.

All three types have an inlet valve and an outlet valve. The inlet valve opens on the suction stroke. The vacuum formed by the piston pulls the paint from the pail. The inlet valve then closes and the outlet valve opens, pumping the paint through the outlet valve into the hose and spray gun. At the end of the compression stroke, the outlet valve closes and the process repeats.

Airless spraying requires high pressures but low volumes. The cylinders are small but the movement is fast. Only a fraction of an ounce of paint is used each time the valve opens and closes. The motion is very fast, pumping hundreds of these little "gulps" every minute. This produces a high volume of paint flow at the nozzle.

Low-cost piston airless spray units are now sold in many hardware stores and lumber yards. The units are good, but they're designed for the average homeowner who only paints occasionally. They're not designed for daily use and high volume production. These units usually apply one gallon of paint in 20 minutes.

A professional painter needs an airless spray rig designed for high volume and daily use. A professional airless units can apply one gallon in three minutes. If you can't afford to buy a professional unit, rent one. Most can be rented for about $35 to $65 per day. Do all your prep work beforehand. You don't want to have the unit sitting idle while you're scraping paint.

Taking Care of Your Airless

Most of the moving parts of the airless are in direct contact with the paint or liquid being used. Since most paints contain abrasive oxides, an airless rig gets heavy wear. That's why the manufacturers of airless equipment advise you to run the equipment at the lowest pressure that will do the job.

Diaphragm units have a diaphragm made of a flexible synthetic material. But this will wear out as soon as, or even sooner than, a piston unit. You must be careful about the paints you're pumping with a diaphragm unit. Certain types of paint, thinner and solvent can destroy a diaphragm very quickly. The manufacturers formulate their diaphragm materials to withstand most paint products, but not all.

Multicomponent Airless Systems

Some paints and spray products require a hardener, catalyst or curing agent to make them set. If the catalyst were added while the paint was in the can, the mass would harden quickly and become useless. With two-component paints you need two sprays working at the same time and in the same place. A two-component spray gun does just that. It sprays both the paint and the hardener at the same time. The droplets mix or combine in the spray pattern and on the surface being sprayed, causing the paint to harden.

A fiberglassing gun can mix three components, resin, catalyst and glass fiber, into the spray pattern. The glass is chopped into small threads when spraying. These threads mix with the liquid and the hardener in the air and on the surface. When the liquid sets, it locks in the threads, which give the coating strength. This process, called *spray lay up,* is a production method used in fiberglassing.

All spray systems are basically the same, whether one, two, or three component. They all have three major parts: a source of liquid or paint, a source of pressure, and a method for spraying the liquid in the form of droplets. Obviously, all three parts of an airless system have to be working correctly to do professional quality painting.

Tips for Airless Systems

Figures 17-3 and 17-4 show typical tip selection charts for airless units. Notice that they show several different pattern widths for each of the same flow rates (gallons per minute of liquid flow or GPM) and orifice sizes. To decide what pattern size you want to use, consider the size of the material or surface you'll usually be painting.

If you're going to be painting large surfaces such

Last 2 digits of standard tip number indicate angle of spray. Complete Twist Tip Assemblies are for 50A, 51, 500, and 700 series guns only.

Std. Tip No.	Orifice	Capacity GPM	Approx. Spray Width 12" Away	Complete Twist Tip Assembly	Twist Tip Nozzle Kit Only
9-1120	.011	.12	4	54-51115	54-61115
9-1130			6	54-51125	54-61125
9-1140			7½	54-51140	54-61140
9-1150			8½	54-51150	54-61150
9-1170			10	54-51165	54-61165
9-1180			11½	54-51180	54-61180
9-1320	.013	.18	5	54-51315	54-61315
9-1330			6	54-51325	54-61325
9-1340			8	54-51340	54-61340
9-1350			9	54-51350	54-61350
9-1360			10½	54-51365	54-61365
9-1380			12½	54-51380	54-61380
9-1500	.015	.23	3½	54-51505	54-61505
9-1510			5½	54-51510	54-61510
9-1520			4	54-51515	54-61515
9-1530			7	54-51525	54-61525
9-1540			8½	54-51540	54-61540
9-1550			10	54-51550	54-61550
9-1560			11	54-51565	54-61565
9-1580			13	54-51580	54-61580
9-1590			15	54-51595	54-61595
9-1670	.016	.28	13	54-51673	54-61673
9-1800	.018	.36	4	54-51805	54-61805
9-1810			5	54-51810	54-61810
9-1820			6	54-51815	54-61815
9-1830			7	54-51825	54-61825
9-1840			10	54-51840	54-61840
9-1850			11	54-51850	54-61850
9-1860			13	54-51865	54-61865
9-1880			15	54-51880	54-61880
9-1890			17	54-51895	54-61895
9-2100	.021	.47	4	54-52105	54-62105
9-2110			5	54-52110	54-62110
9-2120			6½	54-52115	54-62115
9-2130			8	54-52125	54-62125
9-2140			11	54-52140	54-62140
9-2150			12	54-52150	54-62150
9-2160			15	54-52165	54-62165
9-2180			17	54-52180	54-62180
9-2190			19	54-52195	54-62195
9-2610	.026	.72	5	54-52610	54-62610
9-2620			7	54-52615	54-62615
9-2630			9	54-52625	54-62625
9-2640			12	54-52640	54-62640
9-2650			14	54-52650	54-62650
9-2660			16	54-52665	54-62665
9-2680			19	54-52680	54-62680
9-2690			21	54-52695	54-62695
9-3120	.031	1.1	7	54-53115	54-63115
9-3130			9	54-53125	54-63125
9-3140			12	54-53140	54-63140
9-3150			14	54-53150	54-63150
9-3170			16	54-53165	54-63165
9-3180			19	54-53180	54-63180
9-3190			21	54-53195	54-63195
9-3620	.036	1.4	7	54-53615	54-63615
9-3630			9	54-53625	54-63625
9-3640			12	54-53640	54-63640
9-3650			14	54-53650	54-63650
9-3660			16	54-53665	54-63665
9-3680			19	54-53680	54-63680
9-3690			21	54-53695	54-63695
9-4340	.043	2.1	12	54-54340	54-64340
9-4350			14	54-54350	54-64350
9-4360			16	54-54365	54-64365
9-4380			18	54-54380	54-64380
9-5220	.052	2.8	5	54-55215	54-65215
9-5240			12	54-55240	54-65240
9-5250			14	54-55250	54-65250
9-5260			16	54-55265	54-65265
9-5280			19	54-55280	54-65280
9-6260	.062	4.2	16	54-56265	54-66265
9-6280			19	54-56280	54-66280
9-7260	.072	5.7	16	54-57265	54-67265
9-7280			19	54-57280	54-67280
9-7860	.078	7.1	16	54-57865	54-67865
9-7880			19	54-57880	54-67880

*Capacity tabulation based on water at 2000 psi.
To order TWIST-TIP w/o Kit, specify 54-2850 for Models 50 series or 500 series guns or 54-2839 for Model 42 guns. Reference part sheet No. 2043.

Courtesy: Binks, Inc.

Binks tip selection chart
Figure 17-3

Spray tips for types JGB, JGN, PGB and WVA guns

How to order:

First two numbers indicate fan width in inches. — TIP — Part Number Prefix — 1017 — Second two numbers indicate orifice size (.000 inch)

Exemple: A 10" pattern (@ I foot distance from work) is desired with a .017 orifice size for proper material flow.
Order: TIP-1017

Material Guide	Specifications: Pattern Width	Specifications: Orifice Size	Specifications: Flow Capacity* GPM	Tip Part Number
THIN VISCOSITY MATERIALS Lacquers, Stains, Sealers, Enamels, Wash Primers	4" 6" 8"	.007"	.05	TIP-0407 TIP-0607 TIP-0807
	4" 6" 8" 10"	.009"	.07	TIP-0409 TIP-0609 TIP-0809 TIP-1009
	4" 6" 8" 10" 12"	.011"	.10	TIP-0411 TIP-0611 TIP-0811 TIP-1011 TIP-1211
	4" 6" 8" 10" 12"	.013"	.15	TIP-0413 TIP-0613 TIP-0813 TIP-1013 TIP-1213
MEDIUM VISCOSITY MATERIALS Sprayable adhesives, Flat paints, House Paints, Industrial enamels, Wood fillers	6" 8" 10" 12"	.015"	.20	TIP-0615 TIP-0815 TIP-1015 TIP-1215
	8" 10" 12" 14"	.017"	.26	TIP-0817 TIP-1017 TIP-1217 TIP-1417
	8" 10" 12" 14"	.018"	.30	TIP-0818 TIP-1018 TIP-1218 TIP-1418
	8" 10" 12" 14"	.019"	.33	TIP-0819 TIP-1019 TIP-1219 TIP-1419
HEAVY VISCOSITY MATERIALS Block fillers, Coal tar, Epoxies, High build enamels	8" 10" 12" 14"	.021"	.41	TIP-0821 TIP-1021 TIP-1221 TIP-1421
	10" 12" 14" 16"	.023"	.48	TIP-1023 TIP-1223 TIP-1423 TIP-1623
	10" 14" 18"	.026"	.61	TIP-1026 TIP-1426 TIP-1826
	10" 14" 18"	.031"	.91	TIP-1031 TIP-1431 TIP-1831
	12" 18"	.036"	1.20	TIP-1236 TIP-1836
	12" 18"	.043"	1.80	TIP-1243 TIP-1843
	12" 18"	.052"	2.50	TIP-1252 TIP-1852

*Flow Capacity based on 1,500 psi at the gun.

Courtesy: De Vilbiss, Inc.

DeVilbiss airless tips
Figure 17-4

as the sides of a building, buy a tip that will give you a large "fan" or spray pattern. See Figure 17-5. For trim lumber, a 4-inch pattern is fine. Figure 17-6 shows a tip with a narrow pattern. It will do a better job with less work, less waste, less overspray, less cost.

Here's the important point: Each orifice size is available in more than one *fan pattern* size. The orifice size is the diameter of the nozzle opening; the fan pattern size depends on the orifice design. These charts are based on 1,500 PSI and 2,000 PSI pressures. Other charts are available for other pressures, other spray systems, and other manufacturers.

If you don't want to buy two dozen different tips for your gun, Titan Tool of Oakland, NJ, offers six adjustable tips. Orifice sizes range from 0.007 to 0.054 inch. Figure 17-7 shows recommended applications for the six sizes.

Broad fan pattern for covering large areas
Figure 17-5

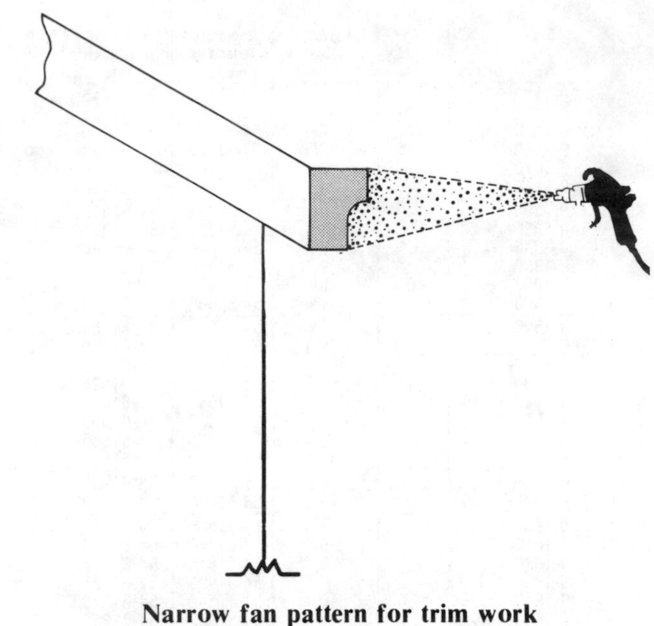

Narrow fan pattern for trim work
Figure 17-6

Remember this about spray tip charts. Most manufacturers show fan pattern sizes and lengths based on a gun that's held 12 inches from the surface being painted. Vary that distance and you vary the fan pattern size. A few manufacturers assume a different spray distance in their charts, as in the DeVilbiss conventional air spray tip chart, Figure 17-8. Read the descriptions carefully before ordering or using the tips.

Controlling the Spray

Both conventional and airless systems produce the same result: paint sprays out the orifice, either pulled out or forced out. A needle valve controlled by the trigger starts and stops the spraying action. Pull the trigger and the end or needle portion of a spring-loaded shaft unseats from an opening to allow liquid or air to pass through. Release the trigger and a spring pushes the needle back into the opening, blocking the flow.

These openings are small, somewhere in the hundredths of an inch range. They can be blocked or clogged easily. To prevent clogging, the incoming air and paint have to be filtered. Most conventional units have a small screen over the fluid intake tube within the canister. Most airless units have a paint filter in the gun handle, a screen filter attached to the paint pickup tube, and in some cases, a filter in the pump. Clean these filters every day or they'll clog or fill up with dirt, cutting your pressure and affecting the spray patterns.

Complete tip number	Type of material to be sprayed	Orifice size range	Fan width
341-015	Lacquer or Stain	.005" - .018"	2" - 14"
341-020	"	.007" - .019"	4" - 20"
341-024	Latex or Enamel	.007" - .026"	2" - 14"
341-028	"	.011" - .026"	2" - 22"
341-041	High Build	.017" - .035"	6" - 24"
341-049	Extremely Heavy	.021" - .054"	6" - 16"

Courtesy: Titan Tool, Inc.

Titan Tool adjustable tips
Figure 17-7

Keeping the Tip Clean

Many tips are supplied with a *tip guard* or *fan* that looks like a small set of wings. This guard is designed to protect the user from accidentally injecting himself with pressurized paint — a serious injury. Be sure paint doesn't build up on the tip guard, as this can alter the spray pattern.

It's essential that the tip be kept clean. If necessary, keep a small container of solvent near your work area. Dip and shake the tip in this solvent frequently. If paint has hardened in the tip, use a remover or liquid sandpaper to clean it. *Don't try to remove dry paint with a metal object.* You're likely to damage the sharp edge of the orifice, ruining it. This can get very expensive very quickly. You've not only lost the tip, but the time it

MODEL SELECTION GUIDE

MATERIAL CLASSES

grouped and numbered according to their spray characteristics

1. Product Finishing Materials
- Automotive, appliance and furniture primers
- Topcoats (acrylics, alkyds, vinyls, epoxies, enamels, lacquers, and water reducibles)

2. Maintenance Paints and Coatings
- Interior and exterior structural primers and enamels (alkyds, chlorinated rubber, epoxy latex, vinyls, and similar materials)

Abrasive Materials
- Glazes, porcelain enamels, etc.

4. Special Purpose Coatings
- roof coatings, block filler, mastics, drywall, etc.

Gun and Nozzle Chart—to order complete gun, add cap and tip to gun model—Example: JGA-502-54FX

Gun Model	Air Cap (number stamped on cap)	Approximate Pattern Length, Gun 8″ from Work	Air Consumption CFM @ PSI	Air Cap & Fluid Tip Characteristics	Fluid Tip			
					Pressure			Suction
					FX .042″	FF .055″	E .070″	EX .070″
1 PRODUCT FINISHING MATERIALS								
AGB-541 JGA-502 JGK-501 MBC-510	30	9″—10″	9 @ 40—11 @ 50	Excellent suction feed cap	•			•
	43	8″— 9″	10 @ 40—12 @ 50	Slightly shorter pattern than #30	•			•
	58	9″—10″	6 @ 40— 8 @ 50	Especially suited for suction feed with limited air supply	•	•		•
	54	8″— 9″	7@40— 8 @ 50	Suited to limited air supply	•	•		
	78	12″—14″	15 @ 40—18 @ 50	Excellent atomization stays clean longer	•	•	•	
	264	14″—16″	17 @ 40—20 @ 50	Very high capacity, fine atomization		•		
	704	8″— 9″	12 @ 40—15 @ 50	Excellent all purpose cap	•	•	•	
	705	8″— 9″	11 @ 40—13 @ 50	Excellent hot spray cap	•	•		
	765	14″—16″	13 @ 40—14 @ 50	Tapered ends for easy overlapping	•	•	•	
	797	13″—16″	15 @ 40—18 @ 50	Exceptionally clean cap—tapered ends	•	•	•	
MGB-501	30	9″—10″	9 @ 40—11 @ 50	Excellent suction feed cap	•	•	•	•
	43	8″— 9″	10 @ 40—12 @ 50	Slightly shorter pattern than #30	•			•
	54	8″— 9″	7 @ 40— 8 @ 50	Suited to limited air supply	•	•		
	704	8″— 9″	12 @ 40—15 @ 50	Excellent all purpose cap	•	•	•	
AGF-506 TGA-510	394	2″— 3″	3 @ 4	Adjustable round or fan pattern	•+			
	944	5″— 6″	4 @ 40	Pressure feed only	•+			
EGA-502	390	2″— 3″□	3 @ 40	Good pattern definition at low fluid flows	•+		•	
	395	3″— 4″□	5 @ 40	Suction feed only				•◆
						D .086″	E .070″	
2 MAINTENANCE PAINTS and COATINGS								
JGA-502 MBC-510 JGK-501 MBE-548 PGC-506	24	9″—12″	12 @ 40—14 @ 50	Large pattern with limited air supply		•	•	
	64	9″—12″	14 @ 40—16 @ 50	Broad pattern—full wet ends		•		
	490△	9″—12″	3 @ 40— 5 @ 50	Long blunt end pattern—not recommended for fast dry materials		•		
	491△	12″—14″	8 @ 40— 9 @ 50	Same as #490 but with different tip		•		
3 ABRASIVE MATERIALS								
AGB-542	64	14″—16″	14 @ 50—17 @ 60	Flows approx. 1000 cc per min.		•		
JGA-5023	67	9″—12″	16 @ 50—19 @ 60	Flows approx. 750 cc per min.			•*	
JGK-5011	704	8″— 9″	15 @ 50—18 @ 60	Flows approx. 400 cc per min.			•	
4 SPECIAL PURPOSE COATINGS								
					¼″	¼″	⅜″	
MBC-516	492M	Variable	13 @ 30—19 @ 50	Fan pattern	•			
PG-504	782M	Variable	21 @ 40—27 @ 60	Round pattern		•		
PG-506	783P	Variable	23 @ 40—28 @ 60	Round pattern			•	

△ Internal mix. *EE tip only. +F tip only (.041″). ◆E tip only. □ @ 4″ from work.

Courtesy: DeVilbiss Inc.

DeVilbiss tip selection guide
Figure 17-8

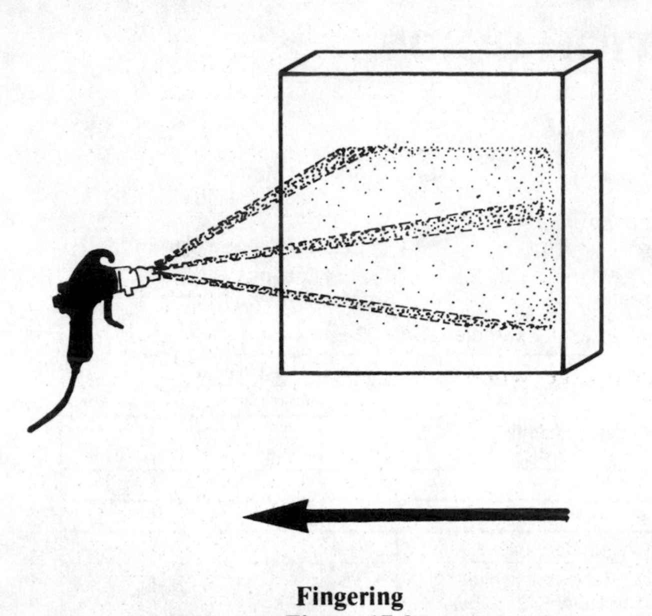

Fingering
Figure 17-9

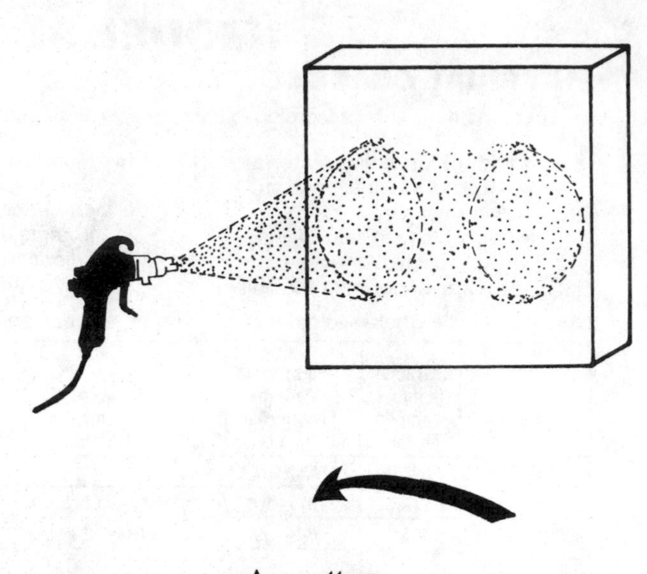

Arc pattern
Figure 17-10

takes to replace it. Use a plastic or wooden toothpick if you have to pick dried paint off.

Selecting the Orifice Size and Shape

Orifice sizes range from 0.007 to 0.110 inch for most applications. Special spray equipment for acoustics, texture coats, and other thick materials can have orifice sizes up to 3/8 inch or larger. The larger or thicker the material, the larger the orifice must be.

The orifice doesn't necessarily have to be round — it can be slotted. A round orifice forms a circular or cone spray pattern. A slotted orifice gives an oblong pattern. The slotted orifice is usually easier to control, but it does have a few drawbacks. There is more atomization at the ends of the slot than in the center. Low fluid pressure will tend to cause fingering. See Figure 17-9. Slotted tips are more prone to clogging or pattern change as the slot fills up and becomes narrower. Liquid tends to dry faster at the end of the slot.

When selecting a spray tip, be sure the tip is tungsten carbide. Paint contains pigments, many of which are oxides and are highly abrasive. A tip made of steel will wear out quickly. As the tip wears, the orifice opens up and the front edge loses its sharpness. A rounded or abraded front edge atomizes paint very poorly.

Droplet Size

I've been talking about the orifice and how it forms the paint into droplets. Just how small are these droplets? The pigments in paint range in size from 0.09 to 8.0 microns. A *micron* is 1/1000th of a millimeter. For an example, it will take 112,007,483 pigment particles 0.3 microns in size to cover a dot 1/8th inch in diameter. That's small. Each droplet of paint has millions of particles of pigment. The droplets themselves range from 100 to 5,000 microns. It takes only 900 droplets to cover that dot. A droplet 100 to 5,000 microns in diameter would be fine for finish spray work. For something like acoustics or texture coats, the droplets can be much larger.

Even though the droplets are small and travel from tip to surface in a fraction of a second, they have a tendency to begin setting up or drying before they're in place. The painter can move closer to the surface to get a "wet coating." But as you get closer to the surface, the concentration of droplets becomes larger. The paint can get so concentrated that it forms drips that run down the surface.

What is a good spray distance? Let's explore that question. It's important.

Tip-to-Surface Distance

Here's a relationship you should remember. If you cut the distance from the tip to the surface in half, you put four times as much paint on the surface. You also reduce the area of the spray pattern by a factor of four.

The opposite is also true. As you move farther away, say from 12 to 24 inches, the spray pattern coverage increases by a factor of four. But now you're spreading out the droplets and the pigments over a large area. They may not cover completely. And since the droplets have farther to travel, they

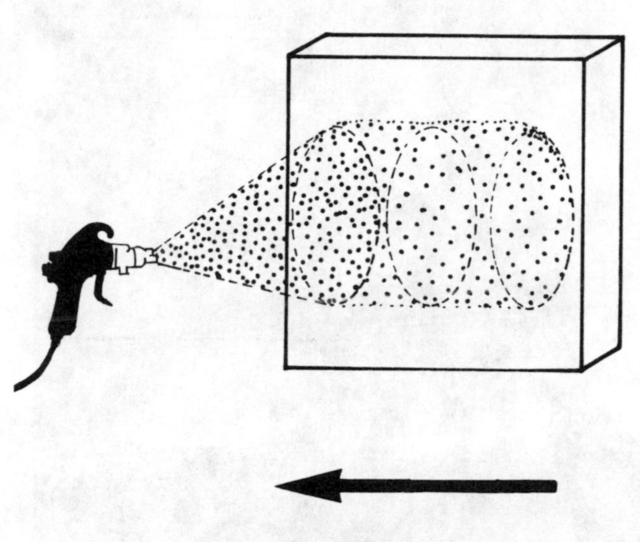

Normal pattern
Figure 17-11

tend to set up or even dry out completely before reaching the surface. The result will be poor paint adhesion.

What's the solution? Watch the tip-to-surface distance at all times. Unfortunately, that's not easy. The normal tendency is to swing the tip in a arc. *Don't do it.* Figure 17-10 shows what happens if you do. Move the tip in a straight line perpendicular to the surface, as shown in Figure 17-11.

Swinging the tip in an arc produces poor coverage, poor pattern, and increases the chances that the coating will run in some places and not stick in others. For professional results, use an even perpendicular stroke at the correct distance.

To further complicate things, most surfaces aren't smooth, at least not from the perspective of a tiny paint droplet. To them the surface is a range of mountains that can block their entry into the valleys below. If you swing the gun in an arc instead spraying straight on, the paint will *shadow* as shown in Figure 17-12. You haven't achieved full coverage when the surface is viewed from a different angle. A straight-on, perpendicular spray pattern will prevent this.

Always look at your completed work from several different directions. Is the coverage or color equal in hue and strength? If not, you're either varying the distance from tip to surface or arcing the tip. It takes practice to lay down a good spray coating.

What happens if the first pass of spraying didn't cover or hide the old surface? Nothing! Just wait a few minutes and shoot it again. Even a third or fourth pass may be necessary. You should be spraying at 10 to 12 inches from the surface.

Why not just get closer and save the trouble of a second coat? Because that's what causes runs and sags in the paint surface. When that happens, you'll have to stop work and clean paint off the surface where the run appeared. That usually destroys the painted surface of the adjacent area.

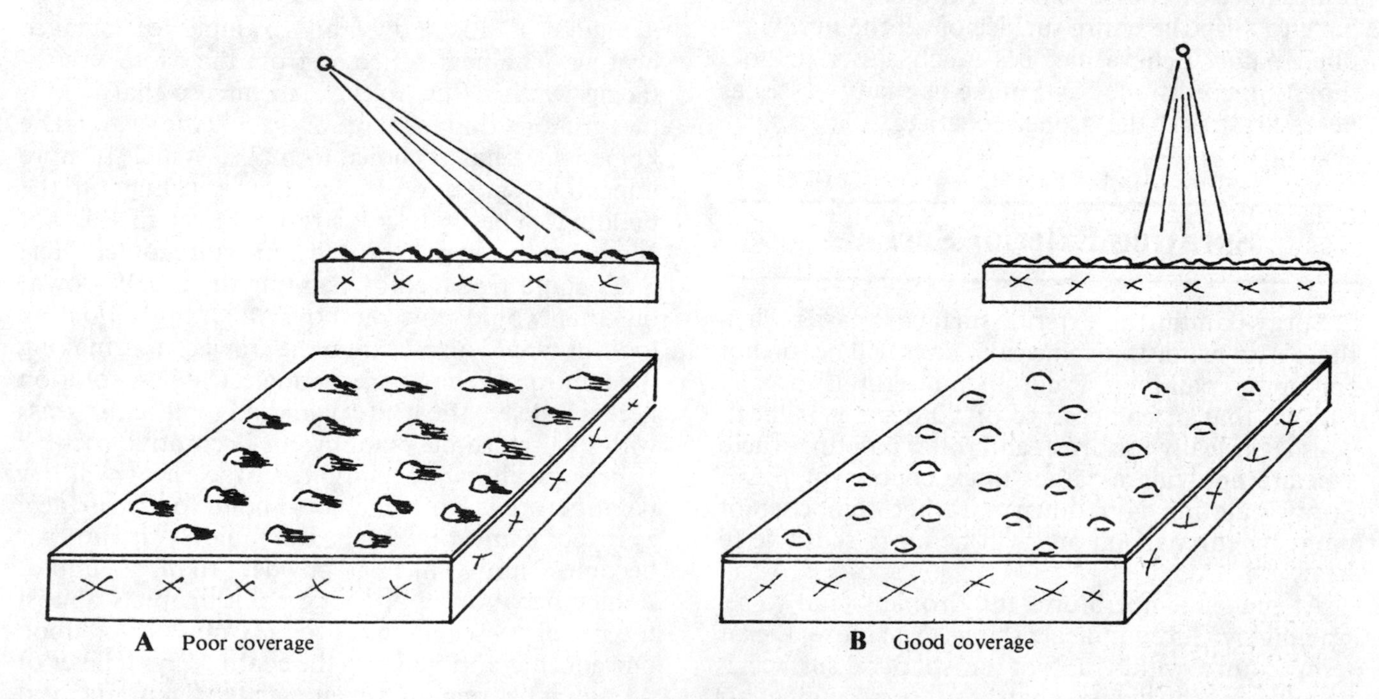

A Poor coverage **B** Good coverage

The cause of shadowing
Figure 17-12

Heat is a common exterior painting problem
Figure 17-13

Even if you're careful, there will be a break or mismatch in the paint coating. This break is very noticeable, of course. To get rid of it, you would have to strip the entire surface of all the newly applied paint. Believe me, it's much easier to hold that 12-inch distance and make as many passes as necessary to get the proper coverage.

Spraying Exterior Surfaces

Spray painting an exterior surface is easier when there's no wind. But some calm days will be too hot for spray painting. Paint manufacturers usually suggest that a temperature of 50 degrees is ideal. This is basically for brush and roller painting where you are applying a rather thick coating of paint. Spray painting lays down a rather thin coat of paint by comparison and will be more sensitive to heat.

At higher temperatures the droplets tend to set up quickly. To complicate things even more, the air temperature within a foot or so of a surface is usually much higher than the surrounding air temperature. Heat may be radiated and reflected from the surface. If the air temperature is 90 degrees, air near the surface may be 130 degrees and the surface itself may be hotter yet.

Figure 17-13 shows an example of surface heating. The heat reflected from the white, south-facing wall of this house is so intense that it kills the grass within a foot or so of the wall. The homeowner has a choice to make: Which is more important, grass or the paint remaining on the building? The light color reflects heat in this hot climate and helps keep the interior cooler. The grass could be saved by painting the block brown, but that would give the house a strange, floating look. It would also heat up the crawl space, making the interior of the house warmer. The best solution here is to leave the white paint and replace the grass with heat-resistant plantings or decorative rock.

Besides killing vegetation, surface heat will dry paint before it has a chance to bond to the surface. Sprayed paint that dries too quickly in hot air becomes more like a powder than a liquid. Remember, you're applying paint droplets almost too small to see. These droplets must layer upon one another and build up the coating. Each layer or droplet must remain wet enough to stick to the next droplet. Paint dried to powder won't stick.

Here's something else to consider. Your spray gun tip is in this extra-hot surface zone all the time it's in use. The tip gets hot, causing it to clog more often. In hot weather you'll waste valuable time cleaning the tip instead of painting.

I suggest you not spray paint in direct sunlight or when the free air temperature is above 75 degrees F. Do your spray painting in the morning after the dew has dried off, or late in the afternoon when it's cooler, or only in the shade.

Heat isn't the only temperature hazard in spray painting. Low temperatures can also cause problems. Cold paint thickens and doesn't atomize as readily. It also takes longer to set up, increasing the length of time required to dry. That means it attracts more dirt and insects. I've found that you can spray paint successfully at temperatures as low as 40 degrees, since sprayed paint dries faster than brushed or rolled paint. Most paint should be "wet" for 30 to 60 minutes after application. This allows the paint to seep into the surface and get a better "bite" or bond.

That summarizes your temperature range for spray painting, 75 degrees as a high and 40 degrees as a low. Some painters find they can spray at higher and lower temperatures. But the more the thermometer exceeds those limits, the greater the risk of a major mistake.

Electrostatic Painting

I said a few paragraphs earlier that powder doesn't stick very well. But there's one situation where that statement isn't true. In electrostatic painting, the paint *is* a powder — a powder that will accept and hold an electrical charge. The item being painted must also hold a charge that's opposite of the charge on the powder. Since opposite electrical charges attract, the charged powder travels to the charged surface, where it temporarily sticks. The painted object is then baked at a temperature high enough to melt the powder into a liquid. This liquid paint flows or spreads slightly and then sets hard.

There's also a form of electrostatic painting that uses a charged liquid paint. The advantage of the powder system is that you can reclaim any powder that doesn't stick and use it again.

The primary advantage of using either electrostatic system is that the electrical charge draws paint into every nook and cranny of the object. Coverage is complete. You can paint irregular or cylindrical objects as well as the inside surface of objects. Just stick the nozzle tip through a hole from the outside and spray. Properly done, there should be very little overspray.

Electrostatic spray painting is normally done in manufacturing plants. But it can also be done in the field, such as painting lockers or playground equipment. It takes a unit that's basically an airless but with a power supply. The only real difference is that the object being painted has to be an electric conductor or has to be able to withstand baking temperatures if you're using powder. Of course, you can't use this system on anything that has electronics or other delicate components that would be damaged by an electrical charge or heat.

Spray Painting Tips and Precautions

If you want your spray equipment to last, cleaning is essential. Spray at least a gallon of solvent through the system after each use. This solvent can be reused many times but will eventually become contaminated. When that happens, discard it and replace it with clean. The cost of new cleaning solution is a small fraction of the cost of replacing the spray unit if you *don't* clean it properly.

What happens if you don't remove all the paint from an airless? Paint will clog the valves, the valves will stick, and the piston will get clogged in the cylinder. If this happens, the only solution is a complete overhaul of the equipment.

The newer units include two buttons that you push before starting up. These buttons manually force the valves open, breaking any paint seal. It helps prevent problems, but thorough cleaning is the best solution.

Here are a few more spraying precautions. Even when the spray pattern is directed on the surface to be painted, there will also be a mist of paint in the surrounding air. This mist is carried by air currents onto everything in the area, sometimes even into other rooms. Remove or cover everything you don't want to paint or clean later on. Wear a dust mask over your mouth and nose, or use a respirator. Dust masks are inexpensive, under a dollar. Respirators cost about $35 but they're essential if you're spraying toxic paints, lacquers, oil/alkyd, urethanes, or liquid plastics.

Always do your spray painting in a well-ventilated area. The mist of paint particles is probably explosive if exposed to open flame. Turn off all pilot lights and extinguish all sources of sparks, like motors or cigarettes.

Finally, never, ***never*** spray an airless into your hand to test the pressure. The paint will be forced into your skin, possibly causing blood poisoning.

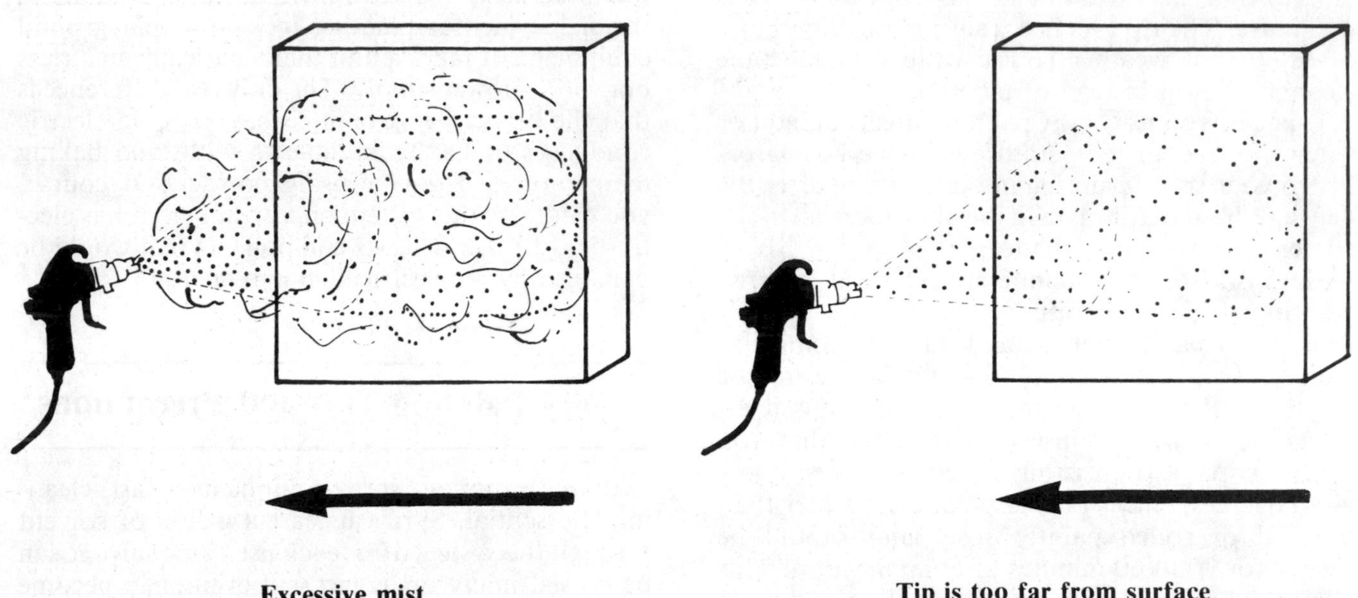

Excessive mist
Figure 17-14

Tip is too far from surface
Figure 17-15

Troubleshooting Guide

I've covered some of the more common spray painting problems and how to prevent them. But there's a whole lot more that can go wrong. The next section is intended to be your troubleshooting guide for airless, conventional and aerosol spray systems. I list symptoms and causes, arranged from most likely to least likely.

Use this section as a reference source. Study it first, so you avoid most of the mistakes in the first place. Then use it to remedy problems if they occur.

Common Spray Problems

Bands of color— The surface looks uneven, as if the paint were applied in horizontal bands. The gun tip is probably the wrong size or improperly adjusted. Change to the correct size or change the adjustment.

Another possible cause is that you're not overlapping from one-third to one-half of the previous pass. Slow down and overlap properly. Try making a fast finish pass in a direction perpendicular to the first coat.

"Bow tie" paint pattern— You're not holding the gun perpendicular to the surface during your arm sweep. Make sure the gun is kept perpendicular to the surface rather than swung in an arc. See Figure 17-10.

Shadowing— This means the paint looks different when viewed from different directions. Look back at Figure 17-12. It's caused by not holding the gun perpendicular to the surface. Surface roughness has kept the paint from filling all voids.

It can also be caused by the paint itself. Some latex paints show a color change due to molecular lineup in the binder. Some paints can *rainbow* from certain angles.

Mist is excessive— You're probably using too much pressure. Reduce the pressure and try again. Figure 17-14 shows the problem.

It's possible that the paint is too thin. Add unthinned paint to thicken the mixture.

If that doesn't work, the tip orifice is probably too small. Choose a tip with a slightly larger orifice.

Overspray is excessive— You're using the wrong tip for the job. Use a tip with a narrower pattern or "fan-out."

Paint doesn't cover or "hide"— The paint you're using may not be able to cover in a single coat. Use a different paint or apply a second coat.

If the paint should be covering in one coat but isn't, the tip may be too far from the surface. Move in a little closer until you get good coverage without runs or sags. Figure 17-15 shows the pattern you'll get if the tip is too far from the surface.

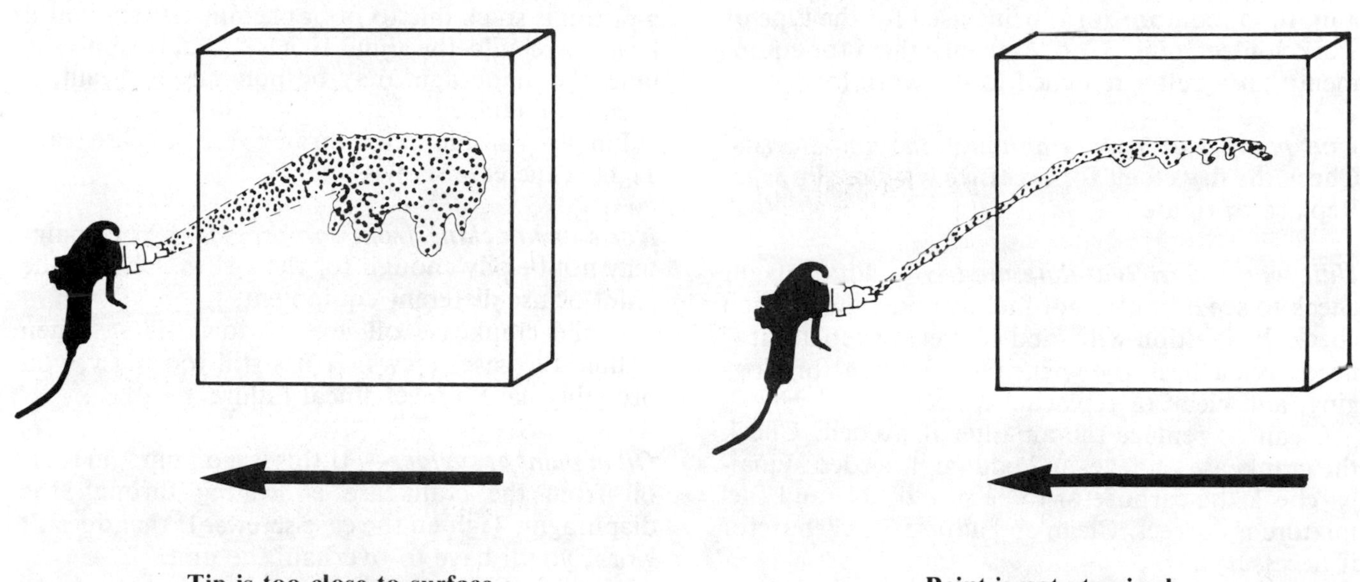

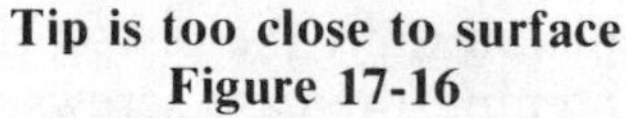

Tip is too close to surface
Figure 17-16

Paint is not atomized
Figure 17-17

Perhaps the surface is contaminated with dirt, oil or grease. Clean it and try again.

Finally, the surface may require a primer. If that's the cause, prime and repaint.

Paint drips, sags or runs— The tip is probably too close to the surface. Move it back a little until the problem stops. Look at Figure 17-16. And read the section on puddling a little farther on.

Paint is sandy or powdery— The tip is too far from the surface. Move closer.

Here's another possibility. The temperature is too high, causing the paint droplets to dry before they reach the surface. Either wait for the temperature to drop or paint in the shade for a while.

Paint isn't atomized— Figure 17-17 shows what happens when the paint doesn't atomize. The most likely cause? The pressure is too low. Increase the pressure until atomization occurs.

It can also happen if the paint is too thick. Thin the paint until it atomizes properly.

The spray tip may be clogged, worn or improperly adjusted. Clean, adjust or replace the tip.

Pattern is distorted— The distortions may include runs or poor coverage. You're using the wrong arm movement, sweeping with the wrist instead of the arm. Use a broad sweep, perpendicular to the surface.

Puddling— Latex paint looks very wet in spots. You're "dumping" or spraying too much paint in one spot, probably because the gun isn't in motion when the spray is triggered, or the arm movement is too slow. Keep a moist roller handy to roll out the excess, then make a fast spray pass to blend in the section rolled.

Spray is erratic— If the spray changes from good to poor during the same operation, lumps in the paint are getting through the filter and blocking the needle valve or orifice. Strain the paint before using it to avoid this problem.

Tips clogs every few minutes— The temperature in the work area is too high. Move to a cooler area or wait until the temperature drops.

Unit blows circuit breakers— The compressor motor is too large and draws too much current for the building's wiring. The wiring in most homes can't handle more than a 1 or 1½ horsepower motor. If the unit is stopped under pressure, it draws a very heavy load when turned on again. If the circuit breakers trip, release the pressure before turning the compressor on again.

Unit requires excessive maintenance or repair— All spray equipment requires service, but keep good records of your repair and maintenance expenses. If unit is defective, your record of repairs may help convince the manufacturer to replace it at his cost.

Most problems are the result of improper cleaning. There's also a possibility that the wrong size gun, tip or compressor is being used for the type of work you're doing. If so, exchange them for equipment that's better matched to the work load.

Unit pump won't work although the unit starts— Check the drive belt for excessive wear or slippage. Replace or tighten.

Unit won't start (gasoline models)— First of all, check to see if you've got fuel. If not, fill up. Then check the ignition wire, and connect or replace it if necessary. Check the spark plug for wear or clogging, and clean or replace it if necessary.

Clean or replace the air filter if needed. Check the crankcase oil level and add oil if needed. Finally, check the carburetor to be sure the air and fuel mixture is correct. Clean and adjust the carburetor if necessary.

Airless Paint Problems

Atomizes only thin materials— The most likely cause is low pressure. Increase the pressure until it atomizes properly.

You could be using the wrong tip orifice. Replace or adjust the tip until it atomizes properly.

Finally, you could be using a unit that isn't adequate for the material you're trying to spray. Most airless units made for home use have a very small piston that doesn't have enough volume displacement or power to handle heavier and thicker paints consistently. Replace it with a professional unit.

Continuous loss of pressure— If the unit is clean and the paint isn't too thick, you probably have a piston seal or diaphragm leak. An overhaul may be required. Most low-cost units made for homeowners are prone to this type of problem. Stick to good, professional-quality equipment.

Fingering— Look back at Figure 17-9. The most likely cause is low pressure. Increase it until you get full coverage.

The paint could be too thick or the filters may be clogged. Thin the paint or clean the filters as appropriate.

No spray— Check to see if the inlet filter is clogged. Clean it if necessary. Then check the in-line filter and the orifice. Clean them as needed. If you're using an adjustable tip, check the adjustment and readjust if needed.

If it's not a tip or filter problem, you may have stuck valves. Newer models have a button that opens the valves manually. On older models, tap on the compressor housing until valves are free.

If this doesn't solve the problem, check to see if a piston is stuck due to poor cleaning. If so, you'll have to rebuild the unit. If it's a diaphragm-type unit, the diaphragm may be punctured. Again, it means a rebuild.

Finally, consider a diaphragm to crankcase leak. Tighten the case screws.

Noise or knocking from compressor— Your paint may not be oily enough for the airless. Change the paint or use different equipment.

If the crankcase oil level is low, fill it. Then tighten the case screws. If it's still too noisy, you probably have a mechanical failure.

Oil in paint at surface— If this is a diaphragm unit, oil from the crankcase is leaking through the diaphragm. Tighten the case screws. If that doesn't work, you'll have to overhaul the unit.

If this is a piston unit, the piston rings are leaking oil. If tightening the case screws doesn't help, rebuild the unit.

Paint film is picking up dirt— The paint spray stream is creating a vacuum that's picking up dirt and mixing it with the paint. Clean or shield the area to be painted.

Paint usage is too high— You'll use more paint when spraying with an airless system because they stop spraying with up to a quart of paint still in the hose lines, can bottom and compressor compartment. Force part of the remaining paint out by pouring the proper thinner into the paint can. Then continue spraying until the paint begins to thin.

Spraying with wide overlaps can also cause excessive paint consumption. Make sure you're not overlapping more than one-third to one-half of the previous pass.

Pressure drops quickly after trigger is activated— The paint may be too thick. Thin to the proper viscosity. Or you could have the wrong tip size or an improper adjustment. Use a smaller orifice if necessary.

The inlet or in-line filter may be clogged. Or the unit may not be powerful enough. Use at least a 3/4 horsepower unit for heavy-bodied paints such as unthinned latex.

Delay between trigger pull and start of spray— This is normal for most airless systems and for bleeder-type conventional systems. Practice a little until you get accustomed to it.

Conventional Spray Problems

Fluid leakage from the air nozzle— The tip is dirty. Clean the tip and the tip area.

It could also be that the needle isn't seated properly. Remove the needle, clean and reinstall. Finally, see if the fluid needle packing screw is overtightened. If so, loosen it.

Fluid leakage from needle packing screw— The packing screw is loose; tighten it. Or the packing material is worn and needs to be replaced.

Fluttering or jerky spray— There's too little paint in the cup or tank. Refill. Or you're holding the gun at too great an angle for the paint to be picked up evenly. Hold the gun level.

The needle packing may be worn. If so, replace it.

Finally, check to see if any of these components are loose: the tip, the needle, the packing screw or the fluid tube. Tighten any loose parts.

Material fails to atomize— Most likely, the air pressure is too low or the paint is too thick. If not, check for a blocked air line or air passages in the gun. Check the air filter.

Material runs or sags— The gun tip is too close to the surface. Move away. You may be sweeping the gun too slowly. If so, sweep your arm more rapidly.

The paint may be too thin. Check it, and add unthinned paint if needed.

Finally, check the adjustment of the needle. It may be open too much.

Mist is excessive— The gun tip is too far from the surface. Move closer.

Spattered spray— The air pressure is too low. Increase the pressure.

The paint may be too thick. Thin it using a viscosimeter.

Check to see if the air cap is dirty. Remove and clean it.

Split spray pattern— The air pressure is too high; adjust it. Or the fan adjustment may be open too far.

Spray pattern is too small— It may be lumpy paint. Remove, filter and reuse.

Check the bleeder screw to see if the fluid pressure is too low.

If the air cap is dirty, remove and clean it.

The last possibility is an incorrect fluid adjustment. Check and readjust.

Aerosol Paint Problems

Color is wrong— There may still be pigment in the bottom of the can because it wasn't shaken enough. Or the can just may have had the wrong cap on it. Check the label for the color.

Coverage is low— The average can of aerosol paint should cover 40 square feet. Check the label for the coverage rate for the brand you're using. Here's another possibility: If you're spraying in a windy area, some of the paint is blowing away before it hits the surface. Move to another area or wait for the wind to die down.

Gloss surface when you're using flat paint— Some of the pigment is still in the bottom of the can. Shake well and try again.

No mixing ball— If you're using polyurethane, the can doesn't need a mixing ball. Other kinds of paint have a mixing ball in the can. If you don't hear it when you shake the can, the ball may be bonded to the bottom of the can. Strike the can sharply on the floor to knock it loose.

No paint coming out— The most likely reason is that the can is empty. Replace it. If there's paint in the can, the nozzle may be clogged. Try cleaning the nozzle: If that doesn't work, return it to the store for a replacement. Minimize clogging by shaking well before using and following all instructions on the label.

Paint appears OK but it peels off easily— The problem is poor surface preparation. There's oil, grease or other surface contamination. Take time to thoroughly clean and prime the surface before applying the final spray coat.

If the surface is extremely hard or glazed, you must provide some "tooth" for proper adhesion. Clean, scuff sand, and clean again before you paint.

Paint doesn't atomize— The nozzle may be partially clogged. Shake the can and try again. If necessary, dip the nozzle in thinner, shake and try it. If it still doesn't work, remove the nozzle and clean it with a wire or pin.

Paint doesn't cover or "hide"— The paint may not have one-coat capability. You'll have to respray. If it should cover with one coat and doesn't, you may be moving your arm too fast.

Slow down and see if it covers.

You may be holding the can too far from the surface. Move closer. Finally, check to see if the surface was properly undercoated. If not, clean, undercoat and spray again.

Paint drips and runs— Most likely, your arm travel is too slow. Speed up a little. Or the can may be too close to the surface. Back up and try again.

The third possibility is that you're overlapping too much. Make sure the overlap isn't more than one-third to one-half of the previous pass. Finally, the surface may be contaminated with oil or grease, preventing proper adhesion. Clean the surface and try again.

Paint looks rough or sandy— The paint isn't properly mixed in the can. Shake the can well. Or, if you're holding the can too far from the surface, move a little closer.

If the temperature of the air, the surface or the paint is too high, it can cause this effect. Paint in a cooler area.

Cement Products and Texturizing

This book is about painting and wall coverings. You know the difference. Paint comes in a can and wall coverings come in a roll — at least most of the time. But there are some wall coverings that are neither paint nor paper. As a professional in the field, you should know about these products. That's the purpose of this chapter.

There was a time in history, between the thirteenth to sixteenth centuries, when some painters had international reputations. These master craftsmen painted on wet plaster using a technique called *fresco*. The most famous of these painters was Michelangelo.

Today you can still see these old frescoes in churches, palaces and public buildings. They're very durable. The paint pigments were mixed with water and applied over the wet plaster. As the plaster dried, the color was absorbed and bonded to the hard surface. But only certain pigments could be used because plaster bleached the darker, brighter colors. Grays, tans, and rust tones were best. Blues, reds, greens, and yellows were used less frequently.

Unfortunately, fresco painting is nearly a lost art in this busy world. But it did give modern man an idea or two. Today, murals similar to the old fresco paintings are made out of easier-to-use materials. And mural tiles, wallpapers, decals, and stencils are used like fresco paintings to bring beauty to plain walls.

Some fresco-like painting is still done. Figure 18-1 shows a mural painted on an exterior wall in Pine Bush, New York. I just wish the artist had painted the meter boxes into the picture. Still, it shows that even a dull building can be made exciting.

Using Stucco

The second idea based on fresco painting was to color an entire plaster wall by mixing pigment into the final plaster coat. This is called stucco. It's a mixture of sand, water, lime, cement, and color pigment. Applied wet and pliant, it dries to a rocklike hardness. Stucco is generally applied over a wire supporting mesh.

Figure 18-2 shows a building being prepared for stucco application. Impregnated kraft paper with wire screen bonded to it is stapled directly to wall studs. The paper will act as a vapor barrier, while the wire supports the stucco.

The stucco is applied in three layers. The bottom or first layer locks the material to the mesh. The second layer helps to build thickness and strength. The final layer, or *dash coat,* has the color and is used to give the surface its final texture. Later in this chapter, I'll show you how to create several of the more popular textures.

Should stucco be painted? Usually not, as it absorbs and expels moisture. Stucco needs to breathe. If it's painted, one of two things can happen: If the

A modern mural
Figure 18-1

Building ready for stucco
Figure 18-2

Crack in stucco caused by structural problem
Figure 18-3

moisture moves as it should, it can cause the paint to peel. Or the paint may trap the moisture in, leading to destruction of the stucco and even the framing lumber within the walls.

This is especially true in parts of the country where weather is more extreme. In the south and southwest, stucco can be painted with a latex paint, since the latex will allow the moisture to escape. But *never* paint stucco with an oil-based paint, or a semi-gloss or full gloss paint.

Repairing Stucco

If you encounter a stucco surface that's faded or in serious need of repair, there are several alternatives. If the owner just wants a color change, painting with latex paint will give good results in the southwest or drier climates. Repainting will cost much less, of course. In the north or northeast, I would suggest a restucco job. The southeast and northwest areas of the United States have more rain, so a restucco will give a more durable repair or color change.

To paint over stucco, first remove any accumulated dirt by hosing or waterblasting. Let the surface dry a day or two and then paint with latex using an airless spray rig.

To restucco, the first step is removing the dash coat with a sandblaster. Then apply the new dash coat with a spray gun designed for applying stucco, or with a brush or trowel. The coating should be 1/8-inch to 1/4-inch thick. Texture it while it's still in a "plastic," workable state.

But be sure to correct any subsurface problems before you paint or restucco. If you don't, the new surface will soon be damaged. The crack in Figure 18-3 was caused by poor drainage at the foundation, causing the underlying cement block to settle. Unless the drainage problem is corrected, the new stucco coat will crack again as the settling continues.

Masonry Paints

Cement-based paints are similar to stucco. Masonry paints are a mix of cement and plastic or synthetic rubber. They act as a waterproof sealer to keep water, even under pressure, from coming through the surface.

They can be tinted with a specially-formulated dry powder tint. You can't get bright colors, though. As with the frescoes, cement bleaches bright or dark pigments.

Before you paint or repaint, remove all dirt, grease, grime, and efflorescence from the surface. Use a solution of muriatic acid and water. The formula is in Chapter 11. But be careful: Muriatic acid can cause skin burns and even blindness. Wear rubber gloves, eye protection, and suitable clothing. The fumes can damage lung tissue, so avoid breathing them.

Seal cracks and holes with a hydraulic cement. Hydraulic cement, unlike other cements, expands instead of shrinking as it cures so it won't fall out of the patches. For best results, wet the surface to be coated or patched to keep the new material from drying too soon at the bonding layer.

Give cement block or brick masonry joints a primer coat before they're painted. The primer can be a paintable masonry sealer paint or a paintable clear waterproofing liquid. Moisture will come through the joints if they're not sealed correctly.

Finally, find and eliminate the source of any water seepage before repainting. Then paint the bottom third of the wall and let it dry for 24 hours before painting the entire wall.

You can use masonry sealer paint on the insides of fish ponds, water holding tanks, and swimming pools made of concrete or masonry. After applying the masonry sealer paint to the interior, allow it to dry for 24 hours. Then fill the tank with water and let it set for another 24 hours before draining. Of course, a swimming pool takes too much water to do this, so hose it down well instead. Why do this? It gets rid of excessive alkalinity which can and will damage the surface as well as pollute the clean usable water.

Make sure you schedule time for this step and notify the property owner that you'll be doing it. And note this: Masonry sealer paints, like most all cement products, can be toxic. Work in a well-ventilated area.

Masonry sealer paints are applied with a brush. A gallon will usually cover about 75 to 100 square feet of surface. You've got to stir the mix frequently to keep the heavier cement particles from settling to the bottom. The dry powder type masonry paints must be used within two hours of adding water or they'll harden in the container and become useless.

Don't use masonry paint on floors. It isn't abrasion resistant. Instead, use specially formulated masonry floor paint.

Masonry Floor Paints

These are special paints formulated for coating concrete, cement, flagstone, brick, and terrazzo floor surfaces. They produce a film that prevents surface dusting, cracking, pock marking, and spalling. The floor paints also act as a barrier against staining from spills or other sources. They penetrate the surface deeply to fill the surface voids. Their biggest advantage, however, is that they contain mineral pigments that stand up to abuse and abrasion.

But don't paint a concrete floor until it's had a year or two to age. As the concrete ages it gives off minerals and salts. If you paint too soon, the sealer won't adhere.

For interior concrete floors, an epoxy or a polyurethane-based material works best. Be sure to give the floor a good paste waxing for maximum protection and shine.

Figure 18-4 shows a set of exterior steps that weren't coated. The damage could have been prevented by painting with an epoxy base paint or a clear waterseal. The only remedy at this point is to set forms and pour a 1-inch layer of acrylic-rich concrete over the old concrete. Then apply a coat of waterseal.

Damage to unprotected cement steps
Figure 18-4

For walkways, patios, driveways and other exterior concrete surfaces, I recommend a water-based epoxy paint or stain. Look at Figure 18-5. Exterior masonry paint can withstand a good deal of abrasion before it requires stripping and repainting. This sidewalk "red zone" has had thousands of people walk on it over the years. It's survived 110-degree heat, 24-degree cold, rain and snow, and scuffing from inattentive drivers. The white reflective street paint has withstood all of the above plus thousand of cars and heavy trucks.

Masonry paint used in street marking
Figure 18-5

Painting Concrete or Masonry Walls

Concrete or concrete block walls should be weathered at least eight weeks before painting. Then paint interior walls with a latex filler paint, regular latex paint or the combination of the two.

Paint exterior concrete block walls with a watersealer paint, a waterseal with a latex top coat, or with a masonry sealer paint. You're sealing the surface against wind-blown rain that can seep in through the walls and ruin the interior walls and the building's contents. For below-grade or ground-level walls, I recommend waterproofing with an asphalt coating.

Painted brick that's flaking and peeling
Figure 18-6

If you have a choice, don't paint brick walls at all. If you must paint over brick, first let it cure for at least a year. Then paint only with paint specially formulated for brick, or with latex. Why? Easy! Brick is porous and must be allowed to breathe. Painting with an oil-based paint will only lead to peeling and repainting. Figure 18-6 shows a good example. It's a section of a painted brick building that's flaking and peeling. Water entered at the roof line and flowed along the mortar lines. It got behind the paint, lifting it off. The solution is to sandblast the entire building, then leave it natural or paint with latex.

I've been recommending latex paint. But here's a note of caution. Some of the newer formulations of latex "skin over" just like an oil-based paint. And that kind of latex will peel off just like oil-based paint. Use exterior latex formulated for use on masonry. But note that latex paint formulated for masonry is not the same as masonry sealer paint.

Plaster

Plaster is a white powder made from gypsum or calcium sulfate. The gypsum is crushed, mixed with water, and then heated. When the water evaporates, a powder is formed. When you add water to that powder, it forms a workable plaster that hardens back into stone like the original rock.

Filler additives such as cattle or goat hair, manila, jute, and wood fibers can be added to give additional strength. Sand, vermiculite, and perlite can give extra resistance to fire and sound transmission.

Wallboard, or gypsum board, is gypsum pressed between heavy paper. It's the most popular interior wall building material in America today. It's inexpensive and easy to work with. Traditional plaster walls aren't common in modern construction because plaster costs more and takes more time to apply and dry than wallboard. Wallboard goes up faster and easier. That's helped keep home prices down.

New Finish Techniques Require New Skills

Modern construction requires efficiency and good management. Material and labor costs are too high and there's too much competition to allow waste and inefficiency. Knowledgeable property owners demand durable materials and good workmanship. The builder is behind the eight ball even before he breaks ground.

When labor costs were lower, laborers could be hired to work for weeks — even months — on each phase of the construction. The accent was on doing it right. It still has to be done right, but now the emphasis is on speed. Older buildings had walls of real plaster and lath. Highly skilled craftsmen worked the plaster into a smooth, paintable surface. Today there are very few skilled plaster craftsmen. The popularity of wallboard has made them practically obsolete.

Walls are now installed in a few days using semiskilled labor. As the saying goes: "Time is money and money is time." To stay competitive and turn a profit, a builder has to put his buildings up fast, using the lowest priced materials that will do the job.

Substituting drywall for plaster reduced construction time. But now even conventional drywall application takes too much time. The taping, spackling, and sanding take days to complete — days in which very little other work can be done.

Texture Coats over Drywall

As usual, when there's a problem, someone finds a solution. In this case, the solution is the hand-held automatic taping machine and spray-on wall textures. The tape is applied fast and in a thin coat, followed by a one-time spackling, quick sanding and spray texture to cover up the inevitable defects. The spray textures, and even the hand-troweled textures, go on quickly and mask most imperfections and poor workmanship. It's just one of the ways a building contractor can help keep new homes affordable. Texture coats aren't new. They were probably used before paints to decorate walls. But the equipment to apply them fast and conveniently is new. Most of the spray-on textures have been developed in the past 20 years.

Textured walls can help you stay competitive. Learn as much as you can about applying and repairing them. For a painting contractor, doing wall texturing is just a natural extension of the basic painting business.

The next pages cover both trowel-on and spray-on wall textures. I'll describe what they are and how you can apply them to both interior and exterior surfaces. I use the term "surfaces" because you're not limited to walls. Ceilings, fences, planters, flatwork, driveways, walks and most any surface where you want to create a permanent design are included.

Wall texturing materials can be paints, plasters, and cements. There are only a few requirements. They must:

- Apply as a plastic
- Remain in a semiplastic state long enough to be properly worked
- Dry to a hard, durable coating
- Be able to be colored

Kinds of Texture Material

The simplest of the texture materials is standard off-the-shelf paint to which size-selected and washed sand or mineral powders are added. The texture material can be purchased in most paint stores in dry form. You just add it to your own paint and apply it with brush, roller, or the proper spray equipment.

Texture Paints

Next are the texture paints themselves. These are specially formulated paints that have thickening agents added to give them body and workability. Thickening ingredients vary but most are talcs, clays, finely ground nut and animal shells, or finely powdered minerals.

There is also a thixotropic agent, which is a silica aerogel. It's a safe, inert substance, used as a filler, that's pure white and very lightweight. One cubic foot of the material weighs about 3.8 pounds. The material is also mixed with fiberglass resin when applied to vertical surfaces.

Texture paint usually contains a plasticizer that makes it more workable, aids adhesion, and improves the final strength of the internal bonds. Texture paints range in consistency from smooth to rough to very rough or sand-grained.

Apply them by brush, roller, or with special spray equipment. You can buy roller covers that create texture designs as you roll the material on. The problem with these is that the design repeats once every revolution of the roller. It takes time and skill to get the desired result. Some texture roller covers are non-repeating and leave a desirable random pattern.

Plaster and Stucco

The next material I'll cover is standard interior wall plaster. It should be used only on walls and ceilings that have been designed to accept plaster — that have wood or wire lath. The plaster is troweled into place and can be textured by any of several common methods. But there are two problems with plaster: One, it doesn't stick very well to smooth surfaces. Second, it shrinks as it ages.

For exterior surfaces, you can use cement or stucco. This can be blown on by machine or hand troweled onto the surface. Wire lath is recommended for proper adhesion.

Taping Compound or Spackle

One of the best and most used materials for interior applications is drywall taping compound (sometimes called *spackle*). Use taping compound, not *finishing* spackle, as it has more body and workability. It sticks to unpainted drywall, doesn't shrink, and can be brushed, troweled, or sprayed in place. And best of all, it's inexpensive. A 50-pound box of premixed spackle costs less than $10, and depending on method of application and texture desired, can cover a lot of area. If you're spraying it, you should be able to cover all the walls in a 1,500 square-foot home with one box. For a heavy texture troweled on, it will cover an average 9 foot by 10 foot room.

Acoustic Ceiling Mix

The final texture coat is for ceilings. Acoustic ceiling mix is a dry spackle mix that has Styrofoam bits or other soft plastic granules added to it. This material is mixed with water to a consistency of thick oatmeal and blown or sprayed in place. The cost is low, under $15 a bag, and one or two bags will cover the ceilings of a small home.

The acoustic mix should be applied over white primer for best adhesion and light reflection. Because it's thick, it will cover a lot of surface imperfections. However, large surface bumps must be flattened before you apply it, as ceiling lighting will accent them.

I don't recommend using acoustic ceiling texture inside closets. The material is easily abraded and the homeowners will forever be picking up pieces of foam. Also, who hangs around in closets admiring the ceiling?

Tools Used for Texturing

Texturing tools for these materials fall into two categories: hand tools and machine application tools. The hand tools are brushes, rollers, trowels and hand-operated pump sprayers. The machine tools are paint sprayers, air-driven hopper guns, and special commercial hopper systems. You need to consider the type of texture desired, the quality and thickness of the material, and the cost of application when deciding which type of equipment to use.

For fast, low-cost commercial or residential application, special purpose hopper equipment is essential. The equipment costs several thousand dollars. But if you want to do this type of work for a living, you have to have it. If you're just doing an occasional texture job, you can rent the equipment from most rental centers for about $60 to $80 a day.

If you own a one horsepower or larger air compressor, you might consider a hand-held hopper gun like the one in Figure 18-7. One source is The Wallboard Tool Company, P.O. Box 20319, 1708 Seabright Avenue, Long Beach, CA 90801. Their model GHG-30 costs about $100. It's easy to use and good for applying dozens of spray textures as well as acoustical ceilings.

If you want to do quality decorator-type texturing, all you need are simple hand tools, time, and imagination. The results depend on your skill with the tools and material.

Hand-held hopper gun
Figure 18-7

Hand-Textured Finishes

To help you compare textured finishes, I applied several examples to a sample board. I troweled the material into place and used the tool in the picture to get the final texture. I suggest experimenting on a 4-foot by 4-foot scrap piece of wallboard before you tackle a complete room or building. It takes time to get the "feel" of the material so you can get results that are consistently professional.

For interior walls, use a commercially available texture paint with any of the methods pictured. For exterior walls, use stucco or a mixture of one part white portland cement, three parts white lime, and three parts water. Textures can be made using combs, brooms, trowels, putty knife, newspaper, brushes, sponges or rollers. Try different designs to see which is most suitable to the decor and the owner's taste.

Texture both hides surface defects and adds to the mood or charm of a room. Below, and on the following pages, are some of the more popular hand textures, and the tools and techniques for creating them. With practice, you can create any of these effects.

Spanish swirl or sponge twist (Figure 18-8)— Each circular pattern should just slightly overlap the preceding pattern. With a little practice, the arm action becomes even and regular

The adobe or stucco look (Figure 18-9)— A 2-inch putty knife makes this an easy texture to apply and an easy texture to duplicate when repairs are required.

Mission stipple (Figure 18-10)— The mission stipple is the finer of the two stipples. It's created with a sponge. A smooth flat coating is applied and then you just dab, dab, dab.

English stipple (Figure 18-11)— This is coarser than the mission stipple because it's done with a stiff brush.

Modern stucco or coarse stipple (Figure 18-12)— Lay down a smooth flat coating. Then just press and lift with the crumpled newspaper.

Trowel texture (Figure 18-13)— Use different trowel sizes to vary the effect. Try to create a random pattern.

Fine texture (Figure 18-14)— Smooth on the material and then roll the surface using a dry roller cover.

Colonial texture (Figure 18-15)— I used a wallpaper paste brush in a short semi-circular pattern.

Sandstone (Figure 18-16)— Here I used a stiff 12-inch dust brush. But any brush will do. Criss-cross at an angle while varying direction and overlapping.

Comb swirl (Figure 18-17)— A regular hair comb will make interesting design effects. You can swirl the comb so it looks like the Spanish swirl or sponge twist.

Comb criss-cross (Figure 18-18)— You can also use a comb to create a criss-cross pattern.

Brush furrow (Figure 18-19)— Use a large wallpaper paste brush in a continuous sweeping motion from wall to wall.

Trowel tooth (Figure 18-20)— Several patterns can be created depending on your arm motion across the surface. The trowel tooth applied in straight horizontal lines is the base or bottom layer for the next texture.

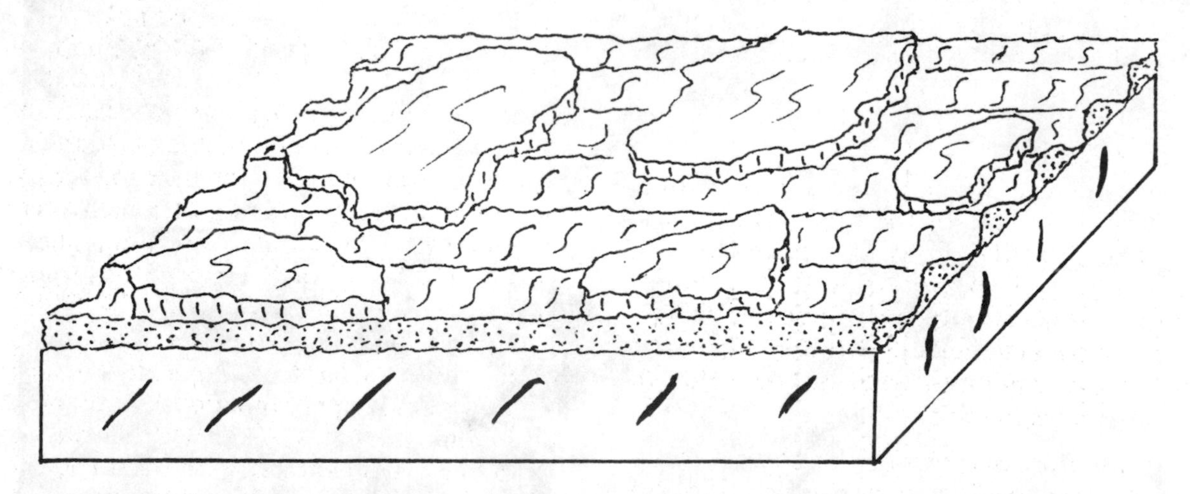

Aged Mediterranean or knock-down (Figure 18-21)— This texture uses a two-step procedure. First the material is spread in a thick coating, 1/8-inch or so, and troweled horizontally across the surface. Use a notched or toothed trowel designed for applying mastics or glues. You're "furrowing" the surface. After this coat is dry, apply a second coat, either sprayed in very large droplets or, better still, hand troweled into place. After these large splotches have set up, they're troweled or "knocked-down" flat. Ideally, you're creating roughly equal areas of "furrowed" and "knocked-down" design.

Limit the Aged Mediterranean texture to exterior work using cement or stucco. It's very coarse. The only way to clean it is by hosing.

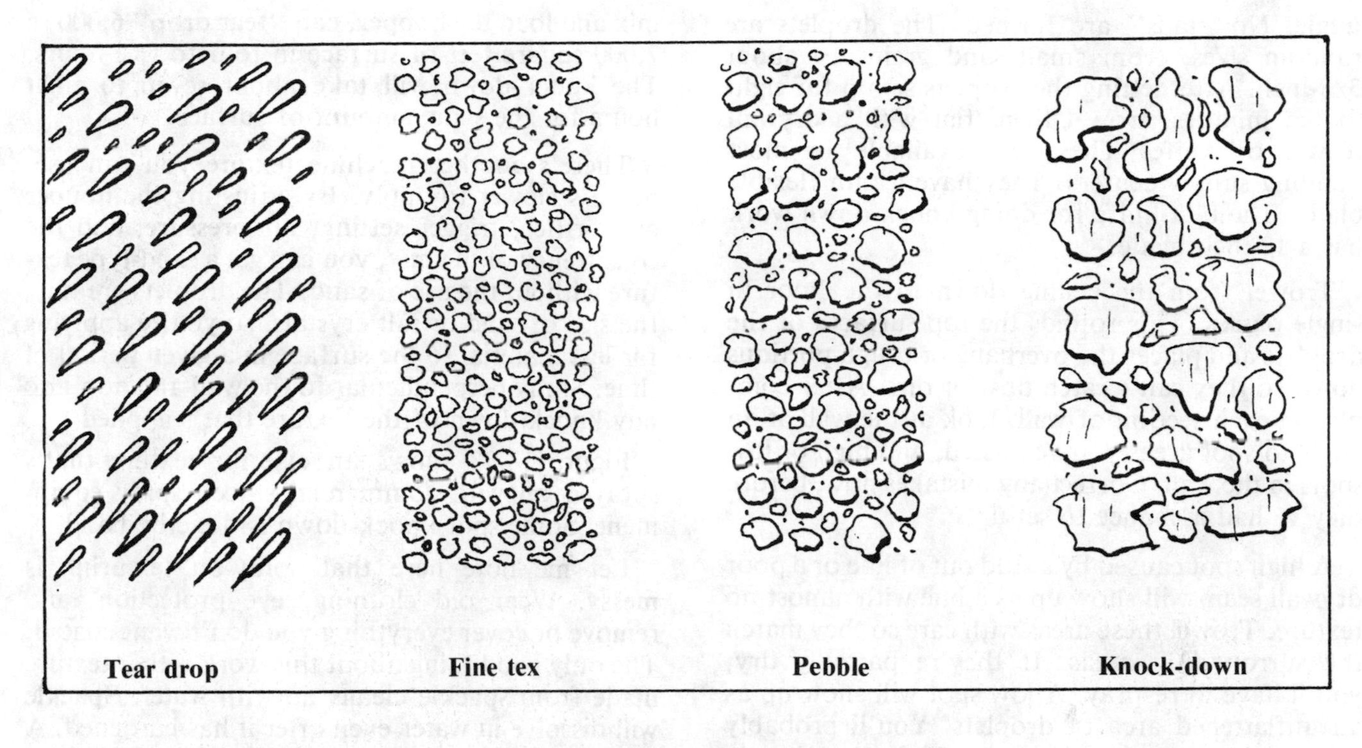

Four common spray textures
Figure 18-22

The only requirements for doing good hand texturing are your patience and desire to do a good job. Perfect your technique before you begin. Take your time, plan your work. Step back occasionally to see how the overall surface looks, and have a little fun.

Spray-on Textures

I don't recommend thinning the paints, texture paints, cements, or stuccos for most texture work. But for many of the spray-on applications of spackle, thinning with water is essential. Thin only enough to get a consistent droplet size that will remain in place long enough to be worked. Thin to a consistency the equipment will handle. But be careful not to overthin.

I like to use the finer textures for kitchen, bath, and laundry rooms as these areas need a scrubbable surface. Choose the medium textures for most interior walls of halls, closets, bedrooms, dining, and living rooms. You can use medium to coarse textures for exterior sidings, walls, fences, and planters.

Figure 18-22 shows only four of the literally dozens of spray patterns created by proper adjustment of air pressure, trigger pull, orifice size, and technique.

The tear drop— This is the machine-sprayed texture that I personally like the least — because it's the hardest for a painter to repair. It's a machine-sprayed tear drop that's not knocked-down, or flattened. This texture is common in the southeastern U.S.

The coating is sprayed in droplets at high speed, similar to spraying paint with an airless spray gun. The spraying motion is a fast side-to-side sweep but all in one direction, usually diagonally across and up the wall. The droplets are thin, like a heavy cream, and "tail" or "tear drop" as they splatter against the wall surface.

If you have to make repairs in a tear drop textured wall, it's hard to duplicate the texture by hand. The only approximation I've found is to apply a thin mix of spackle with a whisk broom. Wave the broom sharply at the surface, letting the thinned spackle form into droplets. Practice this on a piece of scrap, to get the feel, consistency, and proper orientation of the "tails."

The knock-down— The texture I like best is the knock-down, shown in Figure 18-22. The fine tex and pebble textures are just finer versions of the knock-down. Using a hopper gun system, spray spackle *straight* at the wall surface instead of at an angle. No "tails" are formed. The droplets are random sizes, from small sand grains to about 3/8-inch. After letting the droplets set up for eight to ten minutes, trowel them flat with a drywall trowel or knife. These are available in most building supply centers. They have a thin flexible blade. The best knife for doing knock-down work has a 12-inch blade.

Trowel from the ceiling down to the floor in single passes. This rounds the top surfaces of the droplets and places the overhang or rough portions down so they can't catch dust or dirt. After completing each section of wall, look at the wall at an angle to spot areas you've missed, and high or low spots in the wall. Correct any mistakes now, before they've had a chance to set up.

A high spot caused by a stud out of line or a poor drywall seam will show up as a line with almost no texture. Trowel these areas with care so they match the surrounding areas. If they're partially dry, you'll have to respray. A low spot will show up as an unflattened area of droplets. You'll probably have to cross-trowel these areas. If they've already set up, sanding them flat may work.

After the material has set up for a few hours, retrowel in a side-to-side or bottom-to-top motion. This will knock off any overhangs, rough edges, and loose particles missed in the first sweep. Let the area dry for a day, then prime and top coat.

There are a few tricks to learn, but they're not difficult. The first is knowing when to start the knock-down. If the material hasn't set up enough, it'll spread into large, thin smears. If this happens, wait a few more minutes before going ahead.

If the material is allowed to set up too long, it'll be hard to flatten and will flake a lot. Setting time of eight to ten minutes is about right. But time varies with the mix, water content, the wall's absorbency and the temperature of the surface and air.

Be careful to watch for grit (small particles of spackle that have dried) as you do the knock-down. Grit will stick to the drywall knife and scratch the softer material. I keep a towel tied around my pants leg when doing the knock-down. I clean the knife blade on the towel after each sweep of the wall.

Start your knock-down in the same place you started the spray and follow in the same direction. This keeps the drying time consistent from area to area.

Both of these machine-applied textures go on fast. A texture mechanic, with the aid of a helper to mix and load the hopper, can "tear drop" 6,000 to 7,000 square feet of surface in four to five hours. The knock-down will take about seven to eight hours for the same amount of surface.

There's another machine texture you can use, but it's slower to apply. By adjusting the hopper gun orifice, trigger setting, air pressure, and the consistency of the mix, you can get a sand-type texture without the use of sand. The droplets are tiny, the size of sand or salt crystals, so you're applying far less material to the surface in a given period of time. Spray perpendicular to the wall and don't do any knock down of the texture that's applied.

Figure 18-23 shows an exterior texture that's sprayed on with commercial stucco spray equipment. Some spot knock-down is done by hand.

Let me note here that spray-on texturing is messy. Wear old clothing, eye protection, and remove or cover everything you don't want coated. The only good thing about this work is that texture made from spackle cleans up with water. Spackle will dissolve in water even after it has hardened. A big bucket of water and a large soft sponge will be needed to clean overspray. Be sure to clean up any surface defects before painting. After they're painted, mistakes become a major problem.

Ceiling Texture

Wall textures differ from ceiling textures in the method of application. The wall textures are sprayed on first, followed by the primer and then the top coat of paint. Ceiling textures aren't applied until after a sealer or primer coat is in place and dry. Figure 18-24 shows the order of application for both wall and ceiling textures.

Texturing Additives

For superior adhesion of cements and stuccos, I recommend using a product additive that strengthens the surface-to-texture bond. It's a liquid acrylic plastic that you can use in two ways: First, brush it on the surface to be textured and let it dry until it's tacky. It holds the texturing on the wall, and is especially good for old cement or cement block. The second way to use it is to put a little in the water you mix the stucco with. The label tells you how much to use. It makes the stucco more workable. The acrylic plastic is a very good bonding agent. It dries hard and tough, making the texture more durable and weather-resistant.

The product I use is Acryl 60 by Standard Drywall Products. You'll find it at most building supply centers.

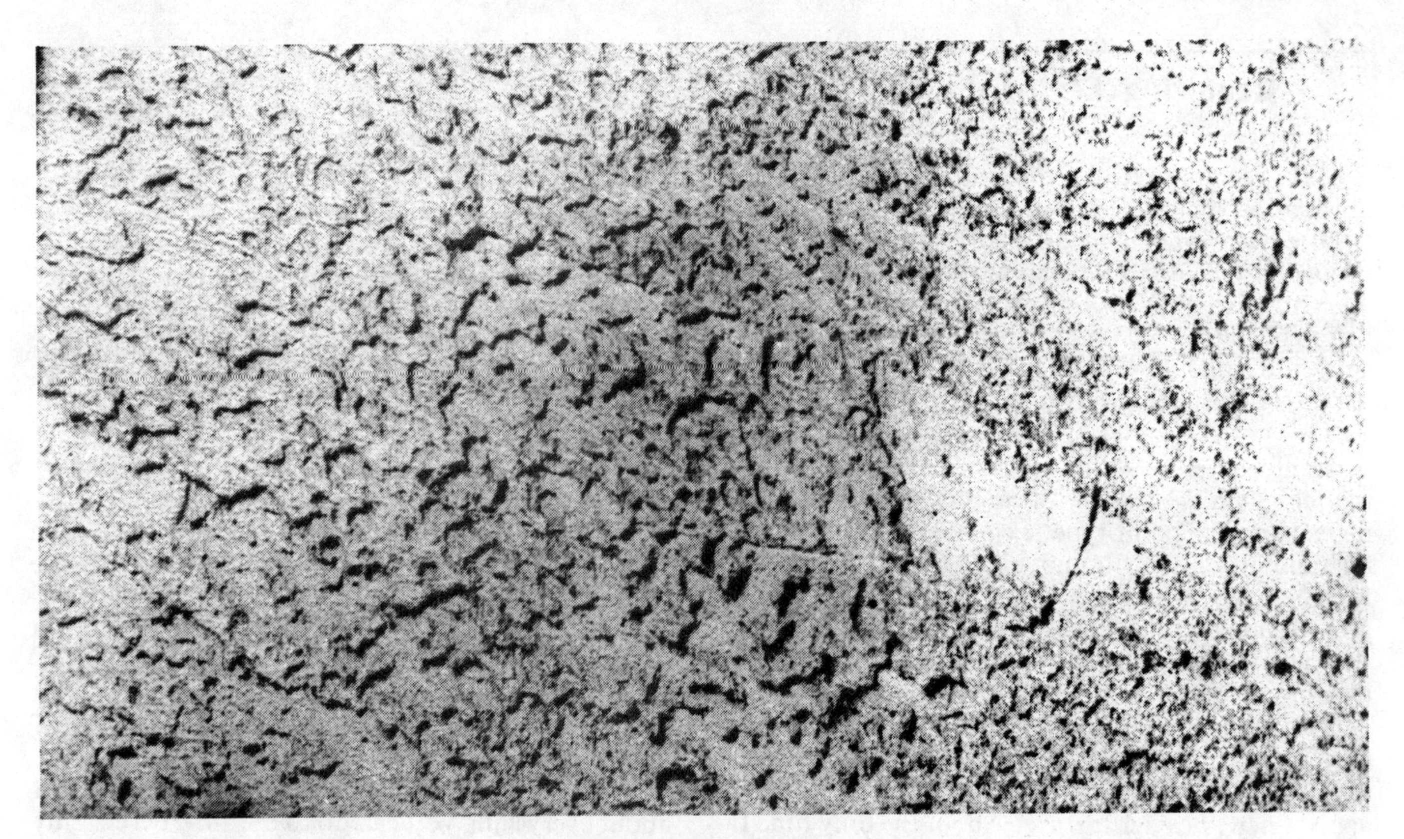

Exterior stucco texture
Figure 18-23

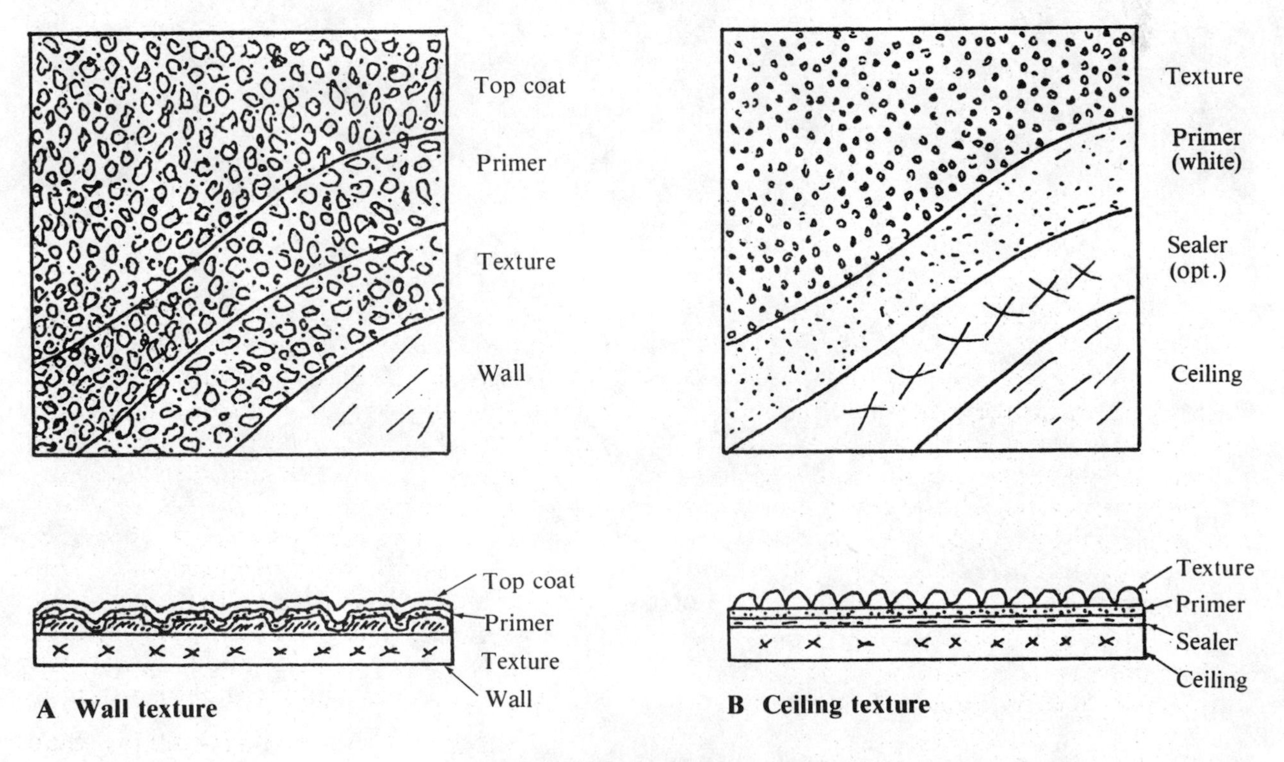

Applying wall and ceiling textures
Figure 18-24

Repairing Textured Surfaces

Make repairs to textured surfaces using the same application method that was used for the original texture. The only real difficulty is matching the original applicator's workmanship. Everyone uses a slightly different tool or procedure or has a little different touch. No two texture finishes will be exactly the same. It'll take practice and a little patience, but all textures can be repaired and matched.

Complete removal of a texture surface is another matter. Don't even try it unless it's absolutely necessary. The paint-type textures can be removed by sanding or with liquid paint removers. If the finish is paint over spackle, it'll take the same treatment followed by a heavy water washing. Remove the cements and stuccos by chipping or by sandblasting.

It would probably be faster, easier, and cheaper to just rewall over the top of the existing coating. If the texture is fairly smooth, you might be able to apply a new, heavier texture. About the only time I would recommend removal of an interior texture is to allow for wallpapering. In most cases, a rough sanding and application of a thick backer paper called *blank stock* should be all you need.

Repairing Textured Walls

Every time a nail comes out of a wall, it leaves a hole. When you patch, sand, and repaint, it will leave a dull spot that sticks out like a sore thumb.

Here's what I do. I use a little bit of patching plaster mixed with paint, not water. For a light traditional skip texture, I apply this mix with my finger, pulling straight out from the wall at the end. This leaves a texture much like the original. Let the plaster dry and then paint.

For larger splatters, use a whisk broom to dab or splatter on the mix. For contemporary or Mediterranean knock-down, use a trowel. Apply the mix in small 1/16 to 1/8-inch balls and flatten with the trowel.

Hand-operated texture guns are available. But it's not worth the $30 to $40 price to fix a few holes. They work well on large surface areas. But you're better off investing in power equipment if you're doing an entire room or house.

Through the first 18 chapters we've covered just about everything you can do to a wall. There's only one left, I suppose. And that's the final topic: alternate wall coverings. That comes next.

chapter nineteen

Alternate Wall Coverings

There are times and places where paint just isn't appropriate. Maybe your residential clients want something different, more stylish or more practical. Maybe your business or industrial clients need fire or chemical resistance. Professional painters who can recommend and install alternate wall coverings have a better chance of expanding their business and making consistent profits.

In fact, the importance of alternate wall coverings may soon become more than just a matter of taste. In 1981 the Environmental Protection Agency District 9 (California) passed *Rule 1113.* Rule 1113 limits the volatile organic compounds (VOC) that can be used in non-flat architectural paints. It reduced the VOC limits to 250 grams per liter of paint, from the current 380 grams per liter. The rule includes varnish, wood stains, preservatives, primers, sealers and enamels.

The EPA gave the paint manufacturers until 1985 to reduce the VOC in their products. The deadline was later extended to 1989. As of mid-1987, there have been virtually no resins or paints developed that can meet this performance level. The industry is experiencing problems in the durability, workability, appearance, and protection coverage of paints that meet or approach the VOC limits of Rule 1113. As a painter, you may soon be limited in the types and quality of product you can purchase and apply. Perhaps alternate wall coverings are the answer.

But what if your only experience is in painting? Should you take a chance and experiment with new materials? There's a risk that you'll do substandard work, hurting both your reputation and your price. Are the potential rewards worth the cost? That's up to you, of course. But let me encourage you to try new materials and techniques. Here are some general principles that I follow:

- If someone else can do it, there's no reason why I can't.
- If I see something in its finished form, I should be able to copy it.
- I can work as hard as the next guy.

In other words, don't stop learning. Keep improving your skills. Stop and look at other jobs, at books, at store displays, anywhere you can get ideas. When you see a finished job that you like, you've found your next challenge. You may lose money on the first few jobs you try. But with experience, your craftsmanship and productivity will improve. Approach the work with a "Can do" attitude and you'll find a way to get the job done. Keep learning and improving your skills — no matter how long you're in the painting business and no matter how experienced you may become.

Fortunately, most manufacturers supply good information on how to use their products. Read and understand the instruction sheets supplied, especially when using a new product. There's usually a good reason for everything in those instructions. After all, the manufacturer wants your job to come out right. If it doesn't, you won't buy the product again. Here's another suggestion: Experiment on your home. There's less pressure if someone doesn't have to pay for your mistakes. And you *will* make a lot of mistakes at first.

No reference manual could explain how to install all the alternate wall cover materials that are now sold. I'm not going to try. Instead, let the manufacturers educate you. Most will supply all the product information you could want. Using your company letterhead, write to those that offer products you could use. Request product catalogs and installation instructions. Most manufacturers will be more than happy to supply all information they can. And believe me, it's an eye-opening experience when you see what's available. What you find at your local hardware or paint store is only the tip of the iceberg.

Master the Basics

The basics of applying wall coverings are about the same, whether you're using paint or something more exotic:

- *Surface preparation* is the same. You need a clean, flat surface.
- *Planning* is the same. You need to know what the finished product should look like, a materials list, and the proper tools.
- *Application and installation* are only slightly different. There are more steps involved, and you're using a solid rather than a liquid.
- *The results* you want are the same: protection, texture, color and a "mood."
- *The skills required* may be similar. You have to know how to use some hand tools, some power tools, and be quality-conscious in your workmanship.
- *The earnings and profits* can be higher. There aren't as many people doing this type of work, so you're more likely to get your asking price.

The success of the job depends on two things: careful planning and taking the time to do the job right. Planning probably includes sketches of the finished job and layout drawings to scale. This helps you plan the quantity and size of materials needed. It'll give you a good idea of how to proceed and of what the finished project should look like.

Workmanship is being a perfectionist. "Good enough" is never good enough. Start it right and do it right. There's no other choice. Skip a step, cut corners, be satisfied with second rate and you're sure to wind up with a job that neither you nor your customer is happy with. A dissatisfied customer never does you any good. He certainly won't recommend you, he may not pay, and if he does, he feels he's been cheated. No professional painter wants a part of that.

The remainder of this chapter describes a wide range of alternate wall cover products. At one time or another in my career I've used each of the materials I'll describe. As you go through these pages, please don't get the impression that I'm plugging any one product. I mention brand names to help you identify the product I'm talking about — and because I'm familiar with that product. I don't own stock in any of them. I call out a brand name to give you a starting point in finding more information. Many companies make similar products. Explore as many as you can. Find the product that best meets your needs.

In the following pages I list the addresses of manufacturers. These addresses were current as of the publication date. But companies, like people, move, change their names and die. If you can't locate a specific company, try this: call 1-800-555-1212. That's the toll-free information operator. Ask for the number of the company you're trying to find. Most major corporations have toll-free 800 numbers you can call for information. If you can't get a phone number, ask your building materials supplier for a current address and phone number.

And here's something else to remember. Some of the suppliers I've mentioned are wholesalers or manufacturers. They may not sell to you directly, or if they do, they can require a minimum order and a credit check. But they'll all send you a list of wholesale or retail suppliers in your area. And they'll all help you select the right materials for your application. After all, they're in business to sell their product.

Now, down to the specifics. There are three types of wall coverings — paint, paneling, and wallpaper — right? Wrong! Actually, there are dozens. Most, unfortunately, are overlooked or only thought of as decorator items. The purpose of a wall covering is to protect the surface: Decorating is really secondary.

For example, what are you going to use on a wall next to a pot-belly stove, in a chemical laboratory, or in a noisy office? Sure there are coatings for

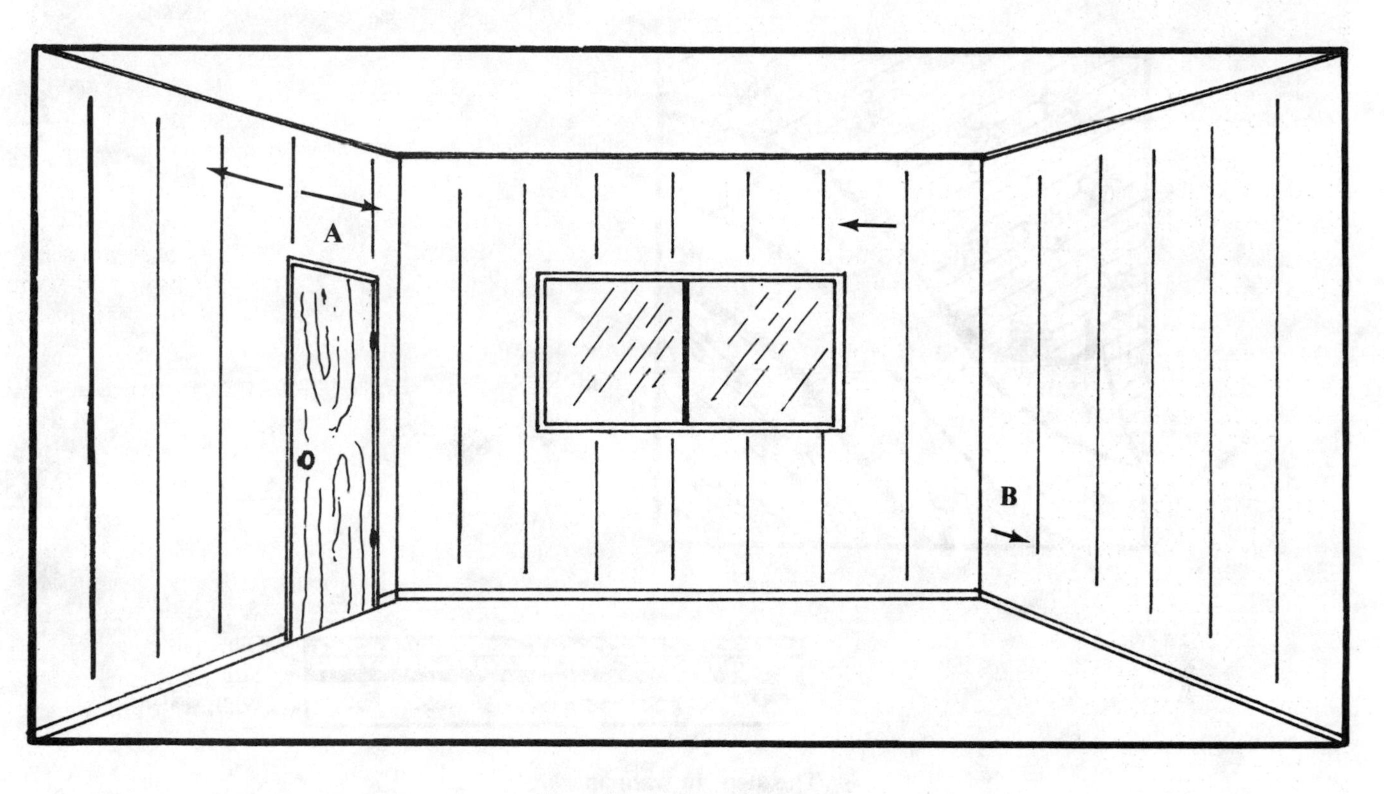

Planning wallpaper layout
Figure 19-1

each of these areas, but in this chapter we'll explore other possibilities.

I'll begin my short course in wall coverings with the most common alternative to paint — wallpaper.

Wallpaper

I won't say much about wallpaper because there are complete books on the subject. The choice of pattern and type is a matter of taste, of course. But you should know enough about the subject to answer questions and give advice on the best type of wallpaper for different areas of the home.

Wallpaper is available in literally hundreds of patterns and styles, including grass cloth, foils, flocks, metallics, and vinyls. Some are easier to hang than others, but the general installation procedures are the same. The first step is deciding where to begin.

Installing Wallpaper

When you walk through the door into a room, what do you see first? The far wall. Next? The walls to the right and left of you. Last? The wall just behind you, the one you entered through. When you're hanging wallpaper, you start with the least visible wall. Your starting (and ending) spot is where there's likely to be a mismatch of pattern or width when you're done. Your job is to make it as unnoticeable and unobjectionable as possible.

Planning the layout— The corner behind the door you entered, or the wall surface over that door, is usually the least noticeable spot in any room. So that's where you begin. From inside the room, facing the door, start your first strip of wallpaper over the knob side of the door, and toward the hinge side.

Look at Figure 19-1. Point A shows where to begin if you're papering the entire room, and the arrows show the direction in which you should work. The first strip goes over the door. The second strip butts to the edge of this paper, extending along the door molding to the floor. The remainder of the strips should work off of the second strip, continuing around the room until you reach the first strip you applied. Here is where the final trimming or mismatch will occur — in the least visible spot in the room.

Some rooms have a natural break-point, like a floor-to-ceiling bookcase or fireplace. In this case, start the first two strips at the sides of the natural

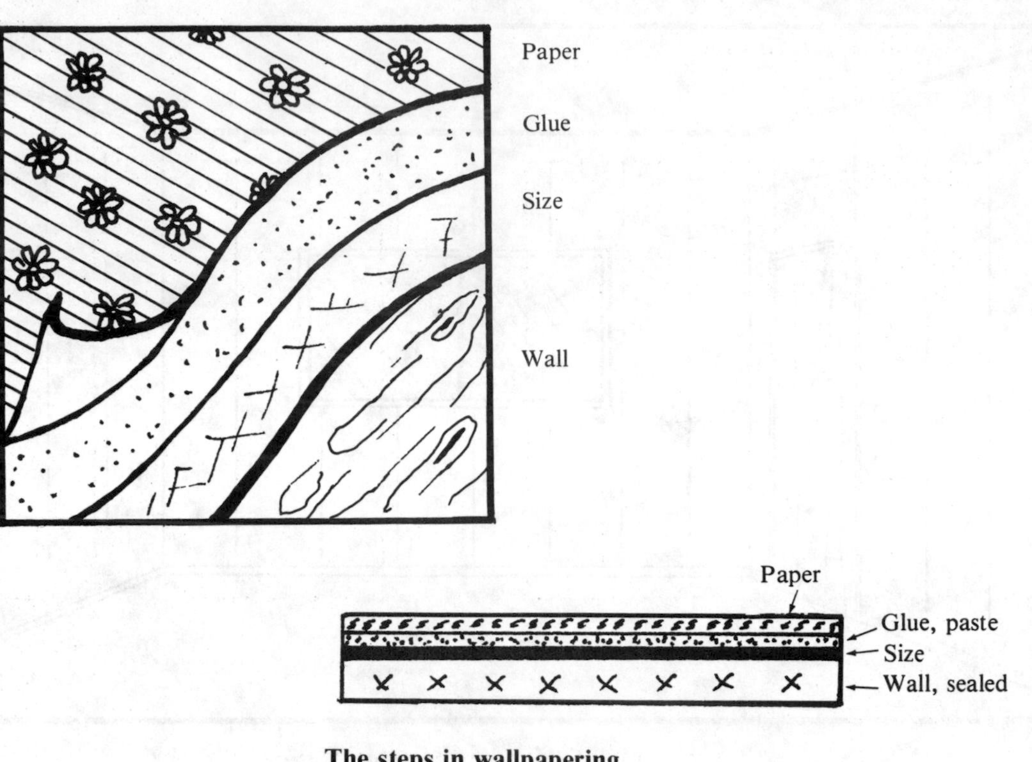

The steps in wallpapering
Figure 19-2

break and then work in both directions until they meet at the door hinge area.

When you're papering only one wall of a room, the natural tendency is to start at the left corner and work toward the right hand corner. But this may not be the best pattern. Where to begin depends on which wall you're papering:

- If the wall to be papered is to your right as you enter the room, start papering at the left corner of that wall and work to the right. See point B in Figure 19-1.

- If the wall is to your left, start at the right corner and work toward the left corner.

- If the wall directly in front of you is being papered, start at the corner you see first when entering the room, usually the one dead ahead of the door.

- If you're papering the wall with the door in it, start at the corner away from the door and work back toward the door.

In each case, you'll see full sheets of paper and a factory-cut edge when you enter the room. The partial sheets you cut during installation won't be noticeable, as they are the last seen by someone entering the room.

But here's something to check before you start hanging. Some wallpapers are designed to be hung from left to right, so make sure your pattern will match up if you hang from right to left. With some papers, it can't be done.

Buying the paper— When you go to buy the paper, buy a little more than you expect to use. As with paint, different wallpaper lot numbers will have slightly different colors. If you run short of paper, getting an exact match may be impossible. You can always return a full roll or bolt of unused paper. But you'll seldom be able to go back to the store and buy more of the same lot number. Use any cut rolls you can't return for patching damaged walls, or for covering wall electrical plates, cabinets, or furniture.

While you're in the store, buy a bunch of super-sharp razor blades. A blade that's sharp enough to cut dry wallpaper isn't necessarily sharp enough to trim wet wallpaper. Hanging paper can be a frustrating job anyway. Don't make it worse by trying to make do with inadequate tools.

Preparing the wall surface— Most wall surfaces need to be sealed before you hang the paper. I use REX sizing. Figure 19-2 shows the layers found in

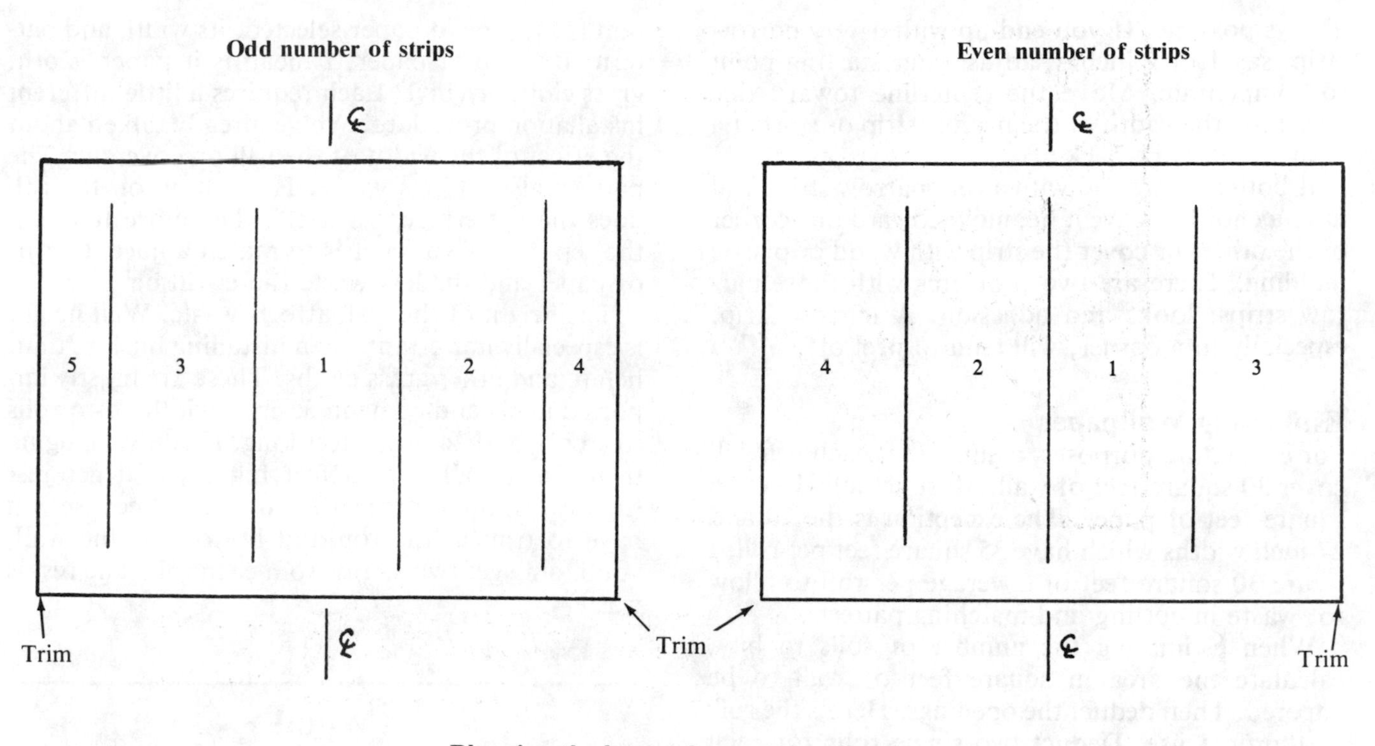

Planning the layout for mural wallpaper
Figure 19-3

properly applied wallpaper. On bare wood or plywood, seal first with 3-pound cut shellac, and let it dry. Then size, let dry again, and finally, paste. This prevents the paste from raising the grain of the wood. On painted wallboard you only have to size, but unpainted wallboard should be painted or sealed first.

Cutting and hanging— Cut each strip of paper 6 to 8 inches longer than the height of the area being covered. The extra length allows for trim overlap at the floor and ceilings. But be sure to check that length before cutting. You may need more than 6 to 8 inches for matching a large pattern.

Wash each strip as soon as it's hung next to the preceding strip and the seams have been rolled flat. (But, of course, you don't wash grass cloths.)

Be careful when hanging vinyls; they have a tendency to stretch when rolled or smoothed. You'll have a good match until the material decides to relax or come back to its normal shape.

Trimming out the room— Boxing or framing the corners and the floor and ceiling lines with molding can be very attractive, and it makes for easier paperhanging. But you'll have added cost for materials and labor. It's something to consider when making your bid.

It might make trim-out easier if you change the width of the paper. Wallpaper is available in 18, 20, 21, 27, 28, and 30-inch widths. But the pattern the homeowner selected may not be available in the width you would like to work with. Not all of them are. If the customer has already bought the paper, you don't have the option of choosing the width. Bid the job accordingly.

Installing Mural Wallpaper

The rules are a little different for hanging mural wallpaper. A mural has a centerline to it. You need to divide the wall at its center and install the paper outwards from this line. If there is an odd number of strips, center the middle strip over the centerline of the wall. If there is an even number of strips, butt the first strip to the centerline of the wall, edge to line. Then work outwards in both directions. Figure 19-3 shows the order.

The trick is to determine which is the center strip. Most kits are numbered, but I recommend assembling the mural on the floor before attempting to glue and match it on the wall. Maybe you can guess how I learned this lesson.

You can use this same procedure for any wallpaper that covers only one wall. The last strips (the one at each corner of the room) should be at least one-third the width of the paper being used, normally no narrower than 10 inches. This isn't always possible. If you end up with a very narrow strip, say 1 or 2 inches, adjust your starting point to compensate. Move the centerline toward one corner by the width of the narrow strip or start at a corner and work back.

If both corners end with a very narrow strip, you have a choice. Leave it be, move toward one corner or the other, or cover the strip with wood or plastic molding. There are two problems with those narrow strips: looks and adhesion. A narrow strip, especially in a corner, will tend to peel off.

Estimating Wallpaper

For estimating purposes, assume that each roll will cover 30 square feet of wall. Most actually have 36 square feet of paper. The exception is the 20 and 27-inch widths which have 35 square feet per roll. I figure 30 square feet of coverage per roll to allow for waste in cutting and matching patterns.

When estimating the number of rolls to buy, calculate the area in square feet of wall to be papered. Then deduct the openings. Here's the rule of thumb I use: Deduct two single rolls for each four openings in a room (windows, fireplace, bookcases, doors, and so on). Divide the net wall area by 30 to find the number of rolls you'll need.

Most wallpaper is sold in what is termed a *bolt*. A bolt is a double or triple length roll. Bolts reduce waste due to pattern matching. The 18, 20, 21, and 27 inch wide papers are packed as double rolls. The 28 and 30-inch wide papers come as triple rolls. Waste can be up to 4 linear feet of paper on single rolls, much less on multiple rolls. For example, if you buy 18-inch wide paper on a 20-foot roll, you get two 8-foot strips and 4 feet of waste. Figure 19-4 shows the coverage for rolls of different widths and lengths.

When you're bidding a wallpaper job, there are several factors you should consider. Most important is the type of paper selected, its width and pattern. By type of paper, I mean is it paper, cloth, grass cloth or vinyl? Each requires a little different installation procedure. We've already talked about the effect of the width of the roll on coverage. The pattern also affects waste. How often on the roll does the pattern repeat itself? The more frequent the repeat, the easier it is to match adjacent strips of paper and the less waste there will be.

The height of the wall affects waste. Wall height is especially important when installing burlap, tulu, hemp, and other grass cloths. These are mostly imported items and come in 36-inch widths, two rolls to a bolt, and 24 linear feet long. If you're hanging them on a wall over 8 feet high, you'll get just slightly less than three strips to a bolt. Because you have to trim at the top and bottom of the wall, you'll only get two strips from each roll. The rest is waste.

Wood

There's little that can match the inherent individuality and beauty of a fine piece of warm, soft wood. Wood is inviting and appealing. It's available in dozens of species, each with its own special look and feel, grain and color. Wood can be left uncoated, clear coated, natural or stained.

You can cover a wall with wood planks, split logs, shingles, panels or veneers, installed with mechanical fasteners or nails and glue. Logs and planking are generally softwoods like pine and fur. If the wood is knotty and light in color, a clear varnish gives it a nice amber tone.

Panels are available in most of the common hardwoods, such as oak, cherry, and walnut. Some panels are available with painted scene overlays.

Width (in inches)	Single roll Length (in ft)	SF	Double roll Length (in ft)	SF	Triple roll Length (in ft)	SF
18	20	30	40	60	60	90
20	21	35	42	70	63	105
21	17	30	34.3	60	51.5	90
27	15.5	35	31	70	46.5	105
28	12.5	30	25.7	60	38.5	90
30	12	30	24	60	36	90
36 (grass cloth)	24	72	48	144	--	--

Note: Allow 5" to 18" for pattern match, floor trim, and ceiling trim.

Coverage estimates for wallpaper
Figure 19-4

Color desired	Paint to use	Ratio of paint to thinner	Treatment
Green	Green resin stain	No thinner	Wipe to effect
Orange	Orange enamel	2 to 1	Wipe to effect
Purple	Blue enamel	3 to 1	Dry for three days,
	Purple enamel	3 to 1	wipe to effect
Pink	Violet enamel	3 to 1	Dry for three days,
	Vermillion paint	3 to 1	wipe to effect
Antique	Flat white enamel	2 coats/no thinner	Dry for three days,
	Thinned white paint		wipe to effect
	with 1 ounce raw amber per pint mixed in		
Red	High gloss red enamel	No thinner	Dry for three days,
	Black enamel	1 to 1	wipe to effect
Clear	Clear boat coat	No thinner	Dry for three days,
	Wood stain-brown	No thinner	wipe to effect
Brown	Brown enamel	2 to 1	Wipe to effect

Color effects for wafer board
Figure 19-5

Most are 4 x 8 feet, 1/8 to 3/8-inch thick, and grooved in the lengthwise direction. The thicker, 1/2-inch panels are generally called siding. But they can be used very effectively on interior walls. They can be applied directly to wall studs with adhesive or nails or both. Paint or stain them just as you would if they were on the exterior.

If your client really wants the ultimate in wood and is willing to pay for it, consider veneers. These thin-sliced panels of wood are bonded to another surface. Many expensive and exotic woods are available. Some home decorating centers sell veneers, but usually in limited quantities. I've found one place that handles almost all veneers and solid plywood, and the tools to work them: Albert Constantine and Sons, Inc., 2050 Eastchester Road, Bronx, NY 10461. If you build furniture, get a copy of this company's catalog.

Wood has one desirable property that most materials don't: It can be colored or stained to give many different effects. If you're interested in wood staining, I suggest that you contact Mohawk Finishing Products, Inc., Route 30 North, Amsterdam, NY 12010. They have just about anything you need to work with wood.

Aspenite or wafer board— These panels are made from large chips or wafers that are glued and pressed into sheets. They're intended as a substitute for plywood and can be used both outdoors and indoors. Painting is recommended to protect the sheets from moisture and water damage. Water can enter behind the wafers at the glue lines and eventually pop the wafers out. This is especially true in freezing weather when wafer board is used as exterior siding. Coat this board with a good quality enamel.

Aspenite or wafer board lends itself to some interesting decorating ideas. By applying contrasting colors in layers, you can make the wafers stand out, creating a marbleizing effect. The color combinations are endless. To start, try some of the combinations shown in Figure 19-5. For example, apply one color, let it dry for three days, then wipe on the second color until you get the effect you're looking for. I recommend practicing on some scrap wood before coating the full sheets.

Installing Paneling

When you're paneling a room, plan the layout before you start, just as we did with the wallpaper.

For new construction, Figure 19-6 shows the proper installation procedure. The wall studs should be spaced no more than 24 inches on center for non-bearing walls, 16 inches on center for a load-bearing wall. The bottom sill plate should be double. For exterior walls, install insulation between studs before hanging the paneling. Interior walls may or may not be insulated. If you're working in an apartment building, insulate the wall between apartments. You also need a vapor barrier, either as part of the insulation or separate (Kraft paper, foil, or polyethylene). For an outside wall, I also suggest that you caulk the sill-to-floor seam to prevent air leaks.

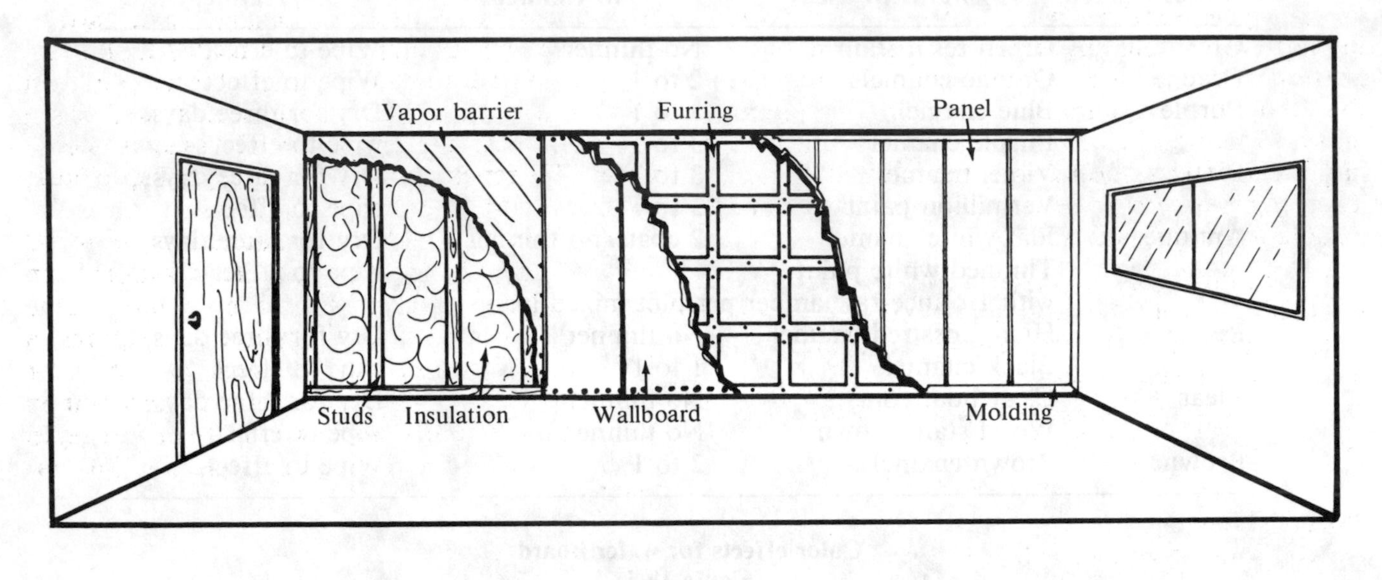

Installing wood paneling
Figure 19-6

The wallboard goes up next. Use water-resistant green board for walls of kitchens, baths, and laundries. If the finish panels don't overlap at the seams, paint the wallboard under where the joints will be with a color that matches the panels. Usually a brown aerosol paint will do fine. All you need is a narrow strip behind the seam.

Then glue and nail furring strips into place. Try to nail to studs. That's easier if you mark the location of the centers of the studs on the floor with a pencil before the vapor barrier is installed. Then it's easy to find the covered studs. As mentioned earlier, paint the furring strips where they may show between panels.

Finally, glue and nail the paneling to the furring. Leave at least a 1/4-inch gap between the top and bottom of the panel and the floor and ceiling joint for expansion. Use colored nails to match the paneling, or countersink all nails with a punch and fill the holes with colored putty. Where possible, nail only in the "V" grooves. This will give a better overall appearance.

Most panels will require some trimming. Be sure to cover the cut edges with molding. Include molding for the corners and floor and ceiling areas in your bid. And don't forget to paint the trim lumber wherever it's sawed and jointed. As a finishing touch, you may want to give the whole surface a good wax coating for protection.

Tile

When someone says *tile,*, you probably think of ceramic tile. And for good reason. It's readily available, accepted by almost everyone, and available in many colors, styles and textures. Its glass-hard surface is resistant to heat, chemicals, stains, and water. The newer tiles have built-in spacers that make them easier to install.

Ceramic tile is available from many sources — in fact some stores sell nothing but ceramic tile. The standard size is 4¼ inches square and trim pieces are available to finish off the job professionally.

Metal Tiles

Metal tile are like the familiar ceramic tiles. They're the same size and install the same way. They are available in smooth, brushed, embossed, flat, semi-gloss and high gloss finishes. Most are aluminum, but colors range from aluminum, copper, brass and silver, to gold. They're a good choice on kitchen walls behind a range or sink and make attractive borders around other wall cover materials.

Mural Tiles

A division of Dal-Tile Corporation, 7834 C.F. Hawn Freeway, Dallas, TX 75217, specializes in custom tile murals. They produce standard 4¼-inch square tile with hand-painted designs. Send them a drawing, picture, or artwork, tell them how large an area you want to cover in tile, and they will duplicate it to your specifications. Each tile has a portion of the picture. When assembled properly, the tile becomes a mural. Contact them for prices and information before sending pictures. Dal-Tile has sales offices throughout the United States. Their general catalog offers decorating ideas and installation information.

These mural tiles are ceramic and therefore are washable, chemical resistant, and fireproof. They're ideal for that extra-special personalized area but can be used like any tile in baths, kitchens, hallways, lobbies, dining rooms, counter tops, and cabinet and furniture inserts.

Mirror Tiles

Do you want to give a small boxy room some size? Just mirror a wall. The pros in new home decorating and sales use this visual trick all the time. Just don't apply mirrors to opposing walls. The continuous reflection will soon become confusing and tiring to the occupants.

Mirror tiles are available in most home decorating centers. They come in 12-inch squares and a variety of other stock sizes. Most tiles can be attached to the wall surface with contact cements. But test a sample with the glue you wish to use. I've seen glues that take off the silvering on the back of the tile. Some mirror tiles must be hung with pressure tape, and you'll find some that have peel and stick coatings.

Here's a word of caution: Don't glue the large size mirrors. They require mechanical fasteners. Since they expand and contract, they'll either crack or peel off the wall if glued. When installing small mirror tiles, don't butt the tiles tightly together. Leave a small space, about 1/32 inch, for expansion.

The mirror tiles are available with designs, from gold and silver veining to frosted or colored pictures. There are also flexible mirror tiles which can actually bend around a curve or a corner. They're made up of hundreds of small mirrors glued to a flexible backing. Flexible mirror tiles of both glass or plastic are available. They come in 1- or 2-foot square sheets in red, green, yellow, blue, and metallics. They can be glued directly to any sound surface including cylindrical and spherical shapes. For information, write to Rohm and Haas Corporation, Independence Mall West, Philadelphia, PA 19105.

Acoustic Tile

Acoustic tile has long been used for ceilings but not for walls because it's easily damaged and hard to clean. The newer tiles are vinyl coated and effectively eliminate some of these problems. You can cover at least the upper half of walls, as well as the ceilings. These tile are commonly used in stereo stores, TV repair shops, TV rooms, halls, offices, factories, and other high noise areas.

Install the tiles with staples, adhesives, or metal and plastic hangers. Acoustic tile is available in 12-inch squares and 2 x 4-foot sheets at most home decorating centers. Wall application requires a hard, solid backing surface such as wood or plaster. On a highly irregular surface, like a bumpy wall or masonry, first apply wood lath or furring to level the surface and provide a backing for stapling.

Installing Tile

All tile, whether metal, glass, ceramic, vinyl, or mirror, is laid out the same way. Look at Figure 19-7. Start by finding the centerline of the surface to be tiled, in both the vertical and horizontal directions. Dry fit your tile outward in a line of single tile in all directions from this center. If you come up short in any direction, move the centerline toward that direction and try again.

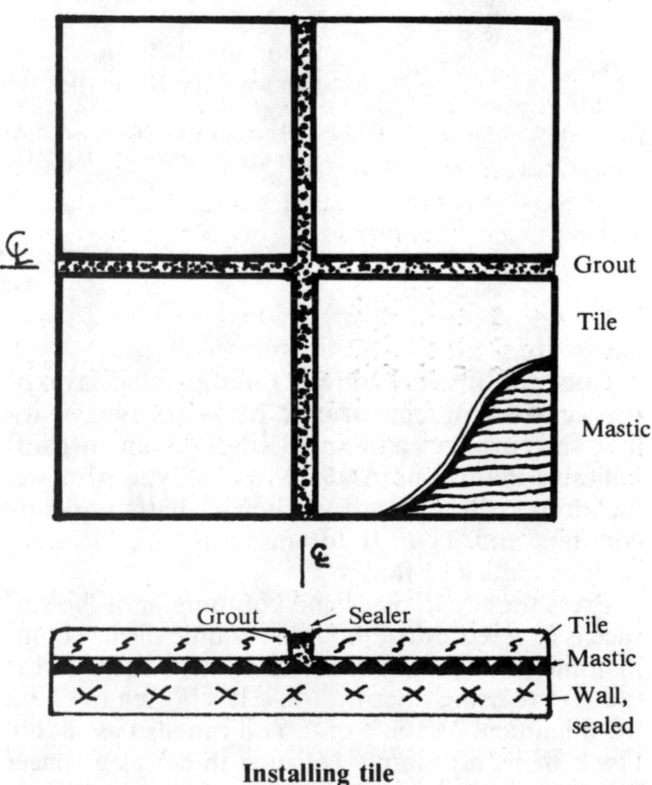

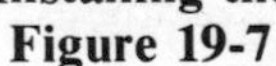
Installing tile
Figure 19-7

Size (in inches)	Type or shape	Number of tiles needed
2¼ x 9	Rectangle	7 to 1 SF
3 x 3	Square	16 to 1 SF
4 x 4	Square	9 to 1 SF
4	Hexagon (6 sides)	9 to 1 SF
4¼ x 4¼	Square	8 to 1 SF
4 x 8	Rectangle	9 to 2 SF
4 x 8	Wedge	10 to 2 SF (approx.)
4¼ x 6	Rectangle	17 to 3 SF
6 x 6	Square	4 to 1 SF
8 x 8	Square	9 to 4 SF
8 x 12	Rectangle	9 to 6 SF
8	Hexagon	9 to 4 SF
8	Octagon (8 sides)	9 to 4 SF
9 x 9	Square	9 to 5 SF
12 x 12	Square	1 to 1 SF
12 x 12	Mosaic sheet**	1 to 1 SF
12 x 24	Mosaic sheet**	1 to 2 SF
17 x 17	4¼'' square tile panel	1 to 2 SF
48 x 96	Tile panel	1 to 32 SF

*Standard sizes

**Mosaic tiles to 2'' x 2'' are mounted on 12'' x 12'' or 12'' x 24'' backer sheets

Standard thicknesses

Wall tile: 3/16''

Floor tile: 3/16''

Heavy duty floor tile: ½''

Standard joint width

Small tile: 3/16''

Large tile: ¼''

Install per ANSI-A108.1 to ANSI-A118.1. Also per CTI, Ceramic Tile Institute technical guide ''TCA Handbook for Ceramic Tile Installation''.

Coverage estimates for ceramic tile
Figure 19-8

Once you find the starting point, apply a layer of mastic with a notched trowel. Make sure you're using the proper adhesive. Choose an organic adhesive that meets ANSI-A136.1, Type I for wet locations such as laundry, kitchen, bath, and sink counters and Type II for moist to dry locations such as halls and floors.

Press the tile in place and continue until the surface is covered. Most modern ceramic tile has built-in joint spacers that make alignment easy. If the tile you're using doesn't, use a level often to check the alignment as you work. You can also use a thin block of wood (about 1/4-inch thick) as a spacer for alignment. For cutting, rent a ceramic tile cutter from the store where you bought the tile. Rates are low and the results will be much better. Figure 19-8 will help you estimate the amount of tile to buy.

For ceramic and some of the metal tiles, the next step is to *grout.* Mix the grout powder with water to form a paste. Rub it into the cracks between the tiles. After the grout has set up, wash the surface with a soft flat sponge and clear water. Repeat several times. Allow the grout to dry once again and then apply *tile grout sealer.* Sealer helps waterproof the joints.

Tile used in areas subject to moisture, including bathrooms, kitchens and laundries, should be ap-

Surface cap

¼" round bead

Cove base

Bullnose cap

¼" round cove

Cove base

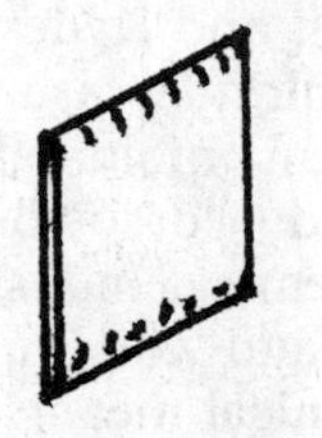
Double bullnose, surface type

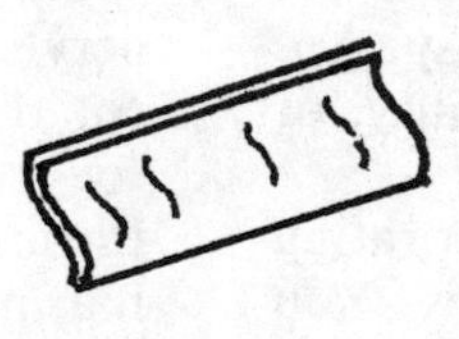
Hospital cap

Cove base

Double bullnose, mortar type

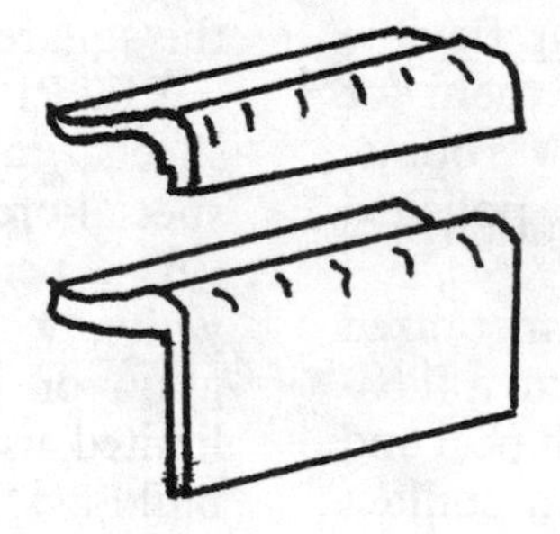
Sink trim set

Cove base

Trim for glazed ceramic tile
Figure 19-9

plied to a waterproof backer. For new construction, *tile backer* or *green board* is recommended. For old construction where green board wasn't used, you can waterproof the wall with enamel paint. Let it dry and scuff sand the paint before applying the tile. Also, although the tile is waterproof, the grout isn't. As the sealer wears off, water can penetrate and destroy the subsurface.

As a service to your customers (and as a sales tool), keep a list of customers and the date installed for each tile job. Every year, call them offering to reseal the grouting. That gets you back in the door, and you may find other jobs that need to be done while you're there. This really isn't a gimmick — grout does need resealing once a year.

Trim for ceramic tile— Figure 19-9 shows profiles for ceramic tile trim pieces. Trim pieces can be used for thinset, glue-on or mortar installation. Many of the edge trims are thicker than the tiles because they're designed for counter top edges. You can use them to trim tile applied to a flat wall if you install a filler board under the tile. It should be thick enough to build out the tile to the height of the trim, usually 1/4 to 3/8-inch. Fasten the filler board securely to the wall with mastic or glue and nails.

Plastic

Plastic used for wall coverings falls into two groups: flexible and rigid. But there's a wide range of variations within these categories. First we'll look at the flexible plastics.

Flexible Plastics

Flexible vinyl is used for upholstery, book covers, and a host of other items. Cemented to a smooth surface, it makes an excellent wall covering. It's resistant to water, most chemicals, and abrasion. The cost is reasonable and application is easy. Apply with vinyl wallpaper paste for strippability, or with contact glue for permanence.

Flexible vinyl is available from home decorating centers, fabric stores, and upholstery shops. You can buy the vinyl only, or vinyl on a cloth backing, in a wide variety of colors and textures. Although there are designer prints available, most flexible vinyls are made to look like leather. Use them for walls in offices, family rooms, dens, TV rooms, and bedrooms. Using these vinyls with polished wood moldings gives a very rich look.

One special form of flexible plastic is metalized polystyrene or metalized Mylar, which form a flexible mirror surface. It comes in sheets, with peel and stick backing or ready to be applied with contact cement. Mirroflex by Heinze America comes in twelve colors and four surface textures. The material is used by both industry and decorators throughout the world. It makes a great decorator accent. But don't use it on surfaces that will get heavy wear.

Another flexible vinyl is *vinyl flooring.* It can be applied to a wall surface only, or it may be part of the floor extended up the wall surface. In wet areas, it's sometimes advisable to *cove* the vinyl, that is, to bring it 3 to 6 inches up all vertical surfaces and cap it with metal or plastic inverted-U channels. The advantages of using vinyl floor covering as a wall covering include its low cost, wide availability, and the range of colors and styles on the market. Standard widths are 6, 9, and 12 feet, and some designs come 15 feet wide. Figure 19-10 shows the square footage of tiles and sheet goods, from 9-inch square tiles to 15-foot wide sheets.

Size	Type	Single	Full pack	Square footage
9" x 9"	Tile	.56 SF	80 tiles	45 SF
12" x 12"	Tile	1 SF	45 tiles	45 SF
6' wide	Sheet		100 LF	600 SF
9' wide	Sheet		100 LF	900 SF
12' wide	Sheet		100 LF	1200 SF
15' wide	Sheet		100 LF	1500 SF

Coverage estimates for flooring material
Figure 19-10

Rigid Plastics

PE (polyethylene), PP (polypropylene), PVC (polyvinyl chloride), and Kydex, a trade name for acrylic-polyvinyl chloride, are rigid plastics used for their light weight, moldability, and resistance to chemicals and corrosion. They're used throughout the chemical industry for wall covering, piping, tanks, and work surfaces. PE must be attached by mechanical means. The others can be applied with contact cements. Figure 19-11 shows the square footage in sheets of various sizes.

PE, PP, and PVC are available in 4 by 8-foot sheets, ranging up to 1/4-inch thick. You can special-order thicker sheets, but they're usually very expensive. PE and PP are translucent clear or white, while PVC may be transparent or white, gray, or brown. These rigid plastic sheets have limited use in homes but are used in commercial buildings to replace stainless and enameled steels. The walls of storage tanks, process areas, freezers, walk-in refrigerators, and even wineries have been covered with rigid plastic sheet materials.

They can be molded by heat or vacuum formed into continuous shapes. They can be welded using a technique called *hot gas welding.* For more information on these materials, write to Kamweld Products Co., 90 Access Road, P. O. Box 91, Norwood, MA 02062.

Kydex is a product designed just for walls that require exceptional protection. The color is molded in so scratches aren't noticeable, and you can get textured Kydex for even more hiding power. Fifteen colors are available, ranging from pastels to brights. Chemical resistance is very good. It will stand up to gasoline, chlorine bleach, most foods, and most acids and bases.

Sheet size (in feet)		Square footage
2 x 2		4
2 x 3		6
2 x 4		8
2 x 6		12
2½ x 5	(some plastics)	12.5
3 x 4		12
3 x 5	(some plastics)	15
3 x 6		18
4 x 4		16
4 x 8		32
4 x 9	(some exterior wood panels)	36
4 x 10		40
4 x 12		48
4 x 16	(some drywall)	64
10 x 12	(some acrylics)	120

Corrugated panels: Metal panels usually 25" wide; Fiberglass panels 26" wide.
Allow minimum 1" for overlap.

Linear feet	Square footage
6	12
8	16
10	20
12	24

Interior panels are available in thicknesses of 3/16", 1/4", 5/16", and 3/8"
Exterior panels 3/8", 1/2", 5/16", 5/8", and 3/4"
Floor panels 3/4", 1 1/4", and 1 3/8"
Drywall panels 3/8", 1/2", and 5/8"
Plastic sheet 1/32", 1/16", 1/8", 1/4", 3/8", 1/2", 5/8", 3/4", 1", 1 1/4", 1 1/2", 1 3/4", 2". Acrylic up to 4 1/2".
Metal sheet .015" to .125". Under .015" is considered "Strip."
Over 3/16" is considered "Plate."

Coverage estimates for sheet goods
Figure 19-11

The material is approved by the ASTM, Underwriters Laboratories, and the building code for most class I, II, and III locations. It's easily worked using standard wood and paneling tools. Contact Rohm and Haas Co., Independence Mall West, Philadelphia, PA 19105 for complete specifications. Ask for PL-10481.

If you want to do commercial wall covering jobs, I suggest you learn about Kydex. Hospitals, offices, laboratories, schools, and most factories are using the material. In homes it's appropriate for baths, kitchens, family rooms, and storage areas. I've used it for cabinet door inserts.

Two other decorative plastics are *acrylic and polystyrene.* You probably know acrylic by its trade name, Plexiglas. It's another Rohm and Haas product. Write for PL-713f and PL-1047K for additional information. Plexiglas is available in several surface textures, almost all colors, and in special formulations. Thicknesses range up to 4½ inches, in sheets from 36 by 48 inches to 120 by 144 inches. You can choose from transparent, translucent or semi-opaque varieties. There are U/V additives that will block or will pass ultraviolet light.

Plexiglas is used as a substitute for glass. It resists breakage, and in the thicker forms, will stop a bullet. Backlighted ceilings and walls often use Plexiglas panels and many lighted signs are made of it. The product has been in use for about 30 years.

But there's a major caution with Plexiglas. Unlike PE, PP, PVC, and Kydex, which tend to be self-extinguishing, acrylic burns like crazy. It will spread a fire because it melts into flaming droplets. Because of this hazard, I don't recommend covering complete walls with acrylic. Use just what you need to get the desired effect. And check with your local building inspector before you use it at all.

Acrylic sheet can be heat formed and solvent welded. But most applications will require mechanical fasteners. Backlighted applications must be removable so the light bulbs can be changed.

Styrene or polystyrene is used the same way as Plexiglas. Many ceiling drop-in light panels are styrene. It's cheap, easily workable with hand tools, and comes in 2 by 4-foot and 2 by 6-foot sheets. The only real drawback I know of is that it's brittle and breaks easily. Be careful in cleaning the back side of the designer series types, the multicolored ones. The color or pattern is usually printed on and some cleaners will remove it.

Laminates

Better known under the trade names *Micarta* and *Formica,* laminates are sheets of paper, plastic, and sometimes metal that are impregnated with or sandwiched between plastic resins under pressure. The resin most used is phenolic. A design is printed on the last sheet just before the top coating is applied. Since most designs are done photographically, many patterns are available.

Panel sizes run from 3 by 8 feet to 5 by 12 feet, with 4 by 8 feet being the most common. You can find almost any color. Textures range from super smooth to rough. Laminates aren't scratchproof, but they're scratch resistant. They're also food, water and chemical resistant. Like most plastics, they will burn. For special applications, you can use anti-static laminates. The cost is somewhat higher than regular laminates, but they're a must in certain process areas, such as where integrated circuits or explosives are manufactured.

Recommended uses for laminates are for counters, cabinets, furniture and the walls of kitchens, baths, laundry rooms, hallways, offices, labs, factories — wherever an easily cleanable surface is required. Laminates can be cut with woodworking tools and are applied with contact cements.

Check with your building supply dealer, home decorating center, cabinet shop or local plastic store for more information. Or you can write to Westinghouse Electric Company, Decorative Micarta, Hampton, SC 29924, or to Formica Corporation, Subsidiary of American Cyanide Company, Wayne, NJ 07470.

Laminates are available with metal top sheets (copper, brass, stainless steel, or chrome) with an epoxy or urethane clear coat. The finishes are bright metal, satin brushed metal, and several embossed designs. The two companies I'm familiar with are the Diller Corporation, Homapal Products Division, 6210 Madison Court, Morton Grove, IL 60053, and Chemetal Corporation, South Norwalk, CT 06854.

A fairly new laminate is manufactured by the Composite Materials Division of Consolidated Aluminum Co., 11960 Westline Industrial Drive, St. Louis, MO 63146. Their trade name is Alucobond. It consists of two electrostatically painted aluminum skins bonded to a polyethylene core. It works like wood but gives the protection of painted metal. The material can be sawed, routed, sheared, rolled, hot gas welded, glued, and painted. It comes in several thicknesses. If you route through one side to within 1/32 inch of the other side, you can bend it by hand to form nice rounded seamless corners.

Alucobond is currently being used as an exterior covering for commercial buildings, as an interior covering for hospital walls, clean-room walls, and other areas where a super-smooth surface is required. Many sign builders use it for making the sign frames. The company also has a stainless steel and polyethylene laminate which is ideal for walls of food processing areas. Both materials are available in single sheets from distributors or from the manufacturer in minimum lots. Sheet sizes range from 4 x 8 feet to 5 x 12 feet. Ask for the manufacturer's very detailed "how to" manual.

There are two other wall coverings I should mention, although they're manufactured in England and not readily available in the United States. One is a wood veneer plastic laminate, the other is a cloth veneer plastic laminate. Both are available from The Diller Corporation. The trade names are *Dillerwood* and *Dillercloth*. Their biggest drawback is that they can't be applied over concrete, plaster, gypsum or wallboard. They require a wood subsurface. They're very resistant to chemicals, meet class I and II fire code requirements, and are wear resistant.

Vinyl coated metal sheet is steel that has been factory covered with a vinyl top layer. The material is used by equipment manufacturers for trim, equipment housings, and panels. It protects like vinyl but has the strength and fire resistance of steel. It comes in many colors and textures, in 4 by 8 foot sheets. If you're doing commercial work, add this to your list of alternate wall coverings.

Metal Sheets

Metal sheets make good wall coverings. They're fireproof, rot proof, insect proof, easily cleaned, and easily painted. The main disadvantage is cost. Metal sheets are available in 3 by 6 and 4 by 8-foot sizes, primed, painted or unpainted, embossed, or designer punched. Surface textures range from silky smooth to wood and leather grained. Materials available are steel, stainless steel, brass, copper and aluminum. Aluminum and steel sheets are the cheapest.

Cold rolled steel (CRS) requires painting. Painted steel is used as a wall material in many commercial stores, offices, and factories. Aluminum is used on exteriors because it doesn't rust like steel. Decorative punched aluminum sheets can be used as a decorative accent on interior walls and as room dividers.

Stainless steel is one of the more desirable metal surfaces because of its corrosion resistance and cleanability. Stainless steel is an alloy of standard steel, (iron), and chromium. The higher grades may also include nickel. The chromium and nickel help resist corrosion. Stainless steel is the preferred wall covering in food handling and chemical handling areas. But note this: Some of the lower grade (301, 302) domestic stainless steels and some of the higher grade (304, 308) foreign stainless steels will rust when sanded or rubbed with kitchen cleaners.

For information on stainless steel types, uses and joining methods, write to the Committee on Stainless Steel Products, American Iron and Steel Institute, 1000 16th Street, NW, Washington, DC 20036.

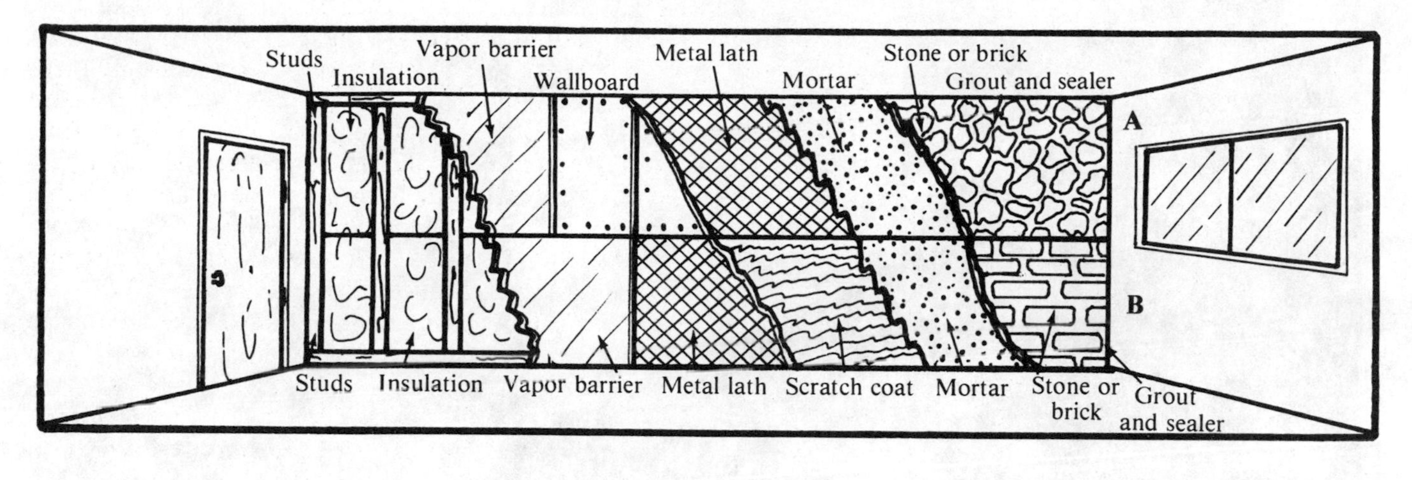

Brick or stone veneer over wallboard or plaster
Figure 19-12

Marble

The ultimate in elegance in wall covering is probably the hardest, coldest material available: *quarried marble* cut in slabs approximately 1/2-inch thick. Sheet size is usually between 2 by 4 and 4 by 4 feet. Marble will outlast most buildings. It can be hand or machine polished to a gloss finish. But it is porous and stains easily. A urethane, epoxy, or wax top coating is required.

Marble is recommended for fireplaces, baths, entryways, and for commercial buildings. Since it's heavy, the wall and floor foundations must be designed to accept the weight. My advice is to leave marble installation to the experts.

A lighter, more colorful, and easily workable material is *cultured marble.* It's imitation marble manufactured from polyester resin with marble or silica powders added. The chief advantage is lower cost. There are two primary disadvantages: It will scratch and it will burn. Many firms make vanity tops and sinks from cultured marble. It's hard to find in sheets for wall covering, but they are available. For more information, write to General Marble Co., 507 Arrow, Rancho Cucamonga, CA 91730.

Brick and Rock

Brick has been used on the exterior of fine homes and buildings for hundreds of years. It resists the elements, doesn't burn, and pleases the eye. But I don't recommend the use of brick on interior walls unless the floor is a concrete slab or the foundation has been formed to carry the weight.

Natural rock, including split fieldstone, riverbed rock, white marble chips and black or red volcanic rock, are available in many areas. Cemented and grouted in place, they provide a natural, rustic look with the added advantage of being fire resistant.

The problem with both brick and natural rock is the weight. But there are brick and rock substitutes that are light in weight and designed for interior wall application.

Brick veneers— These are real brick that are less than an inch thick. They're applied using mastic and grout or mortar. If you plan to use brick veneer behind a stove or near a fireplace, buy materials that meet or exceed ICBO 2113 and UL #209T. Check with your local building inspector. You may have to use asbestos sheet as a backer board.

Brick substitutes— Then there are the brick substitutes, brick-like veneers that are manufactured from cement rather than clays. This product has the advantages of brick but at a lower cost.

No matter what form of brick or rock you're using, interior masonry is usually limited to the lower half of the wall, with paint, wallpaper, or paneling for the top half.

Installing brick or stone veneer— Figure 19-12 shows how to install stone or brick veneer over an existing plaster or wallboard wall or a newly constructed wall. Follow along on the top of the figure, section A, for brick or stone veneer over wallboard. Studs should be on 24'' centers maximum for non-bearing walls and 16'' centers maximum for load-bearing walls. The bottom wall plate should be doubled. Exterior walls should be insulated and interior walls may be. A vapor barrier should be part of the insulation, either foil, Kraft paper or polyethylene. Up to this point, the procedure is the same whether the wall is wallboard or plaster.

Brick embossed metal
Figure 19-13

Still looking at the top half, section A, of Figure 19-12: For a wallboard wall, install the wallboard and then the metal lath. Now you're ready for the mortar and the stone or brick.

Section B shows brick veneer over plaster. After the vapor barrier is up, firmly attach the metal lath to the studs with staples. Use a lath that's at least 1.75 pounds per square foot and overlap it 2 to 4 inches at all joints. Then apply the plaster. In this case, use ASTM Cement Mortar N, mixed according to the section on mortar in Chapter 11. You can tint the mortar to match or contrast with the stone or brick, using colorant formulated for mortar. The scratch coat is mortar used to build out the surface. Use a "plaster comb" or "plaster rake" to scratch the surface so that the next coat will adhere. Then apply mortar with a mason's trowel to a thickness of 1/2 to 3/4 inch. Work on 5 to 10 square foot sections at a time so the mortar doesn't set up before you can apply the stone or brick.

Use a carborundum bladed saw for cutting the stone and brick where necessary. Press the stone or brick into the wet mortar and lightly tap to set. Brick should be spaced 1/2-inch apart, stone 1-inch apart.

Try to keep mortar off the face of the brick or stone. Mortar may leave a stain that's hard to remove completely. If some does get on, don't try to wash it off with water. Let it dry, then use a dry brush to clean it off. After a week or so of drying, remove any residue with a mild detergent wash, and rinse. Finally, grout the seams, strike off if desired, and coat with a sealer when the grout is dry.

To help you estimate, it takes about 3.5 standard clay brick to cover 1 square foot. A 66⅔-pound sack of mortar will cover about 15 square feet at 1/2-inch thick.

Imitation brick and rock— The first imitation brick was just wallpaper with a brick design. It didn't come close to looking like the real thing. Then came stamped sheet metal, shown in Figure 19-13. Yes, the sheet on the right was installed backwards. The paint has peeled because of heat expansion; this spot gets up to 100 degrees F for several hours each day. The paint job was actually done very well, considering that it's about 10 years old.

There's also molded plastic, which approaches brick in realism, but it's not heat resistant.

Then there's imitation rock and stone wall covers. These are made to resemble anything from fieldstone to sandstone. Modern imitations have all the advantages of real rock. They're heat and fire resistant, rot and damage resistant, and have natural beauty and individuality. I recommend them for around fireplaces, the lower half of interior walls, and as an exterior wall and trim treatment. I've even seen them used to build a fence. The contractor used 4 x 4 redwood posts, exterior plywood, and wire lath. Then he cemented the stone in place. It looks like a real stone wall.

The nice part is you don't have to spend hours or days finding and selecting the most desirable rock or brick. Also, imitations are much lighter than the real thing. You can use them on interior wall surfaces without fear of the wall or floor collapsing. They're moderate in price and install just like brick. For more information, write to Stucco Stone Corporation, P.O. Box 270, Napa, CA 94559.

Fabric and Carpet

I think *burlap* should be used more than it has been. It's available in many colors and in various lengths and widths. Most home decorating centers have it in stock or can get it for you. Burlap can be glued directly to the wall surface or be mounted on a fiber or sound board backer sheet. When it's glued or stapled to a backer sheet, it makes an excellent bulletin board wall for an office, shop, kitchen or children's play area. But don't use it in greasy areas; it can't be scrubbed. It's best used on upper wall sections, where there's paneling or veneer on the bottom half of the wall.

We all recognize the luxurious feel of fine carpet. Running carpet from the floor right up the wall can give a rich, plush effect. This treatment is most appropriate for play rooms, TV rooms, and bedrooms. Keep the carpet confined to the lower half of the wall, with paint, wallpaper, or paneling in the upper section. You say you're not convinced? Have you ever been in a custom van? It's like being in a warm, soft cocoon.

Carpet applied to walls should be glued, with no padding under it. A good carpet installer can suggest proper methods for your particular situation. Carpet is available in loop, cut-loop, and plush textures, or it can imitate grass or animal fur. Less expensive carpets are of polyester, nylon, and other synthetics. The most expensive are llama and lamb's wool — both used as wall hangings. They're too expensive to glue down.

Special Purpose Wall Coverings

Fear of asbestos fibers has our schools and government offices around the country spending millions on the removal of asbestos fiber wall and ceiling panels. The truth of the matter is that asbestos isn't dangerous if it's painted. Asbestos fibers cause health problems only when inhaled. If you "lock in" the fibers, there's no health danger. Cement, mortar, or paint will lock in the fibers.

Asbestos fiber-filled cement board has been around for decades. It's common on farms and many urban buildings. It's harmless, easily cleaned, "chew-proof," and fireproof. Many building codes require panels of fiber-filled cement to be used behind furnaces, fireplaces, and wood stoves. It will withstand heat flow and temperatures up to 1,000 degrees F. The cost runs from low to moderate. Cut or drill fiber-filled cement board with carbide tipped blades and bits. Yes, do wear a dust mask or respirator when cutting. And use a suction system or vacuum to pull in the loose fibers. This is the only time the material is dangerous. Cutting releases the fibers.

The life expectancy of the material is about 40 years, longer if it's painted with acrylic latex paints. Asbestos shingles may also be painted. The natural color is a pale gray. The only trade name I currently know of is "APAC" by Turners Building Products, 33778 First Avenue, Mission, B.C. Canada V2V 1H6. Most building supply centers do carry it, but you may have to special order.

One-part polyurethane foam— Imagine a wall covered with that semiflexible, easily cleanable material found in the dash boards of cars. It would give a waterproof, food and chemical resistant, durable surface. The material is polyurethane foam. It comes in quarts, gallons, and 55 gallon drums, and can be brushed, rolled or sprayed in place. The tough skin is resistant to weathering, cracking, and abrasion, and acts as an electrical insulator. It's available in several colors and can be textured to resemble cloth or leather.

Fiberglass

Boat and bathtub manufacturers learned long ago that *fiberglass* is a very tough material. It's strong, water repellent, chemical resistant, and well accepted by the public. I expect that the first com-

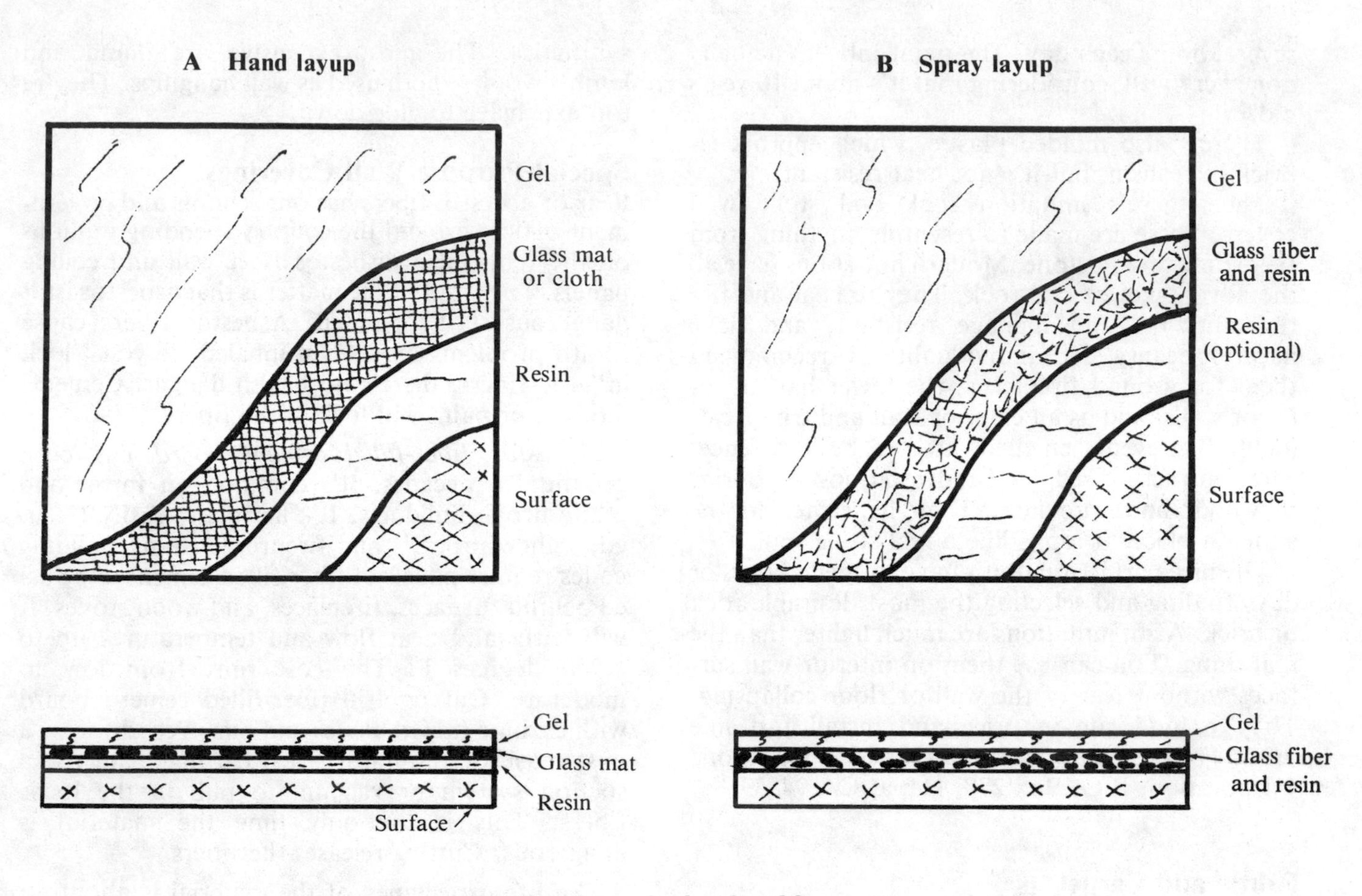

Fiberglass layup by hand and by spray
Figure 19-14

pany to discover a way of applying fiberglass on walls will make a bundle.

Figure 19-14 shows the two ways fiberglass is laid up, by hand and by spray. Epoxy or polyester resin is applied to a surface, glass cloth or matting is pressed into the resin, and a final colored gel coat is applied. The problem is in the curing of the epoxy or polyester. Hardener catalyzes the resin, making the resin warm and runny. On a vertical surface such as a wall, it's hard to control the resin until it sets up. Thickening agents do help, but it's still chancy.

Horizontal surfaces are ideal for laying up fiberglass. The resin smooths out to a nearly impervious, glass-smooth coating. Coating panels, doors, kick-plates, cabinets, handrails, planters, furniture, and window sills are easy because they can all be laid horizontal during application. Like all plastics, fiberglass does scratch. But scratches are less noticeable because the color goes all the way through the gel coat. And colorant can even be added to the base resin.

The cost of applying fiberglass is moderate to high. But if you need durability, color retention, and rot resistance, fiberglass is a good choice. If you want to learn more about fiberglass and its uses, your local Owens/Corning Company representative will be a good source. Or you can write to Owens/Corning Fiberglass Corporation, Fiberglass Tower, Toledo, OH 43659.

For production work, fiberglass can be sprayed. The process requires special equipment that sprays the glass fibers, resin, and hardener simultaneously. They mix in the air and harden on the surface being sprayed. Equipment is available from several companies. The one I'm familiar with is Glas-Craft, 5845 W. 82nd Street, Indianapolis, IN 46278.

Wall Coverings for Sound Control

Cork is available in light or dark colors in sheets, rolls, and tiles. Many of these 12-inch tiles come preglued, with a peel and stick backing. Cork tile is sound deadening, and makes a great bulletin board wall. I'd use it only on upper wall areas, since it's so easily damaged. And use it in well-lighted areas as it tends to cut light levels. Offices, TV rooms, dens, are some of the areas in which cork can be used effectively.

If there's a real noise problem, or an area that can have no reverberation, consider *eggcrate foam.* Available in sheets and 12-inch tiles, it's thick, soft, reasonably inexpensive, and an excellent sound deadener. It only comes in gray or blackish-gray, so again, use it in a well-lighted area. And use it only in air-conditioned rooms. It accumulates dirt and dust very quickly. Recommended uses include sound studios and sound analytical labs, as well as the ceiling of any noisy area. You won't apply much eggcrate foam. But if you spend a career in the painting and wall covering business, you'll have several opportunities to use it to good advantage.

Wall Coverings for Fire Resistance

General Electric Silicone Products Division (Marketing Communications, Waterford, NY 12188) makes some coatings, rubbers, varnishes, greases, and sealants that you should know about. One product, Pensil 851, is an excellent coating for areas exposed to open flame. It can withstand continuous exposure to heat of 600 degrees F, and short-term exposure to as much as 5,000 degrees. It's a silicone, so it resists water, oils, and most chemicals. The product is thick enough to be troweled into place.

It's currently used in commercial buildings as a firestop between floor-to-floor openings. The cost is high and it's not readily available. But think how nice it is to turn any piece of drywall, lumber, or even screening into a firewall in less than 15 minutes. The material sandwiched between two pieces of 3/8-inch drywall creates a fire resistant, vibration resistant, sound barrier.

I can think of many places in the home where Pensil could be used: storage cabinets, garages, heater and boiler rooms, firewalls between rooms or apartments, and electrical equipment housings. Other possible uses include material storage tanks, pipes and valves, vehicle firewalls, and welding areas.

Speaking of storage tanks and pipes, Pocomo Fabricators unit of Patterson-Kelley Co., Harsco Corp., East Stroudsburg, PA 18301 has a product called "Pre-Krete." It's used for corrosion and abrasion resistance in heat exchangers, boilers, water heaters, stacks, ducts, tanks and silos. It's a licensed process, but it could be the niche in the market that you're looking for.

Another licensed process is offered by Neogard Corp., 6900 Maple Ave., P.O. Box 35288, Dallas, TX 75235. It's a urethane foam insulation covered by a rubber elastomeric coating. It's ideal for the roofs of industrial buildings and possibly mobile homes, and may be suitable for the walls of industrial storage buildings.

Trim for a Professional Look

Trim out is always a problem for contractors, whether they're painters, carpet/vinyl or wallboard/panel installers. Most local hardware stores have metal and plastic trim available in 4, 6, and 8-foot lengths in limited shapes, materials, and sizes. But if you're serious about craftsmanship, you need a better source of quality trim in longer lengths and at contractor prices.

Wood trim— Figure 19-15 shows utility wood trim for various applications. It's available in several sizes and lengths, but the usual length is either 10 or 20 feet. Figure 19-15 doesn't show the more decorative trims. Many designs, finishes and colors are available.

Don't be afraid to do some experimenting with trim. Sometimes a combination of several different shapes will make a better looking, more interesting finished job. And here's a helpful hint: Keep a large piece of plywood tacked to the wall of your garage or shop. Each time you use a different trim, glue a small scrap piece to the plywood, labeled as to size, application and cost. This is the way to build a trim reference library for future reference.

Metal and plastic trim— Extruded plastic and metal trim comes in hundreds of shapes, many colors and textures, several different finishes, and sizes to fit panels from 1/16 to 3/4-inch thick. Figure 19-16 shows drywall metal trim for joining inside/outside corners, two or more panels, and finishing cap trim. Most of these trim sections are galvanized steel or aluminum and should be cleaned and scuff-sanded before you paint.

Trim can be installed with nails, screws, adhesives, or a combination of them. They will work on vinyl, wallboard, plastic, metal, and wood panels.

For metal trim, the trade name I'm familiar with is Trimedge, from the William L. Bonnell Company, Inc., 25 Bonnell St., Newnan, GA 30263. Ask for their general catalog. You'll be amazed at how many trim designs are available.

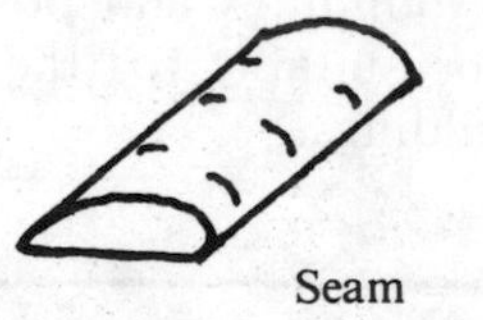
Seam

Stop

Mullion

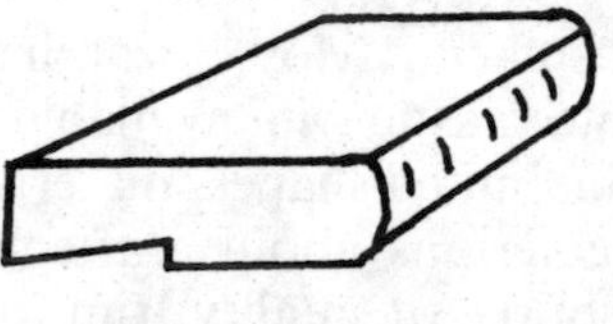
Stool

Stool, window sill

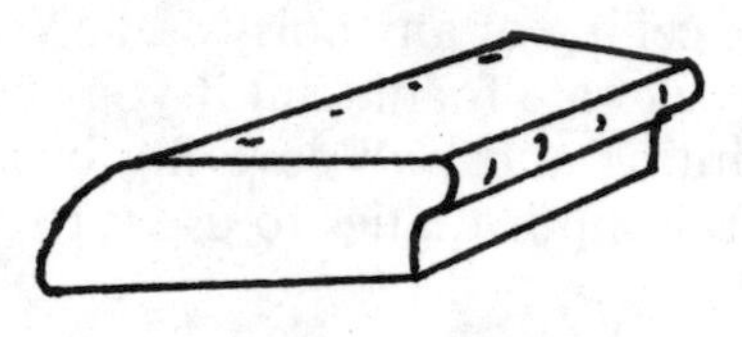
Door sill

Stop or base

Base

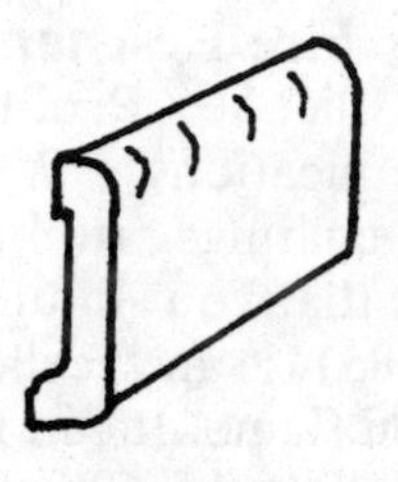
Casting

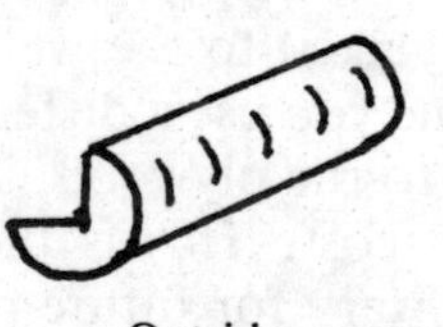
Outside corner

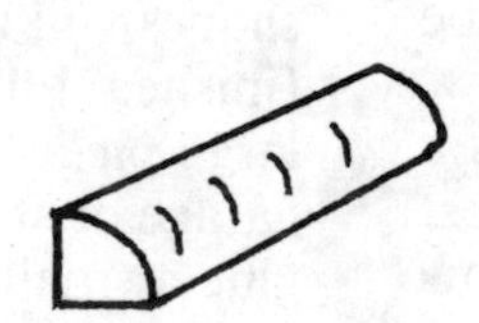
¼ round or base shoe

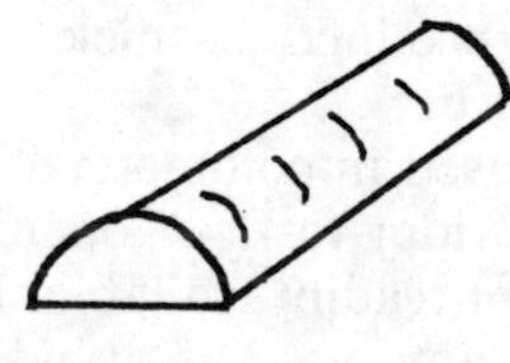
½ round

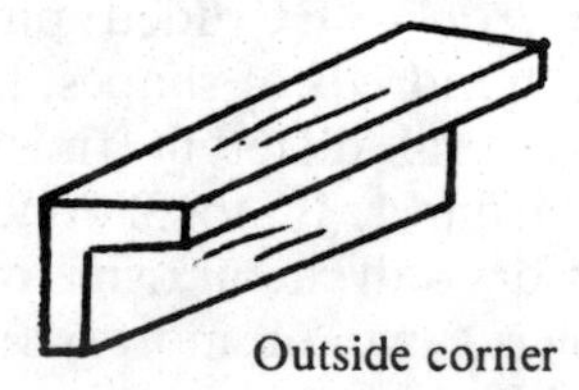
Outside corner

Lattice or lath

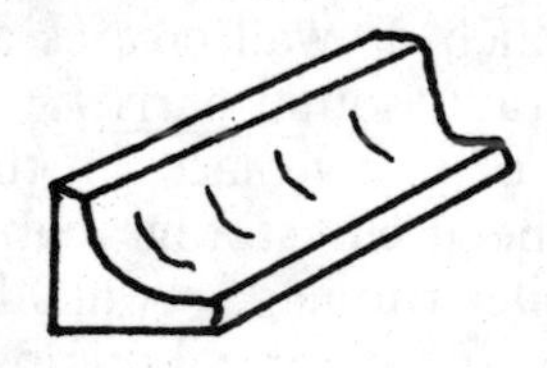
Cove

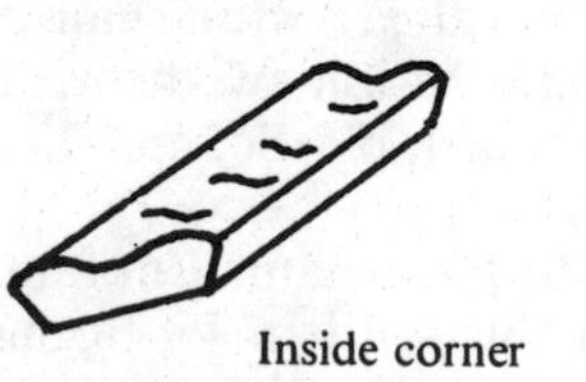
Inside corner

Cap

Drip cap

Utility wood trims
Figure 19-15

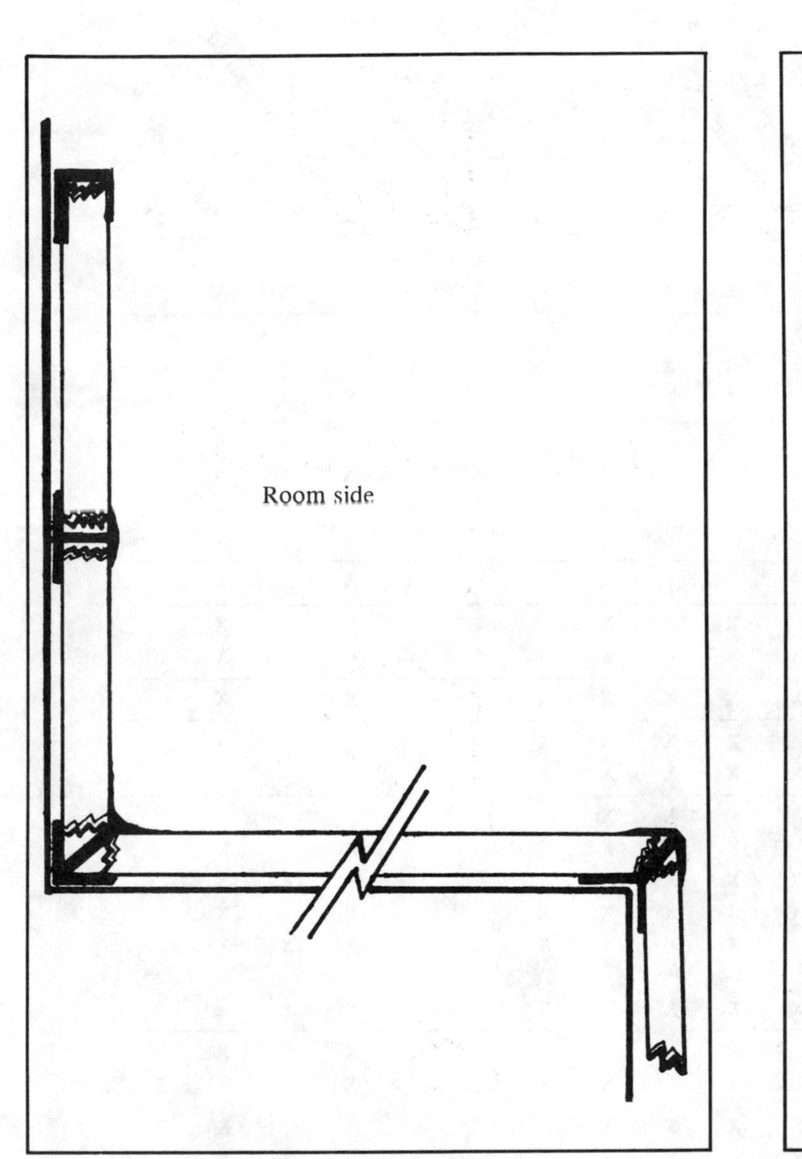

Trim for installing panels
Figure 19-16

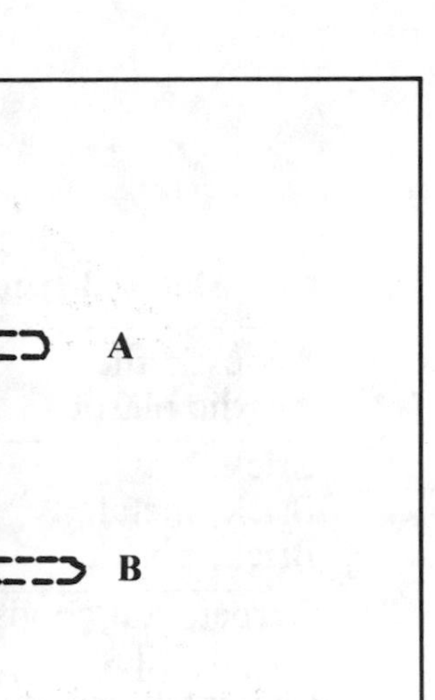

Mounting trim for mirrors
Figure 19-17

Mirror hangers— Figure 19-17 shows mounting trim for mirrors. The mountings labeled A and B are inexpensive, common plastic mirror clips. The C hanger is a U-channel trim. For a picture frame look, you can use a U-channel trim on all four sides. Using A hangers at the top and C at the bottom is the most common method of hanging mirrors. For lightweight mirrors, you can use A or B hangers in the four corners. For medium weight mirrors, I suggest a U-channel at the bottom and B hangers at the top two corners. The B clip does require that the mirror be drilled and a rubber spacer be used. This spacer prevents cracking when the wall and mirror expand and contract. The slight angle I show is deliberate. It helps keep room occupants more comfortable by reflecting window and ceiling light downwards.

Trim work as a separate business— Trim-out work can be a small business of its own. Many general contractors would prefer to sub out their trim work. It takes talent, time, and care to install trim correctly. If you do decide to specialize in this line of work, you'll need your own routers or saws and cutting heads. It's far less expensive to make your own trim than it is to buy ready-made trim. But your selling price can remain the same.

General Installation Tips

For most of the materials we've talked about, you can get detailed instruction sheets at the wholesale or retail outlet where you buy the material. The installation instructions show why and how to do it, and often include a material and tool checklist.

There are usually some things you'll need that

Material type	Chemical resistant	Cleanability good	Fire resistant	Moisture resistant	Rodent resistant	Rot resistant	Reflective to light	Sound absorber	Finish coat required	No finish coat required	Low cost	Medium cost	High cost
Acoustic tile								X		X	X	X	
Acrylic plastic	X			X	X	X				X	X	X	
Brick, clay	X	X	X		X	X				X			X
Brick, metal	X	X	X	X	X	X			X			X	X
Brick, plastic	X	X		X	X	X				X	X		
Carpet, roll goods	X	X						X		X	X	X	
Carpet, tiles	X	X						X		X	X	X	
Cement & cement block		X	X		X	X			X	X	X	X	X
Cork	X							X		X		X	X
Eggcrate foam						X		X		X		X	
Fiberglass	X	X		X	X	X				X	X	X	
Fire-retarding foam	X		X	X		X		X	X			X	X
Glass	X	X	X	X	X	X	X			X	X	X	
Laminates, plastic	X	X		X	X	X	X			X	X	X	
Laminates, metal	X	X	X	X	X	X	X			X			X
Marble, true	X	X	X		X	X				X			X
Marble, plastic	X	X		X	X	X				X		X	
Mirror, glass	X	X	X	X	X	X	X			X	X	X	
Mirror, metal	X	X	X	X	X	X	X			X		X	
Mirror, plastic	X	X		X	X	X	X			X	X	X	
Mirror, tiles (solid or flex)	X	X	X	X	X	X	X			X	X	X	
Plastics, foam	X	X		X		X		X	X	X	X	X	
Plastics, sheet	X	X		X		X	X			X		X	X
Rock, true	X	X	X	X	X	X				X		X	X
Rock, cement formed	X	X	X		X	X			X		X	X	
Rock, plastic formed	X	X		X	X	X				X	X	X	
Stucco		X	X		X	X		X		X	X		
Tile, ceramic	X	X	X	X	X	X	X			X		X	X
Tile, metal	X	X	X	X	X	X	X		X*		X	X	
Tile, plastic	X	X		X	X	X	X			X	X		
Vinyl, cloth & wallpaper	X	X		X		X				X	X	X	
Vinyl, flooring	X	X		X						X	X	X	
Wallpaper, flocks		X		X						X	X	X	
Wallpaper, foils	X	X		X			X			X	X	X	
Wallpaper, grass cloth								X		X		X	X
Wallpaper, murals		X								X	X	X	
Wallpaper, prints		X								X	X	X	
Wood, panels	X	X		X*						X	X	X	

*Usually factory coated

Note: Cost ranges are for material only. Labor is not included. Many items fall into two or more cost ranges due to quality grades. Low cost = under $0.50 per foot. Medium cost = $0.50 to $1.00 per foot. High cost = over $1.00 per foot.

Properties of common wall coverings
Figure 19-18

aren't on the list, however. Plan ahead. Picture each step of the job. Anticipate the problems. It can save you an expensive extra trip to the store.

Here are just a few of the things you might overlook. Always have rags and solvents for cleanup. If you're doing paneling or tile work, you'll need extra-long electrical outlet and switch plate screws, from 1¼ to 1½ inch long. If you're paneling a wall with drapes, you'll need long screws to reinstall the hardware. Remember, you're building the wall out when you add the panels or tiles.

If you're doing insulation and new stud-out, you may need wire and electrical boxes. I keep a few bottles of appliance touch-up paint handy, white and ivory, for touchup on those electrical screws and plates. If you have to remove existing baseboard molding, you'll need color nails and putty sticks for doing touchups.

If you have to drill holes in the tile or mirrors, you'll need a special drill bit designed for this type of drilling.

When purchasing materials, buy at least 5% more than you actually expect to use for the job. It's always easier to return goods than to make several trips for more. Most jobs will come up short, either because you forgot something or because of material miscuts or damage.

To help you decide what materials are best for a particular job, Figure 19-18 shows the advantages and disadvantages of each of the wall coverings we've covered. I haven't listed brick, cement, marble and rock as moisture resistant because they do absorb moisture, even though they resist damage by it.

What Do You Charge for Labor?

What do you charge when you're doing a particular kind of job for the first time and you don't have any history to go by? Consider using this rule of thumb: Total the material costs and multiply by 1.5 to find the labor charge. For example, if the material cost $100, you would charge $100 times 1.5, or $150 for the labor. You'd bid the total job at $250.

After you've kept good records for the first few jobs, then divide the manhours worked into the labor charge to see if you're in the ball park. Are you covering wages, overhead and a reasonable profit? If not, adjust your rates for the next job you bid.

I offer that 1.5 multiplier to use if you're a "start-up" painter, working out of your house or truck. If you're an established paint contractor with an office and much more overhead expenses, use a multiplier of 2.2 to 2.7. And always use the higher rate for small jobs — they're usually very labor intensive.

We've Come a Long Way

I've covered a lot of material in this manual — more than you're likely to absorb in a single reading, and very likely more than most painters will ever use. But I don't expect that you'll read this book and then set it aside. I hope you'll refer to it regularly as the need arises. That's the way to get the most value from these pages.

Let me make two points before sending you on to your next painting job. First, don't ever stop learning about the work you do. No matter how much you know and how much experience you've accumulated, there's still more you don't know and haven't discovered. Study the work of others, read the trade magazines, collect manufacturer's product literature, be alert for new painting references, and, when something goes wrong with your work, figure out what happened.

Every professional should stay current in his or her discipline. That's true in most businesses, of course. But it's especially true in the painting business. Today's paints and coatings are high-tech products — really modern miracles created by major advances in chemistry. As a professional in the industry, you should be aware of new products and how they're used. Anything less and you're doing your customers a disservice.

Finally, remember that you're in business to make money. I feel that every quality-conscious painter should have no trouble making a good living. No painter should depend on shoddy work or inferior materials to stay competitive. The experienced wall covering contractors I know would agree: the most successful pros in this business do the best job possible and charge appropriately. The customer who demands cut rate prices and settles for amateur workmanship will learn the hard way. He'll be back to you soon enough. Make it a practice to give your customers the best you can and pocket the profit you've earned. That's the only way to be fair to yourself and the public you serve.

appendix

Paint Chemistry

The information in this appendix will help you understand the chemistry of paint. I'll admit that many painters will never need to know most of the information in this section. But if you can't figure out why a paint or coating has failed, you'll probably find the answer here.

Let's begin with a brief review of some very basic laws of physics. Stick with me, even if high school science was your downfall. I'll make it as short and painless as possible. You may even find this interesting. Here are a half dozen terms and definitions you'll need to know.

The Definitions

There are two basic ingredients that make up all matter: the atom and the molecule. The *atom* is the smallest particle of a substance that can be identified as being that substance. A *molecule* is the smallest part of an element or compound that retains the chemical identity of the substance. It's a chemically bonded grouping of atoms. The way the atoms are grouped determines what the substance is. *Elements* are the pure substances we find in nature. There are over 100 elements known today. Aluminum, copper, hydrogen, iron, oxygen, silicon, titanium, and zinc are some of the elements that have been identified.

Two or more elements can be combined as a compound or as a mixture. The *compound* is a chemically bonded substance with its own identity. As a rule, it can be further combined with other compounds or substances. Some common compounds are acids, bleaches, bases, and salts. Literally thousands of compounds occur naturally.

A *mixture* is two or more compounds combined with each other to form a different substance. Potassium mixed with nitrates, for example, forms a fertilizer.

The mixture of a solid in a liquid is a *suspension.* The suspension is a mixture where one substance is hanging but undissolved in another substance. An example would be pigment suspended in paint. The liquid or *vehicle* is a compound.

That's some of the vocabulary we'll be using. Now let's go back and look at the basic building block, the atom, in more detail.

The Atom

An individual atom looks something like a bee's nest with bees buzzing around it. The center (or nest, in our analogy) is the *nucleus.* It's made up of two major particles, the *neutron* and the *proton.* The neutron has no electrical charge. The proton carries a positive electrical charge. Spinning around the nucleus are smaller particles (the "bees") called *electrons.* The electrons carry a negative charge. Positive and negative charges attract each other. It's this *electrostatic attraction*

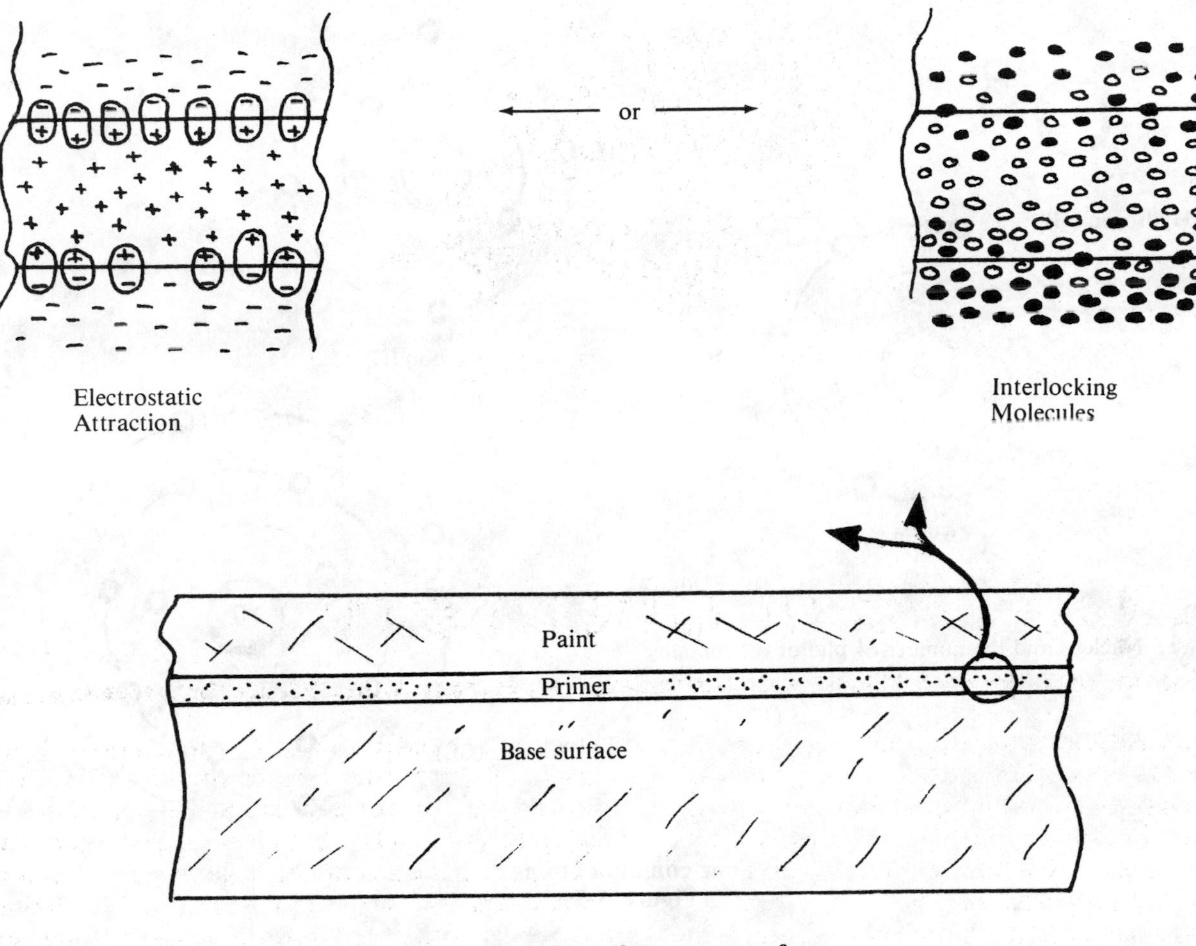

Why paint adheres to a surface
Figure A-1

that keeps the atom together. It's also the reason that atoms can combine chemically. And it's probably one of the reasons why paint sticks to the surface it's applied to. Look at Figure A-1. It shows two reasons why paint adheres to a surface. One is electrostatic attraction. The other is interlocking molecules. Whatever the cause, adhesion is greatly increased if the proper primer is applied between the surface and the paint.

The neutrons normally can't be split off of the atom. The protons and electrons may be. The electrons circle the nucleus in definite orbits called *valence energy levels.* The first or closest orbit to the nucleus can contain no more than two electrons. Each successive orbit can contain no more than eight. When an orbit contains eight electrons, it is full and the atom is said to be *inert.* That means it can't combine with or hold on to another substance by chemical action.

The first orbit, the one containing only two electrons, is the closest to the proton and is tightly bonded by the electrical attraction. The other orbits are further away from the nucleus and are progressively weaker. It is the last or farthest out orbit that can bond two or more same or different atoms together into molecules and compounds. An atom with less than eight electrons in its outer orbit can give up, take in, or share electrons, until it becomes inert.

Most chemical reactions require some form of heat to start the bonding process. The more heat is applied to an atom, the more energy it absorbs and the faster the electrons will move. The atom expands and the electrons move further from the protons, weakening the electrical bond. At some point the electrons will fly free of the atom. At that point, they're free to combine with other atoms, often creating a new substance.

Atomic elements are identified by the number of protons they contain. There can be different numbers of neutrons and electrons, but each element always has a specific number of protons.

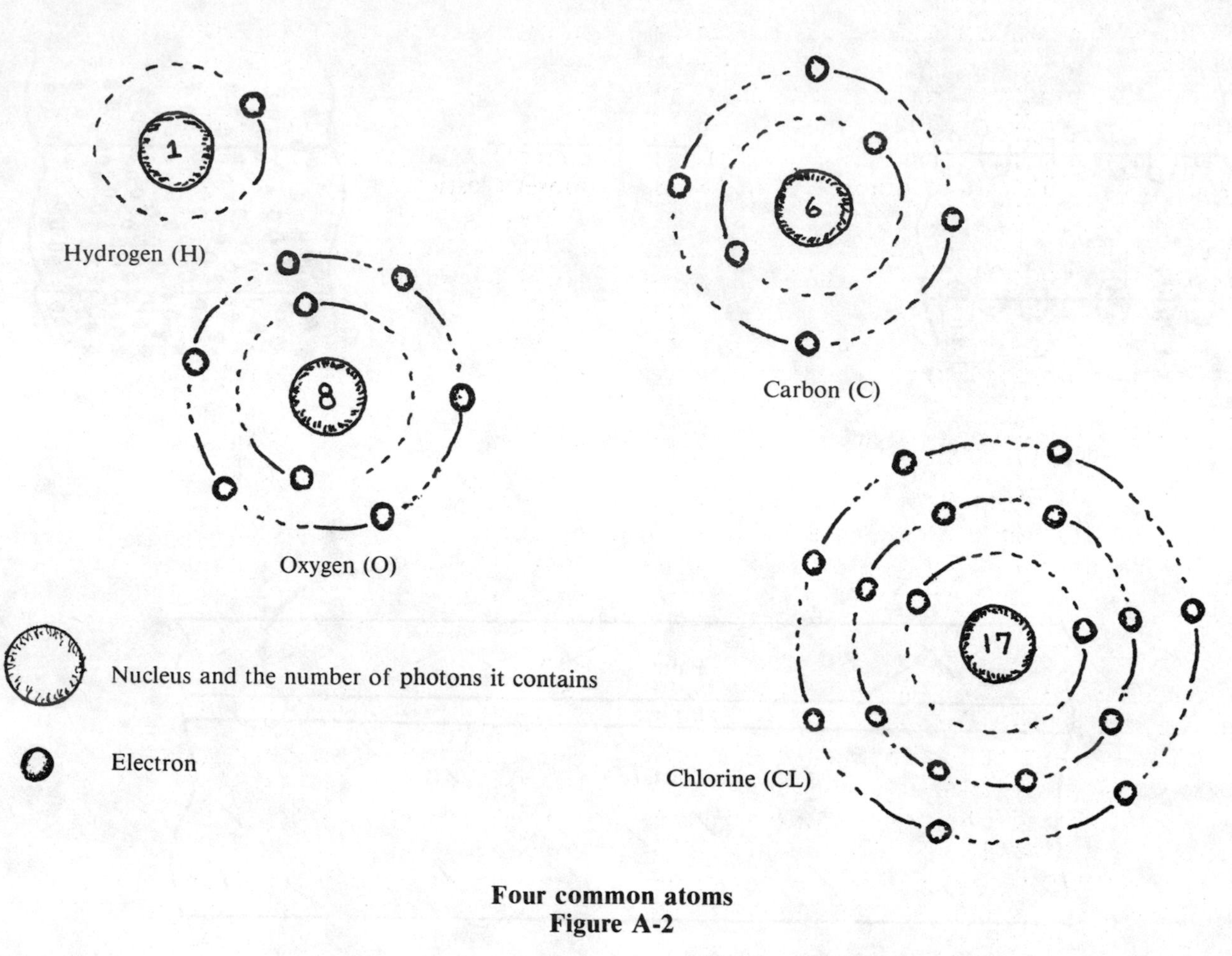

Four common atoms
Figure A-2

The Elements

Hydrogen

The simplest element is hydrogen (Figure A-2). It contains one neutron, one proton, and one electron. Hydrogen is a highly flammable gas.

Oxygen

Another simple (but indispensable) element is oxygen (Figure A-2). Twenty-one percent of our air, all water, most rocks and minerals, and many compounds contain oxygen. It will combine with all known elements except the inert gases. It's required in the processes of oxidation (rusting) and combustion (burning). Nothing will burn in the absence of oxygen.

Carbon

Carbon is found in all living matter. When plants or animals die and decompose or are burned, there will be a residue of carbon. Today we're taking advantage of this. The carbonous remains of plants and animals that were here on earth hundreds of thousands of years ago have formed petroleum. Oil and petroleum are made up primarily of hydrogen and carbon atoms, so they're called hydrocarbons. You may see that word on paint cans. By grouping together molecules of hydrogen and carbon, we form the elementary gases such as ethylene.

The Combinations (Molecules)

One of the simplest molecules is made of one hydrogen and one oxygen atom. This molecule is a very highly flammable gas. If it's bonded to the end of another molecule of select elements, it forms an *alcohol.* Alcohol is a highly volatile and flammable material. It's frequently used in chemical reactions and chemical mixtures in the paint industry to form dyes, pigments and vehicles. It is poisonous and dangerous.

When certain atoms are combined, they form acids and bases. An *acid* is an unstable substance, highly reactive and caustic. That's because it's easy for a hydrogen atom to escape the molecule and combine with other substances. Most acids are poisonous, except for *acetic acid.* Acetic acid is found in vinegar and citrus fruits. The form of acetic acid used for the manufacture of paints, however, is from hydrocarbons found in petroleum. It's very poisonous.

A *base* is the opposite of an acid. It's also reactive and caustic, with a bland, brackish taste. Lye is just one example of a base.

Acids and bases complement each other. One *neutralizes* the other, producing a *salt* in the process. This is the main cause of *efflorescence* on cement or concrete surfaces. Both acids and bases are soluble in water.

Any surface that's cleaned with an acid or base solution should be neutralized with an opposite solution, then rinsed with a lot of clear water. Here's the simple truth: Paint doesn't stick to a salt. Salt is a neutral or inert substance that doesn't adhere to other surfaces.

Water

Water is the chemical combination of two hydrogen and one oxygen atom. Water has several important qualities. Since it's a simple molecule, it's small in size and can easily fill small voids. It doesn't burn; it can be used to quench both thirst and fire; when heated it forms a gas (steam); when cooled it forms a solid (ice); it expands when it changes into a gas or a solid. If water is trapped between a surface and the paint coating that surface, this expansion will blister the paint.

Water is a good solvent, a good lubricant, and a good conductor of heat. Virtually all other matter, given enough time, will dissolve in it. All these things from just three little atoms. A miraculous substance when you think about it.

Plastic

A wide variety of substances spring from the carbon atoms. When carbon atoms are removed from the simple gases and bonded together in a simple chain-like network, they produce a solid substance with unique characteristics. Each time another carbon atom is added to the chain, a new substance with new characteristics is formed. These materials are called *thermoplastics.*

Most of the plastic items in common use are made from thermoplastics. They're also used as bonding agents or *binders* in most modern paints.

Other forms of plastic are created by adding other elements to the carbon atoms. The dissimilar atoms link together the carbon chains and form a substance that has a ladder-type network of molecules in it. Chlorine (Figure A-2) is one such dissimilar atom. When it's added to the vinyl carbon chain, it forms PVC, or *polyvinylchloride.* Poly means many. In this instance, many chains of molecules identified as vinyl are *cross-linked* to one another by the presence of chlorine atoms. Plastics with cross-linked chains are called *thermoset* plastics. Heat transforms the material into a solid. *Epoxy* and *polyester* are just two examples of thermoset plastics.

Plastics have these characteristics: They're flexible, shock resistant and abrasion resistant. They are also resistant to staining, to water penetration, to most chemicals, and are non-corrosive. This combination of characteristics makes them ideal for use as bonding agents in paint.

Motion and Energy: The Photon

If you boil a pot of water, the water will turn to steam and eventually evaporate or disperse into the surrounding air. You've added energy to the process. The more energy you add, the more excited the atoms and molecules become and the more they move about. When they're moving fast enough, they can break away from the material that contains them.

Look at Figure A-3. It illustrates the effect of heat on a carbon atom. When it's heated, the electrons move further from the nucleus until the internal electrical forces can no longer hold the electron in orbit. Then the electron separates and is free to collide or combine with another atom.

Besides the neutrons, protons, and electrons, atoms contain an even smaller particle — the *photon.* Photons are released from atoms when sufficient heat or energy is applied. The faster the photons oscillate or travel within an atom, the more likely they will break free and travel away from it.

Electromagnetic radiation is actually streams of these photons. Visible light is one kind of electromagnetic radiation. *Ultraviolet* or *U/V* radiation is composed of these photons at an energy level that is just beyond our visible senses. As the atom is heated, these particles leave the surface at different heat or energy levels. The speed of travel is what causes the different kinds of electromagnetic radiation — radio waves, X rays, visible light, or ultraviolet waves. It also creates the differences that we perceive as colors.

Photons travel in straight lines from their source to an object. If the object or surface allows these particles to pass through it, it's *transparent.* Glass is an example. A substance that lets some, but not all of these particles, through, is *translucent.* A substance that passes none through is *opaque.* See Figure A-4. A surface that bounces these photons

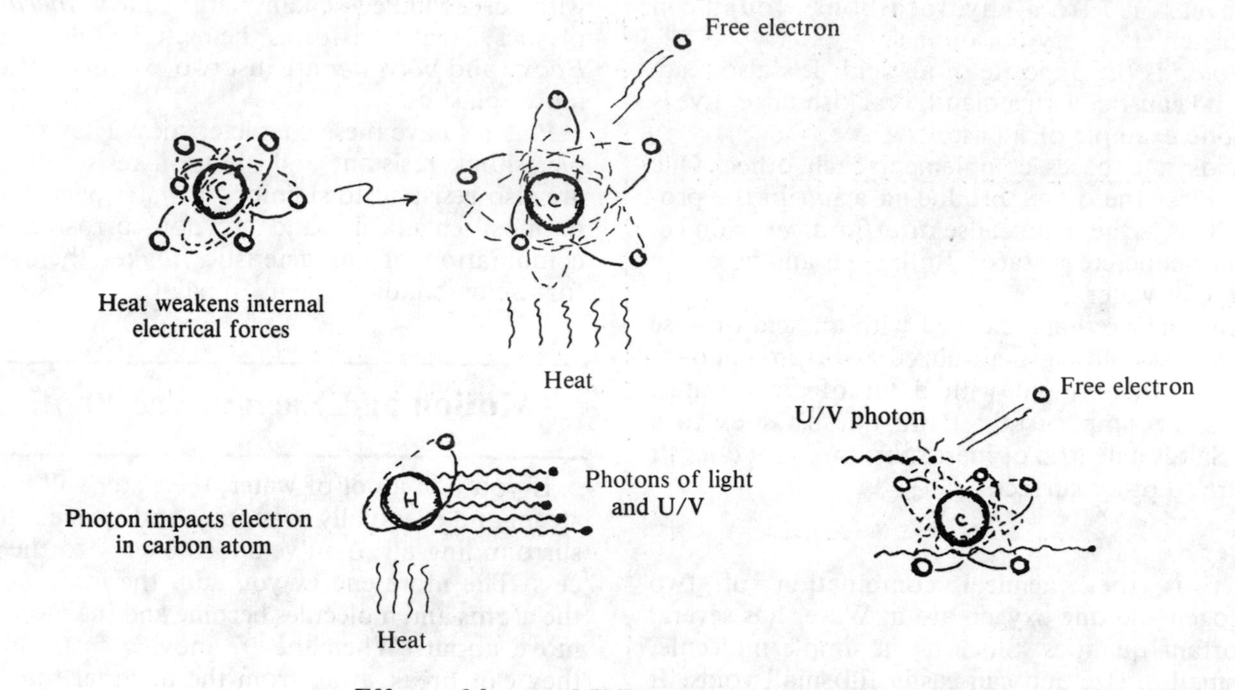

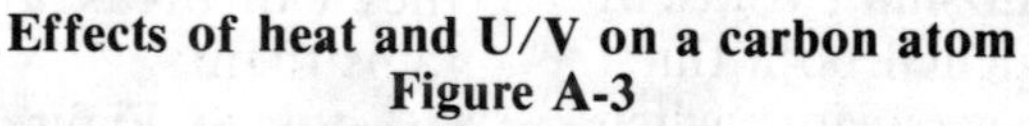

Effects of heat and U/V on a carbon atom
Figure A-3

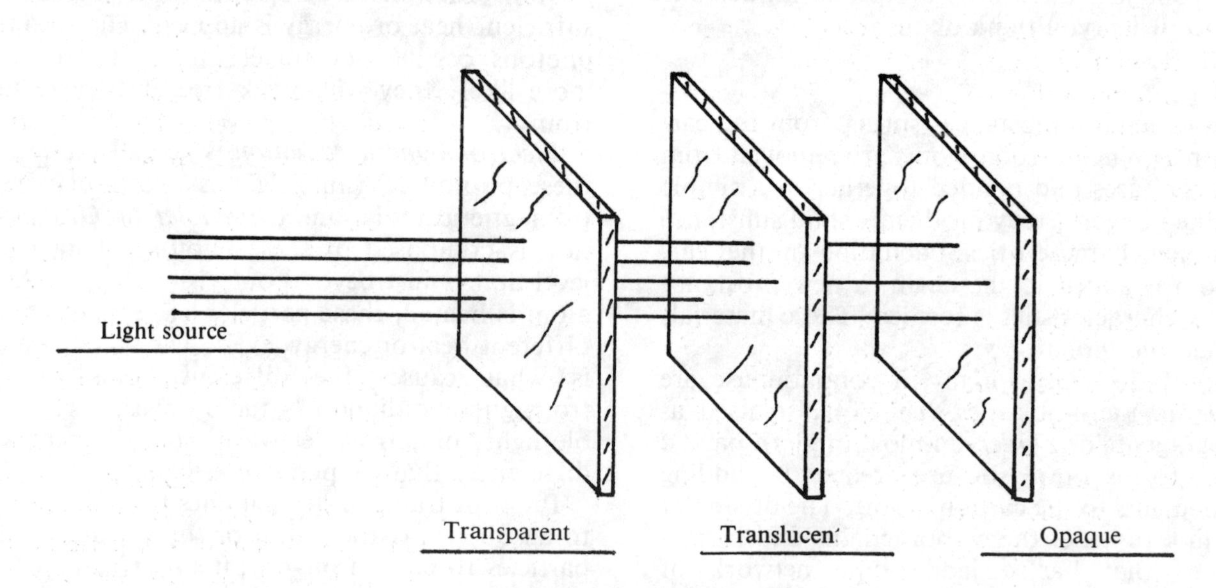

Degrees of transparency
Figure A-4

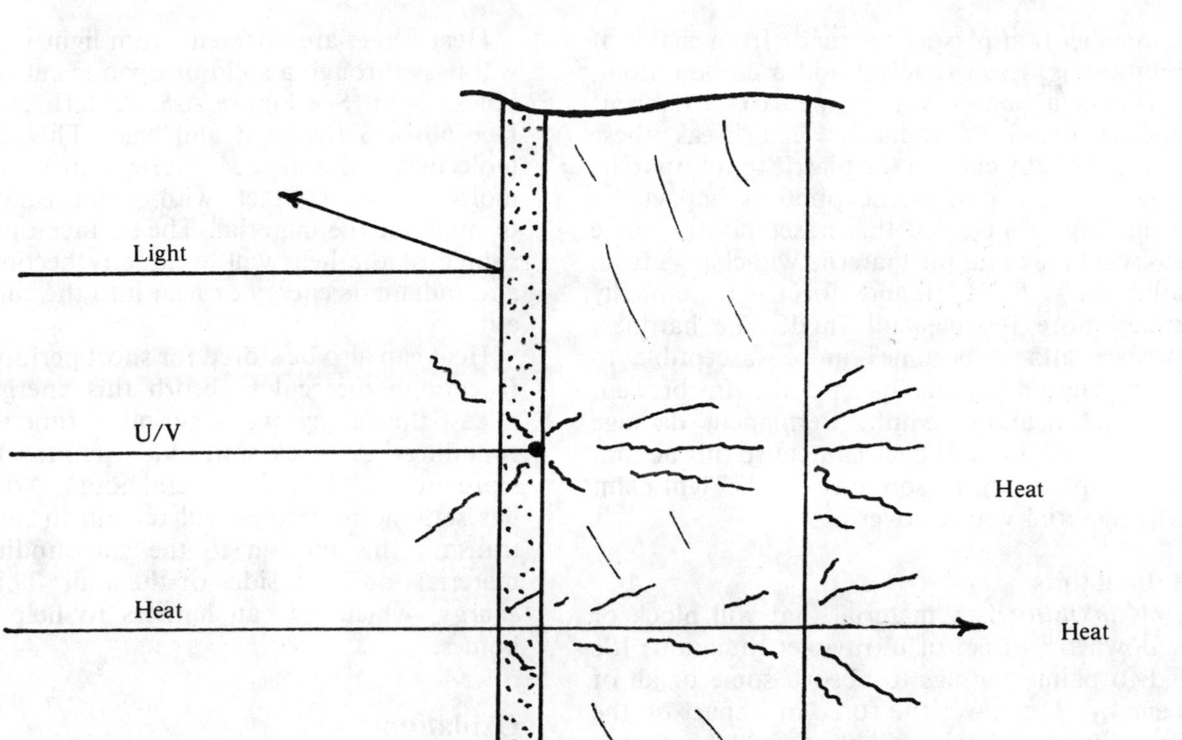

Reactions of light, U/V, and heat striking a surface
Figure A-5

back to their source is called *reflective*. Actually, all that we can see is to some extent reflective. If it wasn't, we couldn't see it. It would be invisible — like air.

One of the laws of nature is that an object in motion will remain in motion until an outside force stops it. Another is that an object in motion carries with it the amount of energy it took to set it in motion. On impact with another object it will give up part or all of this energy. When you throw a ball against the side of a building, your energy puts the ball in motion. The ball will transfer some of this energy to the wall. The remainder of the stored energy will cause the ball to bounce back toward you.

If you shoot a bullet at this wall, the bullet may have enough energy to move the building material aside at the point of impact. If the bullet is moving fast enough, it can force portions of the wall surface to break free and travel away from the spot of impact.

Ultraviolet Radiation

U/V radiation is a stream of photons with enough energy to do the same thing when a surface is intercepted. The photons hit electrons on the surface, knocking some loose from the atoms they're orbiting. Where does this U/V radiation come from? From the sun, which burns millions of tons of hydrogen every minute of every day, sending photons streaming into space.

Look again at Figure A-3. In the bottom half, you see the effect of heat on a hydrogen atom. When it's heated, the hydrogen atom gives off photons of light and ultraviolet radiation. These photons strike the electrons of other atoms, setting the electrons free. When this happens, the atom changes. At the same time, these free electrons can unite with other atoms to form gases, acids, bases, salts, and other matter. They're all destructive to the bond made between the subsurface and a paint coating that's applied to it.

Now look at Figure A-5. It shows the different reactions of visible light, U/V, and heat, when they hit a surface. The visible light tends to reflect from the surface. The U/V radiation will penetrate the paint film and gradually destroy the paint-to-subsurface bond. During this process, heat is generated. Some of this heat will transfer through the material to the opposite surface.

Remember that plastics are made from chains of carbon atoms. Each time we add a carbon atom, we form a new substance of different characteristics. U/V radiation can break these bonds, eventually causing the plastic to return to its original form, a gas. If the photons impact the cross-linking atoms of a thermoset plastic, some bonds will break and the material will change from a solid back to its liquid form. It gradually becomes more flexible and fluid. The hardness disappears and it becomes more susceptible to damage. Once these chains or links are broken, they rarely heal or re-link. Permanent damage results. Both U/V and heat can cause this action. This is an important reason to select the right paint for the material you're covering.

U/V Inhibitors

A *U/V inhibitor* is a material that will block or slow down the effect of ultraviolet radiation. It's added to paint coatings to absorb some or all of this energy. The lower the force of impact on the electron bonds, the less likely they'll be forced from their orbits. That means there's less chance of surface destruction.

Most clear coatings used for exterior surfaces have a slight amber color. This color may be from the base material, but it's probably the U/V inhibitor. A fine suspension of *iron oxide* or some other inhibitor has been added. Iron oxide will absorb or at least slow the U/V photons to a lower energy level. The iron oxide is added in just the correct amount to maintain transparency of the coating and yet slow the U/V destruction of the coating.

Note I said *slow* the destruction. It won't prevent the eventual destruction: Only an opaque material of the proper type can do this. At present there are virtually no materials or coatings that will fully stop U/V destruction over a period of time. Time is the key factor.

Manufacturers formulate the material to last a reasonable length of time for the application and service intended. The silicon paint coating inside the nozzle of a rocket engine only has to last a few minutes during the burn cycle to do its intended job. The paint coating on a building has to last long enough to satisfy the consumer, generally about eight to ten years.

Heat Transfer

Remember, photons release energy when they hit a surface. We perceive this energy as heat. That's why we feel warm when the sun shines on us — and why we get sunburned. The same thing happens to paint coatings. We can use this to our advantage.

Heat waves are different from light waves. Heat will pass through a solid or opaque substance that blocks light. See Figure A-5. A dark colored surface absorbs the light and heat. This excites the molecules in the surface. These in turn excite other molecules in contact with them, and so on, throughout the material. The surface opposite the source of the heat will eventually become excited and radiate its energy or heat into the surrounding air.

Heat can also be stored for short periods of time. If enough molecules absorb this energy, they'll release the energy over a period of time to the surroundings. A block wall that's painted black will store the sun's heat for several hours. After the sun has set, the molecules will remain in motion and transmit this motion to the surrounding air or materials on both sides of the wall. This is *solar energy*, which we can harness to help heat our homes.

Oxidation

Oxygen makes up one-fifth of the air we breath, one-third of the water we drink. It's required to sustain life. But it's also highly reactive and destructive. Oxygen will combine with almost all known elements. When it does, it forms acids, bases, and a host of other compounds. One common compound is *oxide.* The natural combination of oxygen with other substances is called *oxidation.* All substances on earth will eventually oxidize. Some may take hundreds of thousands of years. Others, like aluminum, begin to oxidize in seconds.

Once a material has oxidized, it's hard to bond it to another substance. The oxygen atom and its electrons have filled the valance, or electron rings, of the substance. There's no place left for another material to attach. So oxides are inert materials. That makes them a good protective substance.

When all the surface molecules have oxidized, very little additional oxygen can penetrate to attack the remaining structure. Since oxides are fairly inert and don't readily combine with other substances, they're not a good surface to paint. Clean oxides from a surface if you expect a paint coating to adhere. While oxides can be a problem, they can also be a blessing. Paint manufacturers use oxides in paints and primers because oxides don't readily combine with other elements. This makes them protective and less prone to color fade.

Paint Odor

As paint dries, the vehicle evaporates, releasing molecules into the air. It's these molecules that our nose senses as odor. As the carrier evaporates, the escaping molecules become fewer and the odor diminishes. These fumes are often highly flammable and sometimes very toxic. Ventilation of the area is required to reduce danger to the occupants.

As long as odors are being given off by a paint film, the paint is not dry. It's still in a semi-fluid state and can be damaged.

For most paints, a large part of the carrier evaporates during the first few hours after it's applied. The evaporation is nearly 90% complete within the first 24 hours. The remaining 10% can take several weeks to evaporate. This is because the surface evaporation has allowed the binder to film or skin over. The vehicle now has to escape *through* this film.

The important thing to remember is that there can still be odor weeks after the paint was applied — and the coating may still be soft weeks after being applied. Both conditions are normal for many paints. But the nose gets accustomed to odor very fast. So the odor should become unnoticeable in a few days.

If a paint coating much over a month old has an objectionable odor, there's a problem. It's usually not the paint that's causing the odor. It's what's growing on and consuming the paint, *bacteria* and *fungi.* The odor can also be from an external source like food or chemicals that have penetrated the surface film. Airborne smokes, oils, waxes, and greases can migrate into the surface very easily.

Paint coatings that are formulated with *fungicide, mildewcide,* or *Teflon* can help prevent bacteria and fungus growth. Teflon is Dupont's trade name for polytetrafluoroethylene, a very inert plastic that resists bacteria and fungus growth, water and chemical penetration, chemical attack, and U/V destruction. Very little will stick to the material, making it easy to keep clean. Some of the newer paints are now being formulated with Teflon to make use of these properties.

glossary

A

Abrasive A material such as sandpaper, emery paper, steel wool, or pumice that is used to wear away a surface by friction of rubbing.

Acetone A fast-drying, flammable solvent.

Acid stain A liquid stain made from organic acids that are soluble in water.

Acoustic paint A sound-absorbing coating.

Acrylic A paint in which the vehicle is an acrylic resin.

Acrylic resin Synthetic resin added to or used as a base for latex paints. It gives excellent color retention, weather resistance, and film flexibility to the material.

Adhesion The attraction or bond strength of a coating to a surface.

Air dry A coating that dries due to evaporation of the carrier.

Air entrapment Air bubbles remaining in the paint film.

Airless Spray equipment that uses hydraulic pressure on the liquid paint to force the paint through an orifice, atomizing it.

Alcohol A fast-drying, flammable solvent, commonly ethyl or methyl alcohol.

Alkali A chemical (lye, soda, lime, etc.) that will neutralize an acid. Oil-based paint films can be destroyed by the presence of a strong alkali or alkaline solution.

Alkyd resin A synthetic resin used to enhance hardness, gloss, and impermeability of coatings such as enamels. It is added to the coating or used as the vehicle.

Alligatoring Paint film breakdown that resembles the hide of an alligator. Caused by incorrect paint application.

Aluminum paint A liquid paint in which flakes of aluminum are mixed. Used for heat and flame retardation.

Amalgamate Chemical reconditioning of old paint or lacquer. Also used to remove white rings from lacquer that were caused by heat, water, or alcohol (amalgamator fluid).

Ambient temperature Temperature of one's surroundings.

Amide Epoxy resin curing agent.

Aniline dye A dye, bluish in color, poisonous, made from nitrobenzene.

Arcing Moving a spray gun in an arc rather than parallel to a surface.

Asbestine Trade name for a type of paint filler that is white in color. It is a natural fibrous magnesium silicate.

Asphalt Hydrocarbon product used as a protective coating, used for waterproofing road and roof surfaces. By-product of petroleum refining.

Atomize To break into a mist or droplets. Spray guns atomize paint by forcing the paint through a small orifice under high pressure.

B

Back prime The process of painting the back or unexposed side of material, such as the back side of exterior window shutters.

Backing guide Refers to the weight of the backing paper of abrasive papers. *A* is the lightest: finishing paper. *C* and *D* are medium weight: cabinet paper. *E* and *F* are the heaviest: used primarily for mechanical sanding.

Baking finishes Used primarily on metals. These coatings are baked with infrared, induction heat, or in an oven. Baking speeds up the drying process and produces a hard, smooth finish. Paint is formulated especially for the process.

Ball mill A machine used to mix pigment into the carrier vehicle. It is a large cylindrical drum filled with round balls that fall through the mixture as the drum is rotated.

Barytes A base or extender pigment of barium sulphate, a colorless crystalline insoluble compound. It is opaque to x-rays.

Benzene A toxic, flammable solvent. Its use is restricted.

Binder The carrier or vehicle that binds the pigment together. For example, the oils, varnishes, and emulsions that make up the liquid portions of paint.

Bitumen Another hydrocarbon product. It's used much the same as asphalt.

Bleaching The use of oxalic acid or other bleaching agents to lighten or restore discolored or stained wood to its original color.

Bleeding Paint discoloration caused by leaching of subsurface dyes, acids, stains, or rust.

Blistering Enclosed raised spots on a painted surface resembling a blister. Caused by poor surface prep (usually water or oil under new paint) or poor painting techniques.

Bloom A rainbow-like surface caused by excessive humidity before the painted surface is completely dry. Will disappear on drying.

Blown oil A vegetable oil whose viscosity has been increased by blowing air through it.

Blue lead A rust preventer that's bluish in color. Contains lead sulfide and carbon.

Blushing White discoloration due to excessive humidity; usually affects lacquer-type varnishes.

Body The thickness or viscosity of a fluid.

Boiled oil Oil such as linseed oil that has been boiled so that it becomes a drying oil.

Boxing Act of pouring the paint back and forth from container to container. This improves the consistency by insuring a proper mix of vehicle, pigments, and fillers.

Bridging Property of a paint that fills all cracks, pores, and voids of a surface. Bridging is not desired when painting acoustical ceilings.

Brightwood Non-pigmented clear glossy finishes, such as a marine deck coating.

Bronzing liquid Paint that has particles of bronze powder suspended in it.

Brush hand A painter who is an expert in applying paints with a brush.

Brushability Term given to a paint, varnish, or enamel that indicates the ease or difficulty with which it is applied to a surface by brushing.

Brushing Act of applying paint using a brush.

Brush-in stick Heat-resistant wax stick used for cabinet and furniture touchup. Must be melted at 140 degrees F or 200 degrees F into the surface. Sandable and tintable.

Brushing lacquer A slow-drying lacquer that can be applied to a surface by brushing. Because it dries slowly, it levels and does not show brush marks. Used mainly by the do-it-yourself furniture refinisher.

Bubbling Blisters that form on a newly-painted surface. Caused by trapped air or moisture.

Build or build up The thickness of the coating caused by successive application of coats. May or may not be desirable.

Burnishing Glossy or shiny spots on a painted surface caused by rubbing, washing, wiping, and scrubbing.

Burnt sienna A pigment that's yellow when raw, reddish brown when burnt. Made of iron and manganese oxides.

Burnt umber A pigment, darker in color than sienna, moderate to dark brown when burnt. Made from iron and manganese oxides.

C

Caking When pigment settles and cakes hard in the bottom of the paint can.

Calcium carbonate A natural salt used in lime, cement, and paints as a colorant and an extender. Its common name is chalk.

Casein A protein derived from souring of milk, used to make cheeses, and as an additive to paints, adhesives and plastics.

Cat eye Also called *cat face.* See ***Holiday.***

Catalytic coating A finishing coat that hardens by reaction instead of evaporation. For example, a two-part epoxy where a hardener must be added to obtain the required results.

Caulking compound A pliable and elastic material that can withstand expansion and contraction. Used to fill voids, cracks, and seams to prevent air or water infiltration. May or may not be paintable.

Cellulose acetate Water-insoluble compound made from the chemical reaction of acetic and/or sulfuric acid on cotton or other cellulose materials. Used in manufacture of varnish, and as a paint binder.

Cellulose nitrate An ester of nitric acid made from the chemical reaction of nitric acid on cellulose. Also known as pyroxylin or nitrocellulose. It is used as a paint binder, and in the manufacture of varnish, photographic film, and explosives. It is flammable and soluble in an organic solvent such as alcohol and ether.

Chalking A form of oxidation of paint, usually due to weathering.

Checking Cracks in wood or paint surface that follow the grain of the wood and then crack across the grain.

Chime The chime is the area of the lip or rim of a paint can to which the lid seals.

China clay Fine clay pigment used as an extender. It aids in abrasion resistance.

Cleaner A solvent for cleaning; may be a base, an acid, an alkali or a detergent.

Close-grain wood Hardwoods that when fully dry do not show pores. Cherry, birch, and maple fall into this group. Excellent furniture material.

Clouding Murky, dull, or uneven luster or color. Generally caused by porous subsurface or undercoat.

Cloudy Surface that has not been properly and uniformly coated.

Coal tar pitch Black residue from distilled coal tar.

Coal tar solvent Solvent made from distillation of coal tar: benzene, toluene, xylene, naptha.

Coating in Applying a coat of paint to a surface so that the surface is coated uniformly.

Cohesion A bonding together of two or more items; i.e. paint to surface

Cold-checking Checking caused by cold temperatures.

Cold cracking Cracking caused by cold temperatures.

Cold water paints Casein, glue, or similar materials dissolved in water. Used as a plaster, masonry, or concrete surface coating.

Color-fast Nonfading.

Color person The person who is an expert in the mixing, tinting, and matching of colors.

Color retention Ability to retain original color.

Corrosion inhibitive An additive to paint or primer that aids in the prevention of corrosion or rust. Red lead, zinc chromate, and zinc dust fall into this category.

Corrosion resistant A paint or primer that aids in the prevention of corrosion. An insulator against water vapor and airborne contaminants such as sulphur compounds.

Cracking The first stage of checking; paint cracks with the grain of the wood. It can also be caused by subsurface cracks and splits.

Crawling Uneven surface of paint caused by the paint shrinking during the drying process.

Crazing Minute interlacing cracks that appear on the surface of a varnished or painted object. Can be caused by using the wrong cleaner, detergent or solvent.

Creeping Paint runs together into small droplets because the subsurface is excessively hard, dirty, glossy or waxy.

Cross spraying Spraying in one direction and then again at a right angle to that direction.

Curing Becoming cured, hard, or set.

Curtain Running or sagging of paint into a scallop or curtain-looking design.

D

Dead flat No gloss at all.

Decalcomania (decal) Transfer of pictures or decorations from a backing sheet to a surface.

Delamination Paint layers that did not adhere to each other and have separated. Usually caused by poor surface prep or no primer.

Density Weight per unit volume.

Dew point Temperature at which moisture condenses.

Dipping The act of submerging a surface or article into the coating or paint.

Doctor blade Also called *spreader blade.* A blade used to give a set film thickness during machine coating of panels, coiled steel, etc.

Drag The failure of a paint or coating to slide off the brush or roller evenly and smoothly.

Driers Various metal compounds added to paints and varnishes to hasten the drying action.

Drop One vertical descent of a scaffold.

Drop cloth Cloth made from canvas or plastic used to cover and protect items from paint splash or splatter.

Dry colors Powder-type colors to be mixed with water, alcohol, or mineral spirits to form a paint or stain.

Dry to handle Paint that has dried sufficiently to be handled without being marred.

Dry to recoat Paint that has dried sufficiently to receive the next application.

Dry to touch Paint that has dried enough that light touching will not leave paint on fingers.

Drying oils An oil that when exposed to air will dry to a solid: linseed oil, tung oil, perilla, fish oil, soybean oil.

Drying time Time required for a coating to dry once it's been applied.

Dull finish Just a little gloss, almost dead flat.

Dull rub Mostly a furniture finish, where gloss is rubbed to a mar-free dull shine.

Dumping Applying paint very fast to a surface. The term usually refers to airless spraying.

Dust-free Condition of area used for painting when all dust has been removed; surface condition where dust will no longer stick.

Dye colors Colors that are dissolved, not mixed with water, alcohol, or mineral spirits, to form a paint or stain.

E

Earth pigments Those pigments that are obtained from the earth, including barytes, ochre, chalk, and graphite.

Edging Act of stripping in or painting near the edge of a surface, such as the wall intersection at ceiling, doorway, or window.

Edging stick Specially-formulated sticks used to touch-up or finish raw edges of countertops and cabinetry.

Efflorescence Salt rising to the surface of masonry or plaster, or cement. Paint will not adhere to a surface that contains efflorescence.

Eggshell The sheen or luster of a painted, finished surface that resembles that of an eggshell.

Elasticity Springiness or the ability to be stretched and then return to shape without damage. This property in some paints makes them resistant to dents, chips and scratches.

Emulsion A mixture or suspension of fine particles in a carrier or vehicle. Usually the solids are mixed with one liquid such as oil and then this mixture is suspended or emulsified into a second liquid such as water.

Emulsion paint A paint with the pigment suspended in the vehicle.

Enamel A paint that flows on in a smooth coat and dries to a very hard, glossy finish. Most equipment-coating enamels require baking. Enamels for walls do not.

Epoxy A catalytic coating that is extremely tough, durable, and chemically impervious.

Etching The process of preparing a metal or glass surface by application of an acid. The etched surface that results permits better paint adhesion.

Extender A low hiding-power pigment added to paint to increase coverage. May also be added to bind pigment particles of different shapes and sizes.

F

Fading Reduction in color brightness.

Fanning Movement of spray gun to surface when one's arm moves back and forth in a fan-like motion.

Fat edge A term used to describe a buildup of paint on the edge of an object caused by poor painting techniques.

Feathering The blending of one surface into another. In painting, the act of lifting the brush at the end of a stroke that results in an undetermined edge. In sanding, the blending of old to new surface or coating.

Feel The working quality of a paint: how it spreads, covers, and dries. Painters like to use a paint that *feels* right to them.

Ferrous metal A metal that contains iron. Because it is subject to rust, special primers are required.

Ferrous sulfate A salt used as a pigment in paints and inks.

Field painting Doing your work in the field, at the job site.

Filler An additive or extender to paint to give certain qualities. Also a material used to fill holes, pores, and cracks in a surface before the application of paint.

Film build Progressive recoating to build up in layers the required paint protection. Excessive buildup can lead to paint destruction because different layers of different paints expand and contract at different rates, causing separation.

Film integrity Integrity or continuity of the paint film.

Film thickness Depth or thickness of the coating in millimeters.

Filter Cloth or other media through which one passes a substance in order to purify that substance. Paint should always be filtered before being used in spray equipment.

Fingering Misapplication of paint spray with an airless. Low pressure causes two outside bands of thick paint with very little paint in between. Or fingering due to "dumping" or applying too much paint to a given area, causing the paint to puddle and run. Fingering can also happen with a brush when the brush is poorly cleaned. Fibers stick together in bundles, causing the paint to go on in streaks that resemble fingers.

Fire retardant Paint that aids in extinguishing a fire or prevents flame spread. Modified paint that contains flaked metals and/or nitrogen compounds.

Fish oil Oil from fish, used as a drying agent in rust-preventive primers.

Flaking Paint that breaks off or flakes. May be caused by poor subsurface preparation or poor paint.

Flammability The ease at which a material burns. Many paints are being taken off the market because of their high flammability. Some states require special licenses for the use of highly flammable lacquers.

Flash Color variation due to surface porosity. Paint is sucked into the surface at different rates. Usually caused by poor or no primer or sealer.

Flash point Lowest temperature at which a material will flash or burn explosively if a spark or flame is present. Lacquers have a low flash point, so they're dangerous to use. Many solvents, vehicles, and paints are now banned because of their low flash points.

Flat wall paint A paint, usually interior type, that dries to a flat lusterless finish. Best for hiding surface imperfections. Ideal as a top coat or as an undercoat for more translucent semi-gloss and full gloss paints.

Flatting agent Pigment added to reduce gloss or give a "rubbed" look. Some flatting agents are zinc stearate, calcium, and aluminum.

Floor varnish Varnish formulated with abrasive-resistant pigments, used specifically on floors.

Flowing varnish Varnish formulated to produce a smooth lustrous finish.

Foots The pigments, additives and other solids that settle on the "foot" or bottom of the paint can.

Forced dry Baking the paint to speed the drying process. See *Baking finishes.*

Furane resin A resin obtained from pinewood oil. Flammable, dark in color. Used for paints and the manufacture of nylon.

G

Gel Top coat in fiberglassing. Usually the color coat, and it usually contains a mold-release wax.

Ghosting Nonuniform sheen of paint resulting in a shadowed effect. Usually caused by lack of a primer or sealer, or poor quality ones.

Glazing Also called *antiquing*. Obtained by painting on a second, semi-opaque color over a base color, then wiping off to reveal base color.

Gloss oil Varnish composed of limed rosin and petroleum thinner. Dries to a high shine.

Glossy finish Varnish or enamels that do not contain flatting agents and therefore dry to a hard, glass-like gloss.

Grain The natural look or pattern of wood. Texture due to alignment of the wood fibers.

Grain raising The swelling of wood fibers due to the application of water or water-based paints.

Graininess Paint film that does not dry smooth. It looks as if sand were mixed in.
Graining See *glazing.* Method of making non-wood or poor wood look like real or good wood by application of several coats of different color paints.
Grit number Measurement of the coarseness of abrasives. Refers to number of open spaces per linear inch on standard control screen. #12 is the coarsest. #1200 is the finest.
Ground coat First or prime coat used as a base for all other coats of paint.
Guide coat Coat of paint first applied to a surface, just slightly different in color from the final top coat. Aids in obtaining complete coverage of top coat.
Gum arabic A water-soluble gum used as an adhesive or binder.
Gypsum Calcium sulfate, used as an extender pigment and in the manufacture of wallboard and plaster of paris.

H

Hairline cracks Fine cracks in plaster or paint film.

Hard oil finish Any interior varnish that dries to a moderate or high luster, as if rubbed with oil, and has an extremely hard film.

Hardener Curing agent for epoxies or fiberglass.

Hardwood A term given to wood from non-coniferous trees (trees that do not bear cones). Not a true description, as many hardwoods are softer than so-called softwoods.

Hazing Clouding of the finish. Surface appears to have lost its gloss, or has a smokey appearance.

Hide glue A strong general purpose wood glue made from animal hides. Hide glue is used primarily for furniture but is being replaced by synthetic resin glues.

Hiding power The ability of the paint or coating to mask or cover the subsurface.
Holiday Skips in paint coverage caused by insufficient paint, poor equipment, or poor surface prep.
Hot spray Paint heated to reduce its viscosity so it flows or atomizes more easily.
Hot wall A plastered wall heavy in free lime. Lime is highly destructive to oil-based paints.
House paint Exterior paint formulated to withstand the elements. It is primarily used on houses, barns, fences, and other outbuildings.
Humidity Moisture content of the air.

I

Inert A material that will not react chemically with other ingredients, used in the formulation of paints, varnish, or enamel.
Insulating varnish Varnish formulated so it doesn't conduct electricity. Used to coat terminals, coils, transformers, etc.

Internal mix Exclusive to conventional spray-gun systems, where the air and the paint are mixed in the gun and then exit as a spray.

Iron blue Prussian blue.

Iron phosphate coating Conversion coating, used as a primer/cleaner in industrial equipment painting. It converts a nonpaintable surface to a paintable surface.

Isocyanate resin Salt of isomeric cyanic acid, used as a binder in paints.

J

Journeyman painter An apprentice who has had at least three years of experience and instruction.

K

Kerf The path of a saw blade through a piece of wood that does not result in cutting the wood into two pieces.

Ketone Organic compound or solvent, highly flammable. Acetone is one example.

L

Lacquer Type of paint product that dries by evaporation of solvents, such as cellulose nitrate.

Laitance Milky-white deposit on new concrete, caused by lime and salts rising to the surface with the evaporating water. Paint will not stick to it.

Lampblack Black soot used as a pigment.

Latex Term used for water-based paints. A type of rubber.

Lead carbonate Metal primer or pigment, white lead.

Lead drier Paint additive to enhance drying. Made of lead and an organic acid. Examples: lead linoleate and lead naphthenate.

Lead oxide Lead monoxide and red lead; basically, lead combined with oxygen, used as a pigment and rust preventer.

Leaded zinc Primer made from zinc oxide and lead sulfates.

Leveling Paint film that dries level, so it does not show brush marks. Many paints have additives that aid in leveling.

Linoleum varnish Varnish formulated to be highly elastic and flexible.

Linseed oil Drying oil made from the flax plant. Used as a vehicle for many oil-based paints. "Boiled" linseed oil can be used to protect wood from water damage. Sometimes used as a furniture polish.

Liquid driers Solution of driers in paint or paint thinners.

Liquid wood filler Low viscosity varnish used as a filler coat on open grain woods. Usually contains an extender and may be stained. Acts as a hard, non-porous coat for additional top coats.

Litharge Lead monoxide, used in vitreous enamels.

Lithopone A white pigment of barium sulfate and zinc sulfide.

Livering Paint that has thickened to a liver-like mass.

Long oil A high portion of oil to resin is used, thus it is long in oil.

Long oil varnish A high oil-content varnish with a slower drying time and a tougher, more elastic coating than short oil varnish. Spar varnish is an example.

M

Maintenance painting Repainting to repair or renew on a regular basis.

Marine varnish Varnish formulated to withstand long exposure or immersion in fresh or salt water.

Masking Covering an area not to be painted.

Masking tape Painters tape, pressure-sensitive paper tape which is easily removed after painting. Allows for a clean sharp break between old surface and newly-painted surface areas.

Mildew A growth on damp surfaces produced by fungi.

Mills Machine used in the manufacture of paints.

Mineral spirits Paint thinner. Solvent distilled from petroleum.

Miscible Able to be mixed in any ratio without separation.

Misses Voids or skips in a painted surface.

Mist-coat Very thin spray coat.

Mobility Ease with which the paint flows.

Mottling Speckling.

Mud-cracking Paint or plaster applied too thick; cracks like dried mud.

N

Naphtha Paint thinner from distilled petroleum.

Natural resins Resins from trees. Damars and copals.

Neoprene Synthetic rubber very resistant to oils. Used as a paint modifier.

Nitrocellulose Another term for cellulose nitrate.

Non-volatile Non-evaporative paint solids.

O

Oil-modified urethane Air-drying type of urethane that contains no hardeners.

Oil stain Oil or varnish that has been blended with colorant dyes.

Oleoresin A varnish or solvent made from oil and resins. Turpentine is an example.

Opacity Ability to hide, not see through. A paint with a high opacity value will cover well.

Open-grain wood Wood in which the pores are easily seen: walnut, oak, mahogany.

Orange peel Paint that has dried to a orange peel texture. Generally caused by poor spray painting techniques.

Orange shellac Shellac that has not been bleached. It is amber in color.

Orifice An opening or hole, especially in spray equipment.

Overcoat The top, or finish coat.

Overlap Each pass of the roller, brush, or spray overlaps or covers a portion of the prior applied coating.

Overspray Sprayed paint that doesn't land on the item being painted.

Oxidation Chemical reaction upon exposure to oxygen. Includes rust on metals, chalking of paint.

P

Padding Act of applying stain, lacquer, or polish to a surface with a padding pad, using a pendulum motion to wipe material on a surface.

Padding pad Pad made from trace cloth filled with cotton. Used in padding to apply material to a surface. It is a furniture refinishing term.

Paint Product developed for the beautification and protection of articles. A volatile carrier that has inert ingredients in suspension.

Pass The application of paint in a single line. It takes multiple passes, generally overlapping, to cover an item.

Paste filler Wood filler in paste form.

Patchal stick No-heat putty stick for touch-up work on furniture and cabinets. Colors to match most cabinet lacquers and stains.

Pattern length Spray pattern length, usually governed by arm reach or motion.

Pattern width Spray pattern width, governed by the orifice of the spray tip, the paint, and the distance from the object being painted.

Peeling Paint film that peels off in large segments, generally due to a surface that wasn't scuff-sanded or degreased.

Penetrating finish A finish that sinks into the surface as opposed to settling on the surface.

Permanent colors Tinting colors. Color pigment in an oil base that is paste-like.

Phenolic Resin made from formaldehyde and phenol.

Piano-like finish Rubbed and polished lacquer or varnish. Highest grade of wood finish.

Pick up sags Rebrushing or rerolling wet paint that has begun to sag or curtain.

Pickling Cleaning process for metals, usually by dipping into a mild solvent or acid solution.

Pigment Insoluble, finely ground minerals that are added to the paint carrier. They give the paint its properties of color and opacity.

Pinholing Also called *pitting*. Small holes in the paint film. May be caused by changes in atmospheric conditions during drying cycle, mixing noncompatible materials, or poor surface prep.

Plastic finish The type of finish given to articles coated with varnish, lacquers, vinyls, acrylics, and polyurethanes. Plastic in appearance.

Pockmarks Pits.

Pole gun Extension added to a spray gun to increase reach.

Polishing Act of polishing. Also the shiny spots on surfaces that have been rubbed or scrubbed to a shine.

Polymerization The interlocking of molecules of a substance by chemical reaction. Chief process in the making of plastics and plastic-based paints.

Polyurethane An oil-modified urethane that air-dries without the aid of a hardener.

Polyvinyl acetate (PVA) Drywall primer. A resin that is highly stable, durable and abrasion resistant.

Porous surface A surface that will readily absorb water or liquids.

Poster color Water-based colors used primarily in posters.

Pot life Usable time for two component paints after being mixed.

Powdering Term usually refers to varnish. The varnish through age and exposure crumbles into a powder or dust.

Prime in the spots To apply primer to those areas where the surface has been stripped by mechanical or chemical means.

Primer The first coating to be applied. An undercoat used to clean the surface by an etching action, and then provide a clean surface for the top coat to bond to.

Priming The act of applying primer.

Print free Paint that is sufficiently dry that it will not retain marks if a hand or brush is pressed into it.

Protective life Time interval from application of a coating to the point when the coating no longer protects the surface.

Pumice Fine rubbing compound for finishing varnished or enameled surfaces.

Putty Form of caulking used to set glass into window frames. Generally a mixture of linseed oil and whiting. Can be used as a wood filler.

Putty coat One of the terms given to the final smooth coat of plaster.

Pyroxylin Cellulose nitrate. Base for many paints and lacquers. It's flammable, soluble in ether and alcohol or organic solvents.

Q

Quick drying A material or paint that contains modifiers to aid in drying.

R

Raw oil Oil separated from solvent in solvent extraction process.

Recoat time Minimum waiting time before the next coat of paint may be applied. It depends on the type and brand of paint. The time is usually specified on the label.

Red label Label used to identify materials that have a flashpoint under 80 degrees F. They're highly flammable.

Red lead A drying agent and/or a rust inhibiting pigment. Lead oxide is an example.

Refined shellac Orange or white shellac that has been refined. Wax has been removed.

Remover A liquid or paste formulated to attack and destroy the paint or paint-to-surface bond so that the paint can be removed.

Resin Synthetic or vegetable material used as a base for most paints. Can be translucent or transparent, solid or semi-solid.

Respirator Mask worn to protect one from fumes or dust.

Ride the brush To bear down hard with the brush. Causes premature brush failure and streaking.

Ropey Paint that dries with brush marks or ridges because it lacks levelers or dries too fast because it was applied at a high temperature.

Rosin A pine tree sap resin.

Rottenstone Very fine mineral powder used for obtaining a hand-rubbed effect.

Rubbed finish The finishing of enamel, lacquers, and varnishes by rubbing with oil and a fine abrasive. Produces a low luster gloss.

Rubbing varnish Any varnish that can be rubbed to a low luster finish.

Runs Uneven flow or leveling of paint resulting in sags. Caused by applying too much paint at one spot or spray painting too close to surface.

Rust The corrosion product formed in the oxidation of ferrous metals, mainly iron. You have to remove it before painting because paint won't adhere to it.

S

Sags See *Runs*

Sand down Act of removing gloss from a surface via chemicals or abrasives. It cleans and abrades the surface so that new paint adheres better.

Sand finish Pebble or sand-like texture applied to wall surface.

Sandblast Abrasive cleaning of a surface by blowing sand against it at high velocity.

Sanding sealer Sealer formulated to make sanding easier. It hardens wood fibers so they cut off rather than bending out of the way.

Sap-streak Exposed streak or void in wood surface that is filled with sap. Must be removed, filled in, or coated over prior to painting.

Scale Thin layers of flaking rust. Must be removed before prime or other coating is applied. Special paints are now being formulated to convert scale and rust to paintable surfaces. A light wire brushing is recommended first.

Scuff sand Very light sanding of surface.

Sealer Specially formulated paint that fills pores of surface prior to top coating.

Seedy Also called *sandy.* A newly painted surface with specks of dirt, old paint, or partially dried paint that were transferred with a dirty roller or brush.

Semi-gloss finish Finish that has a low luster sheen. Semi-gloss paints are formulated to give this result.

Set Initial hardening of the paint film before the final hardness is obtained.

Settling Pigment separates and cakes or settles at the bottom of the can.

Set-up A paint film that has set or filmed over and hardened.

Shade Degree of color.

Shadowing Also called *show through.* Prior coating shows through the new coating due to the low pigment content of full gloss enamels or inexpensive paints. May also be due to painting a light color or pure white over a darker color.

Sheen The degree or level of shine the dried paint exhibits.

Shelf life Usable life of product before it deteriorates. Applies to both unopened and opened cans of paint or bags of dry paint mix.

Shellac A coating made from purified lac dissolved in alcohol, often bleached white.

Shiner A glossy spot on an otherwise non-glossy surface. Can be caused by spot priming patched areas, poor wet-edge lapping, or spot painting with poorly mixed or unmatched paint.

Shop primed Articles that have been preprimed by the manufacturer. Garage doors are usually shop primed.

Short paint Refers to a paint that does not exhibit good properties and therefore is *short.*

Silica Ground sand or ground quartz used as an extender.

Silicone Liquid silicone oxides. Paints containing silicone are very slick and resist dirt, graffiti and bacterial growth.

Size A sealer, usually used before application of wallpaper.

Skin The film, or top layer that forms when paint dries.

Skippy A paint that has a lack of vehicle and too heavy a viscosity, making it skip on some surface areas while piling up on others.

Slip under the brush A paint that is slippery or too easy to apply, making it difficult to control application.

Slow-dry A paint or coating that takes more than 24 hours to dry between process steps: repainting, sanding, etc.

Softwood Term given to wood that comes from cone-bearing trees. The wood is not always soft.

Solids The part of the paint that remains on the surface after the vehicle has evaporated. The dried paint film.

Spackling compound Dry power that forms a paste when mixed with water. Used to patch holes, cracks, nicks in plaster or drywall.

Spalling Paint that breaks up into small chips.

Spar varnish Varnish that has been formulated especially for outdoor use.

Specular gloss Mirror-like finish.

Spirit stain Dissolved dye in alcohol.

Spirit varnish Disssolved resin in solvent.

Splitting A form of *alligatoring,* where topcoat was applied over an incompletely dried sealer or prime coat.

Spray Aerosol, spray gun, or airless application of paints.

Spray can The container that contains the pigments and vehicle of paint in a spray system. Term does not refer to airless systems unless the container is an integral part of the system.

Stain Generally a dye rather than a paint, used to change the natural color of wood. Stain is absorbed into the material rather than resting on the surface.

Stand oil Heat-thickened oil.

Stretch The size of the area that the painter can reach without changing location.

Strip To completely remove old finish from the subsurface by mechanical or chemical means.

Stroke Single pass of spray in one direction.

Substrate The subsurface or base surface that is to be painted.

Suction Absorption or porosity of a surface. Generally refers to a plastered wall. See *Flash.*

Surface drying A paint that dries from the outside in. Not a good quality as the paint will skin over, blocking or slowing further drying.

T

Tack Slight stickiness on the surface of the paint before it is set.

Tack rag A chemically-impregnated cloth used to remove dust from a surface just prior to application of coating.

Tacky A paint film that has dried to the point where dust will not stick, but it is not yet set hard.

Tails Another term for "fingering" when spray painting.

Talc Paint extender, white in color, with a slippery feel. Magnesium silicate.

Tank white White paint that gives good coverage on exterior metal. Self-cleaning.

Tar Residue from distillation of coal, wood, shale, peat, or other vegetable or minerals. It's used as waterproofing below ground on walls, pools, and lumber.

Tempera Water paint.

Thinner Paint solvent that is used to thin or regulate the viscosity or consistency of the paint.

Through drying Term used to denote that the paint has dried completely, usually several hours to days after application.

Tie coat A coating formulated to be an intermediate coat between two nonbonding coatings. An intermediate bond coat.

Tint A color that is not pure color: a shade or chroma of the color that is not saturated due to the addition of white.

Tinting Also called *shading.* Act of changing the intensity of a color or the hue of a color by the intermixing of two or more colorants.

Tipping off Smoothing the coating of wet paint by gently brushing the tip of the brush over the surface.

Titanium dioxide White pigment.

Tooth Name given to the ability of a primer to bite into the subsurface and to bond the top coat securely to that subsurface. The holding power of a primer due to a slight roughness of the film.

Top coat The last or final coating of a paint or varnish, sometimes referred to as "overcoat."

Touch up The act of repairing a small spot that has been chipped or otherwise marked. A small quantity of paint that is kept in reserve for future repairs to the painted surface.

Toxic Poisonous.

Trace cloth Special closely woven, soft-textured, absorbent, lint-free cloth used to make padding pads.

Tung oil Oil of the tung tree, used as a drying oil in fine wood-finishing paints.

Turpentine Distilled pine oil, used as a cleaner, solvent or paint thinner for oil-based paints.

Tuscan red Pigment of iron oxide and a lake.

U

Umber See *Burnt umber.*

Undercoat A coating that is not the top or final coating of paint.

Undertone A subdued color; a color that modifies another color. The color of paint viewed through transmitted (reflected) light. It may be caused by using a top coat that's too thin or that's lacking in solids.

Urethane A solvent used in the manufacture of polyurethane.

U.S. gallon Four liquid quarts or eight liquid pints. The U.S. pint is 16 ounces, so the U.S. gallon is 128 ounces. The Canadian, or Imperial gallon, is comprised of eight 20-ounce pints, so the Imperial gallon is 160 ounces.

Useful life Life expectancy of coating before refinishing is required.

V

Value The lightness or darkness of a color when compared to white.

Varnish Type of clear coating.

Varnish stain A varnish coating that has a stain dissolved in it.

Vegetable oils Oils obtained from vegetable growth, as linseed, soybean, hempseed, tung, castor, perilla. Used as drying oils and carriers.

Vehicle The liquid carrier of paint that evaporates upon drying.

Vinyl A synthetic resin used as a base for paints.

Viscosity The thickness or resistance to flow of a paint.

Volatile Term given to a liquid that denotes its ability to evaporate quickly when exposed to air.

W

Wash primer Rust prohibiting paint of thin consistency.

Water blaster Machine that converts low pressure water to high pressure (several thousand PSI); used to clean or remove paint from a surface prior to painting.

Water spotting Spots left behind when water evaporates. Most likely white or brown in color. Caused by dissolved minerals remaining on the surface.

Water stain A dye that is dissolved in water. A durable stain made from water-soluble dyes.

Water-thinned paint Any paint that uses water as a vehicle.

Wet edge The paint edge of a section of wall that remains wet long enough that the next section can be blended into it without lap marks.

Wet film thickness Depth of film in millimeters while paint is still wet.

Wetting characteristics The ability of a paint or primer to flow over and saturate a surface.

White lead A white pigment.

White shellac Shellac that has been subjected to bleaching.

Whiting A white pigment.

Wire brushing Hand process for removal of loose rust, paint, and other contaminants.

Wood filler See *Paste filler.*

Wrinkle finish A novelty finish, obtained by mixing two dissimilar paints together or by using enamel over lacquer.

Z

Zinc chromate Zinc yellow.

Zinc dust Ground zinc metal, gray in color, used in metal primers.

Zinc oxide A white pigment.

Zinc sulfide A white pigment.

Zinc yellow Zinc chromate used in metal primers. It is yellow in color.

index

Other Practical References

Paint Contractor's Manual
How to start and run a profitable paint contracting company: getting set up and organized to handle volume work, avoiding the mistakes most painters make, getting top production from your crews and the most value from your advertising dollar. Shows how to estimate all prep and painting. Loaded with manhour estimates, sample forms, contracts, charts, tables and examples you can use. **224 pages, 8½ x 11, $19.25**

Drywall Contracting
How to do professional quality drywall work, how to plan and estimate each job, and how to start and keep your drywall business thriving. Covers the eight essential steps in making any drywall estimate, how to achieve the six most commonly-used surface treatments, how to work with metal studs, and how to solve and prevent most common drywall problems. **288 pages, 5½ x 8½, $18.25**

Spec Builder's Guide
Explains how to plan and build a home, control your construction costs, and then sell the house at a price that earns a decent return on the time and money you've invested. Includes professional tips to ensure success as a spec builder: how government statistics help you judge the housing market, cutting costs at every opportunity without sacrificing quality, and taking advantage of construction cycles. Every chapter includes checklists, diagrams, charts, figures, and estimating tables. **448 pages, 8½ x 11, $24.00**

Builder's Guide to Accounting Revised
Step-by-step, easy to follow guidelines for setting up and maintaining an efficient record keeping system for your building business. Not a book of theory, this practical, newly-revised guide to all accounting methods shows how to meet state and federal accounting requirements, including new depreciation rules, and explains what the tax reform act of 1986 can mean to your business. Full of charts, diagrams, blank forms, simple directions and examples. **304 pages, 8½ x 11, $17.25**

Contractor's Survival Manual
How to survive hard times in construction and take full advantage of the profitable cycles. Shows what to do when the bills can't be paid, finding money and buying time, transferring debt, and all the alternatives to bankruptcy. Explains how to build profits, avoid problems in zoning and permits, taxes, time-keeping, and payroll. Unconventional advice includes how to invest in inflation, get high appraisals, trade and postpone income, and how to stay hip-deep in profitable work. **160 pages, 8½ x 11, $16.75**

Manual of Professional Remodeling
This is the practical manual of professional remodeling written by an experienced and successful remodeling contractor. Shows how to evaluate a job and avoid 30-minute jobs that take all day, what to fix and what to leave alone, and what to watch for in dealing with subcontractors. Includes chapters on calculating space requirements, repairing structural defects, remodeling kitchens, baths, walls and ceilings, doors and windows, floors, roofs, installing fireplaces and chimneys (including built-ins), skylights, and exterior siding. Includes blank forms, checklists, sample contracts, and proposals you can copy and use. **400 pages, 8½ x 11, $18.75**

How to Sell Remodeling
Proven, effective sales methods for repair and remodeling contractors: finding qualified leads, making the sales call, identifying what your prospects really need, pricing the job, arranging financing, and closing the sale. Explains how to organize and staff a sales team, how to bring in the work to keep your crews busy and your business growing, and much more. Includes blank forms, tables, and charts. **240 pages, 8½ x 11, $17.50**

Remodeler's Handbook
The complete manual of home improvement contracting: Planning the job, estimating costs, doing the work, running your company and making profits. Pages of sample forms, contracts, documents, clear illustrations and examples. Chapters on evaluating the work, rehabilitation, kitchens, bathrooms, adding living area, re-flooring, re-siding, re-roofing, replacing windows and doors, installing new wall and ceiling cover, re-painting, upgrading insulation, combating moisture damage, estimating, selling your services, and bookkeeping for remodelers. **416 pages, 8½ x 11, $18.50**

National Construction Estimator
Current building costs in dollars and cents for residential, commercial and industrial construction. Prices for every commonly used building material, and the proper labor cost associated with installation of the material. Everything figured out to give you the "in place" cost in seconds. Many time-saving rules of thumb, waste and coverage factors and estimating tables are included. **512 pages, 8½ x 11, $18.50. Revised annually.**

Carpentry Estimating
Simple, clear instructions show you how to take off quantities and figure costs for all rough and finish carpentry. Shows how much overhead and profit to include, how to convert piece prices to MBF prices or linear foot prices, and how to use the tables included to quickly estimate manhours. All carpentry is covered: floor joists, exterior and interior walls and finishes, ceiling joists and rafters, stairs, trim, windows, doors, and much more. Includes sample forms, checklists, and the author's factor worksheets to save you time and help prevent errors. **320 pages, 8½ x 11, $25.50**

Handbook of Construction Contracting Vol. 1 & 2
Volume 1: Everything you need to know to start and run your construction business; the pros and cons of each type of contracting, the records you'll need to keep, and how to read and understand house plans and specs to find any problems before the actual work begins. All aspects of construction are covered in detail, including all-weather wood foundations, practical math for the jobsite, and elementary surveying. **416 pages, 8½ x 11, $21.75**

Volume 2: Everything you need to know to keep your construction business profitable; different methods of estimating, keeping and controlling costs, estimating excavation, concrete, masonry, rough carpentry, roof covering, insulation, doors and windows, exterior finish, specialty finishes, scheduling work flow, managing workers, advertising and sales, spec building and land development and selecting the best legal structure for your business. **320 pages, 8½ x 11, $24.75**

Cost Records for Construction Estimating
How to organize and use cost information from jobs just completed to make more accurate estimates in the future. Explains how to keep the cost records you need to reflect the time spent on each part of the job. Shows the best way to track costs for sitework, footing, foundations, framing, interior finish, siding and trim, masonry, and subcontract expense. Provides sample forms. **208 pages, 8½ x 11, $15.75**

Contractor's Year Round Tax Guide
How to set up and run your construction business to minimize taxes: corporate tax strategy and how to use it to your advantage, and what you should be aware of in contracts with others. Covers tax shelters for builders, write-offs and investments that will reduce your taxes, accounting methods that are best for contractors, and what the I.R.S. allows and what it often questions. **192 pages, 8½ x 11, $16.50**

Craftsman Book Company
P. O. Box 6500
Carlsbad, CA 92008

10 Day Money Back GUARANTEE

FREE Select one of these three handy calculators when your check or charge card order totals $100 or more.

☐ E-Z Square ☐ Stairule ☐ Rafterule

Name (Please print clearly) ______

Company ______

Address ______

City ______ State ______ Zip ______

☐ Visa ☐ MasterCard

Expiration date ______

Number ______

Signature (Required on billed orders) ______

- ☐ 17.50 Basic Plumbing with Illustrations
- ☐ 30.00 Berger Building Cost File
- ☐ 11.25 Blueprint Reading for Building Trades
- ☐ 17.25 Builder's Guide to Accounting Revised
- ☐ 11.00 Builder's Guide to Const. Financing
- ☐ 15.50 Builder's Office Manual Revised
- ☐ 14.00 Building Cost Manual
- ☐ 11.75 Building Layout
- ☐ 25.50 Carpentry Estimating
- ☐ 19.75 Carpentry for Residential Construction
- ☐ 17.75 Computers: The Builder's New Tool
- ☐ 10.00 Concrete and Formwork
- ☐ 20.50 Concrete Const. & Estimating
- ☐ 20.00 Const. Estimating Ref. Data
- ☐ 22.00 Construction Superintending
- ☐ 19.25 Construction Surveying & Layout
- ☐ 16.25 Contractor's Guide to the Building Code
- ☐ 16.75 Contractor's Survival Manual
- ☐ 16.50 Contractor's Year-Round Tax Guide
- ☐ 15.75 Cost Records for Const. Estimating
- ☐ 9.50 Dial-A-Length Rafterule
- ☐ 18.25 Drywall Contracting
- ☐ 13.75 Electrical Blueprint Reading
- ☐ 25.00 Electrical Construction Estimator
- ☐ 19.00 Estimating Electrical Construction
- ☐ 14.00 Estimating Home Building Costs
- ☐ 17.25 Estimating Plumbing Costs
- ☐ 21.50 Estimating Tables for Home Building
- ☐ 22.75 Excavation & Grading Handbook Revised
- ☐ 9.25 E-Z Square
- ☐ 10.50 Finish Carpentry
- ☐ 21.75 Handbook of Construction Contracting Vol. 1
- ☐ 24.75 Handbook of Construction Contracting Vol. 2
- ☐ 14.75 Handbook of Modern Electrical Wiring
- ☐ 15.00 Home Wiring: Improvement, Extension, Repairs
- ☐ 17.50 How to Sell Remodeling
- ☐ 24.50 HVAC Contracting
- ☐ 17.00 Manual of Electrical Contracting
- ☐ 18.75 Manual of Professional Remodeling
- ☐ 13.50 Masonry & Concrete Construction
- ☐ 18.50 National Construction Estimator
- ☐ 23.75 Operating the Tractor-Loader-Backhoe
- ☐ 19.25 Paint Contractor's Manual
- ☐ 21.25 Painter's Handbook
- ☐ 23.50 Pipe & Excavation Contracting
- ☐ 13.00 Planning and Designing Plumbing Systems
- ☐ 21.00 Plumber's Exam Preparation Guide
- ☐ 18.00 Plumber's Handbook Revised
- ☐ 18.25 Process & Industrial Pipe Estimating
- ☐ 12.25 Rafter Length Manual
- ☐ 18.50 Remodeler's Handbook
- ☐ 11.50 Residential Electrical Design
- ☐ 18.25 Residential Wiring
- ☐ 19.50 Roof Framing
- ☐ 9.25 Roofers Handbook
- ☐ 16.00 Rough Carpentry
- ☐ 24.00 Spec Builder's Guide
- ☐ 13.75 Stair Builder's Handbook
- ☐ 7.50 Visual Stairule
- ☐ 11.25 Wood-Frame House Construction

pnt card

10 Day Money Back GUARANTEE

FREE Select one of these three handy calculators when your check or charge card order totals $100 or more.

☐ E-Z Square ☐ Stairule ☐ Rafterule

Name (Please print clearly) ______

Company ______

Address ______

City ______ State ______ Zip ______

☐ Visa ☐ MasterCard

Expiration date ______

Number ______

Signature (Required on billed orders) ______

- ☐ 17.50 Basic Plumbing with Illustrations
- ☐ 30.00 Berger Building Cost File
- ☐ 11.25 Blueprint Reading for Building Trades
- ☐ 17.25 Builder's Guide to Accounting Revised
- ☐ 11.00 Builder's Guide to Const. Financing
- ☐ 15.50 Builder's Office Manual Revised
- ☐ 14.00 Building Cost Manual
- ☐ 11.75 Building Layout
- ☐ 25.50 Carpentry Estimating
- ☐ 19.75 Carpentry for Residential Construction
- ☐ 17.75 Computers: The Builder's New Tool
- ☐ 10.00 Concrete and Formwork
- ☐ 20.50 Concrete Const. & Estimating
- ☐ 20.00 Const. Estimating Ref. Data
- ☐ 22.00 Construction Superintending
- ☐ 19.25 Construction Surveying & Layout
- ☐ 16.25 Contractor's Guide to the Building Code
- ☐ 16.75 Contractor's Survival Manual
- ☐ 16.50 Contractor's Year-Round Tax Guide
- ☐ 15.75 Cost Records for Const. Estimating
- ☐ 9.50 Dial-A-Length Rafterule
- ☐ 18.25 Drywall Contracting
- ☐ 13.75 Electrical Blueprint Reading
- ☐ 25.00 Electrical Construction Estimator
- ☐ 19.00 Estimating Electrical Construction
- ☐ 14.00 Estimating Home Building Costs
- ☐ 17.25 Estimating Plumbing Costs
- ☐ 21.50 Estimating Tables for Home Building
- ☐ 22.75 Excavation & Grading Handbook Revised
- ☐ 9.25 E-Z Square
- ☐ 10.50 Finish Carpentry
- ☐ 21.75 Handbook of Construction Contracting Vol. 1
- ☐ 24.75 Handbook of Construction Contracting Vol. 2
- ☐ 14.75 Handbook of Modern Electrical Wiring
- ☐ 15.00 Home Wiring: Improvement, Extension, Repairs
- ☐ 17.50 How to Sell Remodeling
- ☐ 24.50 HVAC Contracting
- ☐ 17.00 Manual of Electrical Contracting
- ☐ 18.75 Manual of Professional Remodeling
- ☐ 13.50 Masonry & Concrete Construction
- ☐ 18.50 National Construction Estimator
- ☐ 23.75 Operating the Tractor-Loader-Backhoe
- ☐ 19.25 Paint Contractor's Manual
- ☐ 21.25 Painter's Handbook
- ☐ 23.50 Pipe & Excavation Contracting
- ☐ 13.00 Planning and Designing Plumbing Systems
- ☐ 21.00 Plumber's Exam Preparation Guide
- ☐ 18.00 Plumber's Handbook Revised
- ☐ 18.25 Process & Industrial Pipe Estimating
- ☐ 12.25 Rafter Length Manual
- ☐ 18.50 Remodeler's Handbook
- ☐ 11.50 Residential Electrical Design
- ☐ 18.25 Residential Wiring
- ☐ 19.50 Roof Framing
- ☐ 9.25 Roofers Handbook
- ☐ 16.00 Rough Carpentry
- ☐ 24.00 Spec Builder's Guide
- ☐ 13.75 Stair Builder's Handbook
- ☐ 7.50 Visual Stairule
- ☐ 11.25 Wood-Frame House Construction

Charge Card Phone Orders — Call (619) 438-7828

pnt card

10 Day Money Back GUARANTEE

FREE Select one of these three handy calculators when your check or charge card order totals $100 or more.

☐ E-Z Square ☐ Stairule ☐ Rafterule

Name (Please print clearly) ______

Company ______

Address ______

City ______ State ______ Zip ______

☐ Visa ☐ MasterCard

Expiration date ______

Number ______

Signature (Required on billed orders) ______

In a Hurry? Call (619) 438-7828
Orders taken on Visa or MasterCard

- ☐ 17.50 Basic Plumbing with Illustrations
- ☐ 30.00 Berger Building Cost File
- ☐ 11.25 Blueprint Reading for Building Trades
- ☐ 17.25 Builder's Guide to Accounting Revised
- ☐ 11.00 Builder's Guide to Const. Financing
- ☐ 15.50 Builder's Office Manual Revised
- ☐ 14.00 Building Cost Manual
- ☐ 11.75 Building Layout
- ☐ 25.50 Carpentry Estimating
- ☐ 19.75 Carpentry for Residential Construction
- ☐ 17.75 Computers: The Builder's New Tool
- ☐ 10.00 Concrete and Formwork
- ☐ 20.50 Concrete Const. & Estimating
- ☐ 20.00 Const. Estimating Ref. Data
- ☐ 22.00 Construction Superintending
- ☐ 19.25 Construction Surveying & Layout
- ☐ 16.25 Contractor's Guide to the Building Code
- ☐ 16.75 Contractor's Survival Manual
- ☐ 16.50 Contractor's Year-Round Tax Guide
- ☐ 15.75 Cost Records for Const. Estimating
- ☐ 9.50 Dial-A-Length Rafterule
- ☐ 18.25 Drywall Contracting
- ☐ 13.75 Electrical Blueprint Reading
- ☐ 25.00 Electrical Construction Estimator
- ☐ 19.00 Estimating Electrical Construction
- ☐ 14.00 Estimating Home Building Costs
- ☐ 17.25 Estimating Plumbing Costs
- ☐ 21.50 Estimating Tables for Home Building
- ☐ 22.75 Excavation & Grading Handbook Revised
- ☐ 9.25 E-Z Square
- ☐ 10.50 Finish Carpentry
- ☐ 21.75 Handbook of Construction Contracting Vol. 1
- ☐ 24.75 Handbook of Construction Contracting Vol. 2
- ☐ 14.75 Handbook of Modern Electrical Wiring
- ☐ 15.00 Home Wiring: Improvement, Extension, Repairs
- ☐ 17.50 How to Sell Remodeling
- ☐ 24.50 HVAC Contracting
- ☐ 17.00 Manual of Electrical Contracting
- ☐ 18.75 Manual of Professional Remodeling
- ☐ 13.50 Masonry & Concrete Construction
- ☐ 18.50 National Construction Estimator
- ☐ 23.75 Operating the Tractor-Loader-Backhoe
- ☐ 19.25 Paint Contractor's Manual
- ☐ 21.25 Painter's Handbook
- ☐ 23.50 Pipe & Excavation Contracting
- ☐ 13.00 Planning and Designing Plumbing Systems
- ☐ 21.00 Plumber's Exam Preparation Guide
- ☐ 18.00 Plumber's Handbook Revised
- ☐ 18.25 Process & Industrial Pipe Estimating
- ☐ 12.25 Rafter Length Manual
- ☐ 18.50 Remodeler's Handbook
- ☐ 11.50 Residential Electrical Design
- ☐ 18.25 Residential Wiring
- ☐ 19.50 Roof Framing
- ☐ 9.25 Roofers Handbook
- ☐ 16.00 Rough Carpentry
- ☐ 24.00 Spec Builder's Guide
- ☐ 13.75 Stair Builder's Handbook
- ☐ 7.50 Visual Stairule
- ☐ 11.25 Wood-Frame House Construction

pnt card

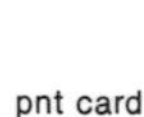

Craftsman Book Company, 6058 Corte Del Cedro, P. O. Box 6500, Carlsbad, CA 92008

NO POSTAGE
NECESSARY
IF MAILED
IN THE
UNITED STATES

BUSINESS REPLY MAIL
FIRST CLASS PERMIT NO. 271 CARLSBAD, CA

POSTAGE WILL BE PAID BY ADDRESSEE

Craftsman Book Company
6058 Corte Del Cedro
P. O. Box 6500
Carlsbad, CA 92008—9974

NO POSTAGE
NECESSARY
IF MAILED
IN THE
UNITED STATES

BUSINESS REPLY MAIL
FIRST CLASS PERMIT NO. 271 CARLSBAD, CA

POSTAGE WILL BE PAID BY ADDRESSEE

Craftsman Book Company
6058 Corte Del Cedro
P. O. Box 6500
Carlsbad, CA 92008—9974

NO POSTAGE
NECESSARY
IF MAILED
IN THE
UNITED STATES

BUSINESS REPLY MAIL
FIRST CLASS PERMIT NO. 271 CARLSBAD, CA

POSTAGE WILL BE PAID BY ADDRESSEE

Craftsman Book Company
6058 Corte Del Cedro
P. O. Box 6500
Carlsbad, CA 92008—9974